❯The Project
MANAGER'S GUIDE
TO MASTERING
AGILE

>The Project MANAGER'S GUIDE TO MASTERING AGILE

Principles and Practices for an Adaptive Approach

Second Edition

Charles G. Cobb

Agile Project Management Academy
https://agileprojectmanagementacademy.com/

WILEY

Library of Congress Cataloging-in-Publication Data

Names: Cobb, Charles G., 1945- author.
Title: The project manager's guide to mastering agile : principles and
  practices for an adaptive approach / Charles Grinnell Cobb.
Description: Second edition. | Hoboken, New Jersey : Wiley, [2023] |
  Includes bibliographical references and index.
Identifiers: LCCN 2022047014 (print) | LCCN 2022047015 (ebook) | ISBN
  9781119931355 (paperback) | ISBN 9781119931362 (adobe pdf) | ISBN
  9781119931379 (epub)
Subjects: LCSH: Agile software development. | Computer
  software—Development—Management.
Classification: LCC QA76.76.D47 C63 2023 (print) | LCC QA76.76.D47
  (ebook) | DDC 005.1—dc23/eng/20220930
LC record available at https://lccn.loc.gov/2022047014
LC ebook record available at https://lccn.loc.gov/2022047015

Cover Design: Wiley
Cover Images: © Den Rise/Shutterstock

SKY10075375_051624

CONTENTS

Part 3 Agile Project Management Planning and Management

14 Adaptive Planning 241

15 Agile Planning Practices and Tools 255

16 Agile Stakeholder Management and Agile Contracts 271

29 Case Study: General Dynamics, UK 499

30 Agile Hardware Development 513

PREFACE

BACKGROUND

THE PROJECT MANAGEMENT PROFESSION is going through rapid and profound changes due to the widespread influence of Agile:

It is becoming very apparent that the classical plan-driven approach to project management that has been the predominantly accepted way of doing project management for a long time is no longer the only way to do project management:

- Rather than force-fitting all projects to a classical plan-driven project management approach, it is much better to fit the approach to the nature of the project.

- It's particularly important to develop an adaptive approach for projects that have a high level of uncertainty and/or where creativity and innovation are more important than planning and control to achieve predictability.

Those changes are likely to dramatically change the role of project managers in many environments as we have known them, raise the bar for the entire project management profession, and perhaps even eliminate the role of some Project Managers as we have known them.

From an Agile perspective, there have also been some equally significant changes:

- Agile and Scrum have grown over the years from a focus on small, single-team projects to much larger and more complex enterprise-level projects requiring multiple teams.

- That has made it evident that scaling Agile for that kind of project requires some kind of overall management framework which might include some kind of project/program management.

In both of these environments, there is a recognition that well-defined and prescriptive "cookbook" approaches are no longer effective for dealing with the complexity of these challenges. Instead, there is a need to focus on defining principles that need to be interpreted in the context of a given situation:

- In an Agile environment, both the Scaled Agile Framework and Disciplined Agile Delivery have moved away from relatively well-defined frameworks to a more flexible, principles-based approach.

- In a classical plan-driven project management environment, PMBOK® version 7 has moved away from previous versions of PMBOK® that attempted to define a checklist of things to do in almost every conceivable project management situation to a less well-defined principles-based approach.

The movement to a principles-based approach in both of these environments will require a lot more judgment and skill for determining and implementing the right approach for a particular project.

It is critical for Project Managers and the Project Management Profession, as a whole, to be proactive, anticipate the most likely impact of these challenges, and adapt accordingly.

It is also important for the Agile community to recognize the need to scale an Agile approach for managing large, complex enterprise-level projects.

This raises a number of questions including:

- What is the role of project management in an Agile project?

- Are classical project management principles and practices in conflict with Agile principles and practices?

- What needs to be done to extend Agile principles and practices to larger and more complex enterprise-level projects requiring multiple teams?

- How does a typical Project Manager shape his or her career to move in a more Agile direction?

Those are the needs and challenges that this book is intended to address. This book should be of value to both project managers and Agile professionals to develop a more integrated approach.

LEARNING OBJECTIVES

The following is a summary of what I believe are the most important steps in the journey toward becoming an Agile Project Manager (not necessarily in this order):

1. Develop new ways of thinking and begin to see Agile principles and practices in a new light as complementary rather than competitive with classical project management practices.
2. Gain an understanding of the fundamentals of Agile practices and learn the principles behind the Agile practices at a deeper level in order to understand why they make sense and how they can be adapted as necessary to fit a given situation.
3. Learn how to go beyond the classical notion of plan-driven Project Management and develop an adaptive approach to Project Management that blends both Agile and classical plan-driven Project Management principles and practices in the right proportions to fit a given project and business environment.

4. Understand the potential roles that an Agile Project Manager can play and begin to reshape Project Management skills around those roles.

5. Learn some of the challenges of scaling Agile to an enterprise level and develop experience in applying these concepts in large, complex, enterprise-level environments.

Relationship to My Online Agile Project Management Training Courses

I have successfully developed an online training curriculum in Agile Project Management that is currently offered on three different platforms with over 175,000 students. Anyone who has taken any of those courses should see a lot of similarity between the material in this book and the material in my online Agile Project Management training courses.

Summary of Changes in the Second Edition

Many of the current trends that are going on in the project management community now have validated the original direction of the book when it was originally published in 2015. As a result, the changes required in the second edition are not radical. Here's a summary of the most important areas of change:

1. More detail on Agile Project Management Planning and Management: One of the professors currently teaching a course based on the book wanted to see more detail on Agile Project Management Planning and Management; so I have added six new chapters on that in Chapters 12 through 17.

2. Less detail on Agile Project Management Tools: The original edition of the book included some detailed material on Agile Project Management tools. Since that time, there have been many changes in that area, and it is apparent that the area of Agile Project Management tools will continue to evolve significantly. For that reason, I have limited the material in this edition of the book to a general, high-level discussion of the capabilities of Agile Project Management tools without going into specifics on any particular tool.

3. Revisions to enterprise-level frameworks: There have been a number of significant changes in the two enterprise-level Agile frameworks that are covered in the book (Scaled Agile Framework and Disciplined Agile framework) and both of those chapters needed considerable changes.

4. Additional case studies: This edition of the book includes two new chapters on case studies. One is on "Agile Hardware Development" and includes material on the Agile implementation at Tesla and the other is on "Non-software Case Studies" to show how to use Agile outside of a software development environment for common projects.

HOW THIS BOOK IS ORGANIZED

Agile Project Management is an art that will take time for anyone to develop and master. There's a concept from martial arts called *shu-ha-ri* that is very appropriate here. It outlines the stages of proficiency someone goes through to develop mastery of martial arts techniques. The same concept can be applied to Agile Project Management:

- **"Shu"**: In the "shu" stage, the student learns to do things more-or-less mechanically, "by the book," without significantly deviating from the accepted rules and practices and without improvising any new techniques. This stage is equivalent to a new inexperienced project manager following PMBOK or other accepted practices like you would follow recipes in a cookbook without necessarily adapting those practices to fit the situation.

- **"Ha"**: In the "ha" stage, the student begins to understand the principles at a deeper level and learns how to improvise and break free from rigidly accepted practices, but it's important to go through the "shu" stage and gain mastery of the foundational principles before you start improvising—improvisation without knowledge is just amateurish experimentation.

- **"Ri"**: Finally, in the "ri" stage, the student gets to the highest level of mastery and is able to develop his/her own principles and practices as necessary.

The way the book is organized follows the *shu-ha-ri* approach to learning. The initial chapters of the book start out with a very basic understanding of the "mechanics" of Agile and learning how to do it "by the book." That is equivalent to the "shu" level of training.

The book will go deeper into the principles behind Agile and why they make sense. It is essential to understand the principles at a deeper level before moving on to the "ha" level and know how to customize an approach to fit a given situation.

The final goal is to move to the master level or "ri" level where you will learn to go beyond current ways of implementing both Agile and plan-driven project management approaches and learn how to blend them together as needed to fit a given project and business environment. That goal will come from actual practice in implementing these ideas in real world situations; however, it is hoped that the information in this book and the case studies that are included will help Project Managers move rapidly in that direction.

The book is organized into Parts as follows.

Part 1 – Fundamentals of Agile

The first step in learning to become an Agile Project Manager is to learn the fundamentals of Agile, which includes not only the mechanics of how an Agile project based on Scrum works, but also understanding the principles behind it at a deeper level so that you can go beyond just implementing it "by the book."

Part 2 – Agile Project Management Overview

Agile is causing us to broaden our vision of what a *Project Manager* is and that will have a dramatic impact on the potential roles that a Project Manager can play in an Agile environment. In fact, the role of a Project Manager at a team level in a typical Agile/Scrum project is undefined. That will cause us to rethink many of the things we have taken for granted about Project Management for a long time to develop a broader vision of what an *Agile Project Manager* is.

Part 3 – Agile Project Management Planning and Management

Part 2 provided an overview of Agile Project Management. In this Part, we will go into much more detail on Agile Project Management planning and management practices including:

- Hybrid Agile Models

- Value-Driven Delivery

- Adaptive Planning

- Agile Planning Practices and Tools

- Agile Stakeholder Management and Agile Contracts

- Distributed Project Management in Agile

Part 4 – Making Agile Work for a Business

There are many precedents for successful implementation of Agile principles and practices at a project team level; however, extending the Agile principles and practices to large-scale enterprise implementations and integrating with a business environment can be very difficult and introduces a number of new challenges, which include:

- Large, complex projects that are commonly found at an enterprise level may require some reinterpretation and adaptation of Agile principles and practices as well as blending those principles and practices with classical, plan-driven project management principles and practices in the right proportions.

- Integrating Agile principles and practices with higher levels of management typically found at an enterprise level, such as project portfolio management and overall business management can be difficult. However, if an Agile implementation is limited to a development process only and does not address integration with these higher-level processes, it is not likely to be effective and may result in failure.

This Part of the book is intended to address these topics and provide an understanding of the key considerations that need to be addressed for:

- scaling an Agile approach for multiple teams and for larger, more complex enterprise-level projects

- integrating an Agile development approach with a business environment

- planning and implementing an enterprise-level Agile transformation.

Part 5 – Enterprise-Level Agile Frameworks

Putting together a complete, top-to-bottom, enterprise-level Agile solution can be a very challenging task, especially when some of the pieces are not designed to fit together.

To simplify the design of an enterprise-level Agile implementation, it is useful to have some predefined frameworks that can be modified to fit a given business environment, rather than having to start from scratch to design an overall management approach.

Three frameworks are discussed in this Part:

- Scaled Agile Framework (SAFe®=) (Dean Leffingwell)

- Managed Agile Development framework (Chuck Cobb)

- Disciplined Agile Delivery framework (Scott Ambler).

Part 6 – Case Studies

In any book of this nature, it's always useful to go beyond theory and concepts and show how companies have actually put these ideas into practice in the real-world. Of course, there is no canned approach that works for all companies—each of these case studies is different and shows how a different approach may be needed in different situations. It also includes a chapter on "Not-So-Successful" case studies, which shows some of the problems that can develop in an Agile implementation.

Part 7 – Appendices

The appendices to the book include additional supplementary information:

- Additional Reading List

- Glossary of Terms

- Example Project/Program Charter

- Suggested Course Outline for a graduate-level course to accompany this book

ACKNOWLEDGMENTS

I USED A VERY AGILE approach for writing this book as well as my previous books. It was a team effort of a number of people who worked with me collaboratively as the book was being written to provide feedback and inputs. I particularly want to thank the following people for their contributions to the original edition of this book:

FIRST EDITION

- Erik Gottesman, Director General Management at Sapient—Erik is a significant thought leader in this area. He played a huge role in helping me develop my two previous books on Agile Project Management and provided some good advice and input on this book as well.

- Dr. Michael Hurst, PMO Director at Harvard Pilgrim Health Care—Michael has played a significant role in providing input and advice for both this book and my last book and he also played a key role in providing a case study on Harvard Pilgrim Health Care that is included in this book.

- Andrew Bone, IT Program/PMO director—Andrew did a thorough review of the entire book, provided a number of good comments and inputs, and also sponsored a presentation on the book with the Long Island, New York, PMI Chapter.

- Liza Wood, Senior Production Manager at Warner Bros. Games—Liza also did a thorough review of the entire book on behalf of the PMI Agile Community of Practice and provided a very large number of excellent comments.

- Several companies generously shared case studies with the results of successful Agile implementations:

  - Michael Hurst, Director PMO, Harvard Pilgrim Health Care—Michael and Harvard Pilgrim shared the results of a very large and successful enterprise-level Agile transformation effort of more than 200 projects.
  - Stephanie Stewart(now Davis), Director of Agile Leadership at Valpak—Stephanie and Valpak shared the results of an enterprise-level implementation of the Scaled Agile Framework at Valpak.
  - Nigel Edwards, Program Manager at General Dynamics, UK—Nigel shared the results of a very large and complex, Agile fixed-price government contracting effort.

SECOND EDITION

I would particularly like to thank the following individuals who made significant contributions towards completing the second edition of this book:

- Stephan Wohlfahrt, Director, Corporate Project Management at Bosch—Stephan has played a significant thought leadership role at Bosch in championing a modern Agile Project Management approach throughout the whole company for all Bosch Project Managers worldwide. He did a very detailed review of the entire second edition of the book and provided numerous helpful comments and suggestions.

- Dr. Winston Gonzalez, D.M., SPC is an Instructor and Subject Matter Expert for the School of Continuing Studies at Georgetown University and is also on the Corporate Faculty at Harrisburg University. He teaches a course based on this book and made numerous valuable contributions and suggestions for its content.

- Joe Justice, Chairman of the Board, Agile Business Institute graciously allowed me to summarize some of his very excellent materials on hardware development practices; and, in particular, on his work with Tesla to implement those ideas. Joe is a real thought leader in this area and has made numerous contributions to implementing Agile in a hardware development environment.

I would like to also thank the following individuals who took the time to review an early draft of the second edition of this book and provided very helpful feedback, comments, and suggestions: Kiron D. Bondale, PMP, PMI-ACP, PSM II, ICP-ACC, PMI-RMP, DASSM, DAC, DAVSC, Senior Consultant, World Class Productivity Inc., and Dr. Monica Kay, PMP, Adjunct Professor, Morgan State University.

① Introduction to Agile Project Management

OVER THE PAST 20 TO 25 YEARS, there has been a rapid and dramatic adoption of Agile methodologies and this trend has significantly accelerated in the last few years:

The "15th State of Agile Report," published by Digital.ai, comments:

- This year's findings indicate significant growth in Agile adoption within software development teams, increasing from 37% in 2020 to 86% in 2021.[1]

Business Wire (a Berkshire Hathaway company) comments:

- Driven by the global pandemic, Agile adoption rates double in non-IT lines of business with continued, strong adoption in software development.

- More than 90 percent of respondents say their company practices Agile, with most saying either the majority of or even all company teams have adopted Agile practices.

- Rapid Agile adoption fuels an increase in adoption of other trends, including DevOps transformation and value stream management (VSM) initiatives for more than two-thirds of organizations.

- Post-pandemic, a vast majority of IT respondents expect to permanently work remotely, making Agile adoption critical to driving collaboration and success across a globally distributed workforce.[2]

These statistics indicate that Agile is not a fad, it is having a significant impact on the way projects are managed, it's definitely here to stay, and it is significantly accelerating in recent years. This trend has a significant impact on the career direction of project managers who have come from a classical plan-driven project management background since there is no formal role for a project manager at the team level in an Agile project.

> Note: Throughout this book, we will focus heavily on Agile as a project management approach because that is where it has the most important impact for project managers; however, it's important to realize that Agile is really a much broader way of thinking and is not limited to projects. It is also not specifically limited to software projects although that is where it is most heavily used.

THE CHASM IN PROJECT MANAGEMENT PHILOSOPHIES

Despite this rapid and sustained proliferation of Agile, there is still somewhat of a chasm between the Agile and classical plan-driven project management communities. When I published the first edition of this book in 2015, that chasm was large:

- There had been only a limited amount of progress at that time on developing a more integrated project management approach that embraces both Agile and classical plan-driven project management principles and practices.

- Many project managers had been heavily indoctrinated into a classical plan-driven project management approach and seemed to see Agile and classical plan-driven project management principles and practices as competitive approaches that conflict with each other, and they essentially treated them as two separate and independent domains of knowledge.

- Considerable polarization between these two communities was based on some part on myths, stereotypes, and misconceptions about what *Agile* and *project management* are that existed at that time.

Since that time, that chasm has narrowed considerably but there is still a gap that needs to be closed. Many people still seem to think that there is a binary and mutually exclusive choice between "Agile" and "Waterfall." The ideal goal would be to have a seamless integration of project management approaches from heavily plan-driven (Waterfall) at one extreme to heavily adaptive (Agile) at the other extreme with lots of alternatives between those two extremes as shown in Figure 1.1.

That leaves many project managers in a conundrum to try and figure out how these two very different approaches to project management can be integrated together. A major goal of this book is:

- to help project managers understand the impact of Agile on the project management profession

- to broaden and expand their project management skills as needed to develop a more integrated approach to adapt to this new environment.

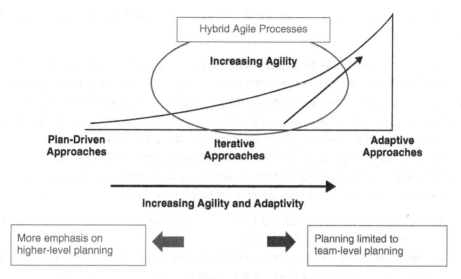

FIGURE 1.1   Spectrum of plan-driven and adaptive approaches

What's Driving These Changes?

In a classical plan-driven project management environment, a project was deemed to be successful if it delivered well-defined requirements on-time and within the approved budget. In today's world:

- There is a much higher level of uncertainty, which, in many cases, makes it very difficult, if not impossible, to document firm and well-defined requirements prior to the start of the project. In that environment, a much more flexible and adaptive approach is needed to further define and elaborate the detailed project requirements as the project is in progress.

- We also live in a very competitive environment where there is a much greater need for creativity and innovation to maximize the business value of the solution. An overemphasis on planning and control can stifle creativity and innovation. In fact, there have been many projects that have delivered well-defined requirements and met their cost and schedule goals but failed to deliver an acceptable level of business value.

This does not mean that a classical plan-driven project management is obsolete and no longer useful, but we need to recognize that it does have limitations and fit the project management approach to the nature of the project rather than force-fitting all projects to a classical plan-driven approach.

The important thing to consider is "value." What is the "value" that the project is intended to produce? Meeting a cost and schedule goal for delivering well-defined requirements certainly has some value in many situations, but it is not necessarily the most important (or only) value in a given project. The choice of an appropriate methodology for a project will depend on a number of factors:

1. **The level of uncertainty in the project:** A project that has a higher level of uncertainty in the requirements would naturally lean more towards a more flexible and adaptive approach.
2. **The level of training and sophistication of the project team:** It takes a considerable amount of skill and judgment to use an Agile approach successfully and it should not be attempted unless the team has been properly trained in Agile, the primary roles of Scrum Master and Product Owner are in place, and the necessary tools to support the project are also in place.
3. **The relationship with the customer of the project:** A classical plan-driven project is typically based on somewhat of a contractual relationship with the customer:

   ■ The customer expects the project to be delivered as defined in the requirements within the approved cost and schedule goals.
   ■ The customer does not need to be heavily involved in the implementation of the project until it is time to approve the final deliverables.

   An Agile approach requires a much more collaborative approach with the customer. The customer needs to share responsibility for the successful completion of the project team by:

   ■ taking an active role in the project to provide feedback and inputs on incremental results
   ■ further defining and elaborating requirements as the project is in progress.

This means that there is no longer only one way to do project management and it takes a considerable amount of skill and organizational maturity to fit the most appropriate project management approach to the nature of the project:

■ Not only is it important that the individuals responsible for product development are trained and skilled in an Agile approach.

■ In addition, the businesspeople who are required to take an active role in the process need to understand how the process works.

■ The overall organization needs to be committed to whatever level of organizational change that may be needed to make it successful.

The Impact on the Project Management Profession

This isn't just a matter of getting another certification—it can require a major shift in thinking for many traditional project managers that will take time and experience to develop. The Project Management Institute (PMI) has created the PMI-ACP® (Agile Certified Practitioner) certification, which has been very successful and is a great step in the right direction—but it doesn't go far enough, in my opinion.

- It doesn't test whether a project manager knows how to blend Agile and classical project management principles and practices in the right proportions to fit a given situation, and that is the real challenge that many project managers face.

- PMI-ACP is also not designed around a specific Agile role as many other Agile certifications are and the role that an Agile Project Manager might play is still somewhat undefined.

A lot of the polarization that has existed between the Agile and classical plan-driven project management communities has been rooted in some well-established stereotypes of what a *project manager* is that are based on how typical projects have been managed in the past. The role of a project manager has been so strongly associated with someone who plans and manages projects using classical plan-driven project management approaches that many people cannot conceive of any other image of a project manager. It's time to develop a new vision of what an *Agile Project Manager* is that goes beyond all those traditional stereotypes and fully integrates *Agile* within the overall portfolio of project management principles and practices.

It feels very similar to an evolution that took place when I worked in the quality management profession in the early 1990s. Up until that time, the primary emphasis in quality management had been on *quality control*, and inspection, and the image of a *quality manager* was heavily based on that role:

- The predominant quality management approach was based on final inspection of products prior to shipping them to the customer and rejecting any that didn't meet quality standards. It's easy to see how that approach was inefficient, because it resulted in a lot of unnecessary rework to correct problems after the fact, and it also wasn't that effective because any inspection approach is based on sampling, and it is impractical to do a 100% sample. For that reason, it can result in mediocre quality.

- A far better approach was to go upstream in the process and eliminate defects at the source by designing the process to be inherently more reliable and freer of defects, and build quality into the inherent design of the products. That didn't mean that the prior emphasis on quality control and inspection was obsolete and eliminated; it was just not the *only* way to manage quality and wasn't the most effective approach in all situations.

That was a gut-wrenching change for many in the quality management profession—instead of being in control of quality and being the gatekeeper with the inspection process, a good quality manager needed to become more of a coach and a consultant to influence others to build quality into the way they did their work. This changed the nature of the work dramatically for many in the quality management profession and eliminated a number of traditional quality management roles that were based on the old quality control and inspection approach. The similarity to the changes going on in the project management profession should be apparent:

- To be successful in more uncertain environments, project managers need to be able to take an adaptive approach that is appropriate to the level of uncertainty in the project and integrate quality into the process rather than relying on final acceptance testing at the end of the project to validate the product that is being produced.

■ They also need to give up some of the control that has become associated with the project management profession—in some cases, they may need to become more of a coach and a consultant to influence others rather than being in absolute control of a project.

This can dramatically change the role of a project manager. In some situations, the role of a project manager as we've known it may no longer exist. For example, at a team level in an Agile project, you probably won't find anyone with a title of *Project Manager* because the project management functions have been absorbed into other roles and are done very differently. That doesn't mean that *project management* is no longer important, but it may cause us to dramatically rethink what project management is in a much broader context than the way we might have thought about it in the past.

THE EVOLUTION OF AGILE AND WATERFALL

You will often hear people make a comparison between Agile and Waterfall. Many of those discussions are polarized and position them as competitive approaches. Here's an example:[3]

> According to the 2012 CHAOS report, Agile succeeds three times more often than Waterfall. Because the use of Agile methodologies helps companies work more efficiently and deliver winning results, Agile adoption is constantly increasing.

While that statement is generally true, it's an oversimplification. There are at least two problems with that kind of statement:

1. It makes it sound like there are only two binary, mutually exclusive choices: Agile and Waterfall.
2. The meaning of the words *Agile* and *Waterfall* are typically not well-defined and are used very loosely.

For those reasons, I prefer to avoid comparing Agile to Waterfall because it tends to be a very polarized discussion—I prefer to take a more objective approach that is based on a comparison between a plan-driven and an adaptive (value-driven) approach to project management. So, let's first define both *Agile* and *Waterfall*, and then compare the two approaches.

Definition of Waterfall

The word *Waterfall* actually has a very specific meaning, but that's often not how the word is really used:

> The Waterfall model is a popular version of the systems development life cycle model for software engineering. Often considered the classic approach to the systems development life cycle, the Waterfall model describes a development method that is linear and sequential. Waterfall development has distinct goals for each phase of development. Imagine a Waterfall on the cliff of a steep mountain. Once the water has flowed over the edge of the cliff and has begun its journey down the side of the mountain, it cannot turn back. It is the same with Waterfall development. Once a phase of development is completed, the development proceeds to the next phase and there is no turning back.[4]

Another aspect to the Waterfall model is that it is plan-driven; it attempts to define and document detailed requirements and a plan for the entire project prior to starting the project.

- When someone makes a statement comparing Waterfall to Agile, the word *Waterfall* is often used very loosely to refer to any kind of plan-driven methodology, and that's not really a very accurate and meaningful comparison.

- In some other comparisons like this, the word *Waterfall* refers to a general style of project management that obsessively emphasizes predictability and control over agility, and that's just bad project management. The Waterfall model will be discussed in more detail in Chapter 2.

Definition of Agile

Officially, Agile is defined by the principles and values of the Agile Manifesto of 2001 which will be discussed in Chapter 2. Agile is also an umbrella term used by many Agile practitioners to refer to different methods and frameworks that are based on adaptive, experimental, and extreme programming practices that have emerged since the mid-to-late 1990s.[5]

From a general perspective, Agile is a flexible and adaptive approach for developing and optimizing solutions in an uncertain environment. It is both incremental and iterative:

- "Incremental" means that the solution is broken up into "chunks" that are developed and tested individually and might also be released individually rather than waiting for the entire solution to be developed, tested, and released as a whole.

- "Iterative" means that the solution is progressively optimized and refined based on user feedback and inputs to maximize the value of the solution to the users.

It is particularly well suited for an environment with a high level of uncertainty because the process can start with only a high-level view of the project goals and requirements and those goals and requirements can be further elaborated and refined as the project is in progress. Of course, that does not mean that all Agile projects start with only a high-level view of the project goals and requirements. That will vary from one project to the next depending on the level of uncertainty in the project.

In actual practice, the meaning of the word *Agile* in this kind of comparison is also somewhat elusive because it has taken on some very strong connotations in actual usage. At a project level, the word *Agile* has frequently taken on a specific connotation associated with using the Scrum methodology on software development projects.

> Scrum is an Agile software development framework based on multiple small teams working in an intensive and interdependent manner. The term is named for the scrum (or scrummage) formation in rugby, which is used to restart the game after an event that causes play to stop, such as an infringement. Scrum employs real-time decision-making processes based on actual events and information.

That definition has evolved over the years as Scrum has become somewhat of a de-facto standard for Agile projects; however, the original definition of *Agile* conceived in the *Manifesto for Agile Software Development*,[6] published in 2001, was much broader than that. Better known as the Agile Manifesto, it laid out some simple and general principles and values that can apply to any kind of project (not just software development) (see Chapter 2).

Comparison of Predictive (Plan-Driven) and Adaptive (Value-Driven) Approaches

Traditional, classical plan-driven project management is a style of project management that is applied to projects where the requirements and plan for completing the project can be defined to a large extent prior to implementing the project. The emphasis in this style of project management is on predictability, and for that reason, the PMI calls it "predictive." However, *plan-driven* is a relative term, and you won't find many projects that start out with an absolutely rigid plan that is not expected to change at all. This style of project management is often loosely called "Waterfall."

In contrast, an adaptive (or value-driven) style of project management starts the implementation of a project with a less well-defined plan of how the project will be implemented and recognizes that the requirements and plan for the project are expected to evolve as the project progresses. *Adaptive* is also a relative term; you won't find many projects that have no plan whatsoever of how the project will be done.

The important point is that the terms *predictive (plan-driven)* and *adaptive (value-driven)* are relative—they are not discrete, binary, mutually exclusive alternatives. They should imply a continuous range of approaches with different levels of upfront planning. Table 1.1 shows a comparison of the two approaches.

TABLE 1.1   Comparison of approaches

	Classical project management approach	Hybrid	Agile approach
Project management approach	Plan-driven, predictive: The emphasis is on planning and predicting costs and schedules for projects with well-defined requirements	Blend of both	Value-driven, adaptive - The emphasis is on maximizing the value of the solution in an uncertain environment with uncertain requirements
Project management responsibility	Typically, a project manager is responsible for managing the overall project to meet approved cost and schedule goals		There may be no single "Project Manager" at the team level. Project management responsibility is typically distributed (see Chapter 17)
Project environment	Best suited for projects with a lower level of uncertainty where some level of predictability of costs and schedules is important		Best suited for projects with a higher level of uncertainty where some level of flexibility and adaptivity is needed to define the solution
Requirements management	Detailed requirements are defined prior to the beginning of the project		Detailed requirements are further defined and elaborated as the project is in progress
Change management	Changes in scope are controlled in order to maintain control over cost and schedule estimates		Changes are encouraged in order to support flexibility and adaptivity
Customer relationship	Contractual based on well-defined requirements		Collaborative based on a spirit of trust and partnership
Development management	Typically, Waterfall or an equivalent SDLC with controlled sequential phases		Typically, Scrum or an equivalent incremental and iterative approach
Solution delivery	The entire solution is tested and delivered all-at-once at the end of the project		The solution is tested and delivered incrementally at the end of each sprint and release
Testing	Testing is typically done sequentially by a separate and independent QA organization		Testing is integrated into the development effort and is done concurrently with development

Which Approach Is Better?

You will often hear people say an Agile approach is inherently better than a Waterfall approach. That is often a very biased statement and it's not necessarily accurate. Saying "Agile is better than Waterfall," is like saying, "A car is better than a boat."

■ Agile and Waterfall are different kinds of approaches designed for different kinds of projects.

■ The problem is not so much that Waterfall or Agile are inherently good or bad; the problem comes about when those methodologies are misused, and people try to use a single methodology (whatever it might be) for all projects.

■ Using a "one-size-fits-all" strategy to applying either Waterfall (plan-driven) or Agile (adaptive) approaches to all projects is not likely to yield optimum results.

In my opinion, being able to objectively understand the difference between a plan-driven approach and a more adaptive (value-driven) approach—as well as the principles behind those approaches at a deeper level—is probably one of the most important skills an Agile Project Manager needs to have. An Agile Project Manager needs to recognize the following:

■ **There is a broad range of alternative approaches between being plan-driven and being adaptive (value-driven):** The Agile Project Manager must choose the right level of upfront planning to be applied to a project, based on the level of uncertainty and other factors in the project.

■ **It takes some skill to make the right choice:** There is nothing inherently wrong with either of these approaches (adaptive or plan-driven). The problem comes about when people try to force-fit a project into one of these approaches rather than selecting and tailoring the approach to fit the project. For example, if I were to set out to try to find a cure for cancer (which has a high level of uncertainty) and I attempted to apply a highly plan-driven approach to that project, the results would probably be dismal.

The important point is that a heavily plan-driven approach (what some loosely refer to as Waterfall) is not the only way to successfully manage a project. In many projects, a good approach is to use an adaptive (value-driven) approach to start the design effort without fully defined and detailed requirements and perhaps prototype something quickly. Then, user feedback can be added to further refine the design as the project progresses. With a more adaptive (value-driven) approach:

■ **The elements of the approach are much more concurrent than sequential:** Instead of doing the entire design and then turning it over to quality assurance (QA) for testing, the design is done in small chunks called *iterations* or *sprints* that are typically two to four weeks long. During that time, developers and testers work collaboratively to design and test the software during each sprint.

■ **The customer also provides detailed inputs on the design during each sprint:** The customer accepts the results of each sprint at the end of each two- to four-week period rather than waiting for user acceptance testing (UAT) at the end of the project. That has the advantage of finding and resolving any problems quickly and early in the project.

Advantages of an Adaptive (Value-Driven) Approach

One primary advantage of a more adaptive approach is that the project startup is accelerated because less time is spent upfront in attempting to define detailed requirements and to develop a detailed plan to fulfill those requirements. In addition, engaging the user more directly in the design process is more likely to produce an outcome that provides the necessary business value and really meets the user needs.

An adaptive (value-driven) approach has a higher probability of maximizing the business value to the customer because:

- The customer is directly engaged with the design team as the project progresses.

- The project might be delivered incrementally rather than waiting for the entire solution to be complete.

- Control is somewhat decentralized, allowing people to make decisions at the appropriate time and level.[7]

Disadvantages of an Adaptive (Value-Driven) Approach

However, an adaptive (value-driven) approach is worse for predictability and control because the customer can make changes as the project progresses. In an Agile project, change is the norm rather than the exception. Several other disadvantages include:

- An adaptive (value-driven) approach typically requires a higher level of judgment and skill to implement and manage successfully and that requires training and coaching.

- It also requires the customer to actively participate in the project as it is in progress in a collaborative spirit of partnership.

The Best of Both Worlds

This is not an "all-or-nothing" proposition to have either total predictability and control or no control at all. There are many ways to achieve the right balance of control versus agility. For example, prior to the start of a project, the high-level requirements might be defined and stabilized, and then only the more detailed requirements need to be further elaborated as the project progresses.

THE EVOLUTION OF THE PROJECT MANAGEMENT PROFESSION

Many of the techniques associated with project management that are in use today haven't changed significantly since the 1950s and 1960s. I believe that we are on the verge of a major transformation of the project management profession that will cause us to redefine project management in a much broader context that includes both Agile and classical plan-driven project management.

The Early History of Project Management

In order to understand this transition and to put it in perspective, it is useful to understand how the project management profession has evolved over the years and how we got to where we are today.

- Project management has been practiced for many years in one way or another—I'm sure that there was some kind of "project management" approach to building the great pyramids of Egypt or the Great Wall of China or other similar large efforts many years ago, but it probably wasn't even thought of as project management in those days.

- They didn't have Gantt charts and Pert charts and other sophisticated project planning and management tools, because those things weren't even invented until the twentieth century.

The Industrial Revolution created the need for a more disciplined approach to project management, and a well-defined body of knowledge associated with project management began to evolve:

> In the late nineteenth century, in the United States, large-scale government projects were the impetus for making important decisions that became the basis for project management methodology such as the transcontinental railroad, which began construction in the 1860s. Suddenly, business leaders found themselves faced with the daunting task of organizing the manual labor of thousands of workers and the processing and assembly of unprecedented quantities of raw material.
>
> Near the turn of the century, Frederick Taylor began his detailed studies of work. He applied scientific reasoning to work by showing that labor can be analyzed and improved by focusing on its elementary parts that introduced the concept of working more efficiently, rather than working harder and longer.[8]
>
> Taylor's associate, Henry Gantt, studied in great detail the order of operations in work and is most famous for developing the Gantt chart in the 1910s.[9]

World War II brought about the need for more large-scale project management for organizing very large projects like the Manhattan Project; however, it wasn't until the 1950s and 1960s, that it became apparent that a much more well-defined body of knowledge and a disciplined approach were needed to successfully manage some of the large and complex projects that were evolving at that time, which led to the following:

- The Program Evaluation and Review Technique or PERT was developed by Booz-Allen Hamilton as part of the US Navy's (in conjunction with the Lockheed Corporation) Polaris missile submarine program.

- The Critical Path Method (CPM) was developed in a joint venture between DuPont Corporation and Remington Rand Corporation for managing plant maintenance projects.

- The Project Management Institute (PMI) was founded in 1969.

Many people probably assume that the project management profession is now reaching a stage of maturity and stabilizing, but I believe that we have only seen the beginning, and project management, as we've known it, will continue to grow in entirely new directions.

Transformation of the Project Management Profession

Sometimes we get so immersed in day-to-day activities that we don't take time to step back and see some fundamental changes that are going on around us.

- It seems clear to me that the project management profession, as we know it, is going to go through such a major transformation. The exact nature of that transformation isn't completely clear as it is still evolving; however, it does seem likely that it will cause us to rethink many of the things we have taken for granted in the project management profession for a long time in a much broader perspective.

- It feels very similar to the evolution that has taken place in other technology areas and disciplines. For example, there is a strong similarity to the evolution from classical physics to modern physics.

> By the close of the nineteenth century, the study of physics was widely thought to be essentially complete, with the exception of only a few "loose ends"—minor unsolved problems to be dealt with.[10]

Up until that time, the study of physics had been heavily dominated by Newtonian physics, which defines some fundamental laws of how the universe behaves such as Newton's laws of motion. These fundamental laws were taken for granted in the world of physics for many years, even though we didn't fully understand why things in the universe behaved as they did.

As modern physics has evolved in the twentieth century, based on quantum mechanics and relativity, we began to develop a deeper understanding of the real dynamics behind these laws, and we began to understand that the universe is not as simple and deterministic as we might have thought it was.

The transition from classical Newtonian physics to a more complete and more dynamic model based on quantum mechanics provided a deeper understanding of the forces and principles behind those laws, as well as the limitations in those laws and when and where they are really applicable. That deeper understanding didn't invalidate the laws of Newtonian physics in most situations—"on an 'everyday' scale; that is, situations in which energies are large enough to permit one to neglect quantum effects, but small enough to neglect relativistic effects."[11]

The similarity to the transition in the project management profession should be apparent—we're moving from a world in which we had the impression that the behavior of the universe was highly predictable and controllable and totally subject to some well-defined rules to a world that is much more dynamic, much more probabilistic, and much less predictable.

What's Driving This Change, and Why Now?

You might ask, "Why is it becoming so essential for the project management profession to make a change at this particular point in time?" There are several major factors that will force us to rethink the concept of project management:

1. **The nature of projects is changing:** The modern concepts of project management were developed as result of big projects like the transcontinental railroad. Today, we have new industries and a much broader range of projects such as web development, e-commerce, large IT projects, etc., which weren't common before the mid-1990s. It is becoming increasingly apparent that applying a "one-size-fits-all" approach to such a broad range of projects will not have optimum results.

2. **Technology is rapidly changing:** Figure 1.2 shows how the adoption rate of new technologies has changed over the past century. Project management approaches that worked in the 1950s and 1960s must be reexamined to adapt to the current fast pace of technology adoption.

3. **The general level of uncertainty in the world is increasing.**

Uncertainties impact the way we live and work and the way we manage our work, including how we manage our personal lives and how we collaborate with others across boundaries. For example, the number of uncertainties of our times, which affect us at our workplaces and at home, include the economic crisis, the pandemics, the wars, and the ongoing conflicts among world-nation leaders.

As Ann Deaton (2018) would imply: we live and work in a VUCA world requiring VUCA tools, where VUCA is an acronym that refers to a Volatile, Uncertain, Complex, and Ambiguous business world. The term VUCA was originally used in the U.S. Army War College in 1990s to describe how military leaders manage to operate under different circumstances.[12]

A similar transformation took place in the quality management profession in the 1980s and early 1990s.

- At that time, the Japanese auto industry was demonstrating huge improvements in quality of products that made conventional quality management methods based primarily on quality control and inspection very inadequate.

- They forced people to rethink the whole strategy and approach for doing quality management. Without the leadership of people like W. Edwards Deming and the significant improvements in quality that were demonstrated in the automotive industry, the transformation of quality management might never have happened.

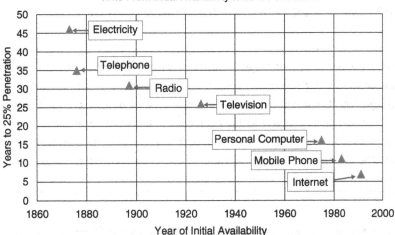

FIGURE 1.2   Adoption rates of new technologies

Source: Singularity.com

- What started primarily in the automotive industry has now become a more modern approach to quality management that is widely used in all industries.

    A similar thing is happening with Agile and classical plan-driven project management today:

- The leadership of W. Edwards Deming in establishing a Total Quality Management (TQM) philosophy can be compared to the thought leadership behind the Agile Manifesto in 2001.

- The broad-based adoption of TQM starting in the automotive industry and eventually spreading to many other industries can be compared to how Agile has started out in software development today and is now beginning to spread to other areas.

    Other researchers have come to a similar conclusion regarding this; Manfred Saynisch published his findings of a research project in *Project Management Journal*.

> Traditional Project Management . . . is based mainly on a mono-causal, non-dynamic, linear structure and a discrete view of human nature and societies and their perceptions knowledge and actions. It works on the basis of reductionist thinking and on the Cartesian/Newtonian concept of causality (the mechanistic science).[13]

The article proposes a new approach to project management called "PM2" where classical plan-driven project management will play an active and important role but will be "extended to consider dynamic, nonlinear, multi-causal structures and processes as well as the principles of self-organization, evolution, and networking." The article goes on to say:

> For an effective attainment of project goals at a defined finishing point in time, we need the linear processes and Cartesian causality and the Newtonian logic from classical project management. But evolutionary and self-organizational-based management methods are necessary to master complex and uncertain situations on the way to the defined finishing point in time for a project. A well-balanced interaction of traditional project management and the evolutionary and self-organizational principles is the message of the Project Management Second Order.[14]

I agree with that view—we are on the verge of a new generation of project management that will cause us to rethink many of the accepted notions of what "project management is." It requires blending classical plan-driven project management principles and practices with a much more empirical and evolutionary approach to deal with the uncertain, dynamic, and fast-paced environment we live in today.

The key message is to fit the project management approach to the nature of the project, as shown in Figure 1.3. That takes a lot more skill, but it definitely can be done.

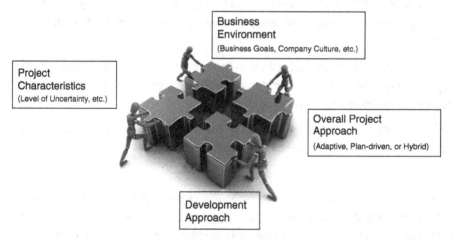

FIGURE 1.3   Fitting the project management approach to the project.

Source: Carsten Reisinger/Adobe Stock.

AGILE PROJECT MANAGEMENT BENEFITS

I am a strong believer in Agile, and there are some very significant benefits of an Agile approach in many situations. However, many proponents of Agile oversell the benefits and sometimes position Agile as a panacea that should be used for all projects.

- The real benefit to a typical project manager of developing an Agile project management approach is not in throwing away any notion of using a plan-driven approach, converting to Agile, and using a totally Agile approach for all projects.

- Rather, the benefit results from recognizing that a classical plan-driven approach is not the best way to manage all projects and thus learning to blend adaptive/Agile (value-driven) and plan-driven principles and practices in the right proportions to fit a given situation.

Even if a project manager never uses a *fully* Agile approach, I believe that knowledge of Agile concepts and principles will make him/her a better project manager. It's really a matter of learning a broader range of approaches (adding more tools to your toolbox) and developing a more adaptive project management approach (developing more skill in using those tools). In my previous books, I used the analogy of a project manager as a "cook" and the project manager as a "chef" (with credit to Bob Wysocki):

- A good *cook* might have the ability to create some very good meals, but those dishes might be limited to a repertoire of standard dishes, and knowledge of how to prepare those meals might be primarily based on following some predefined recipes out of a cookbook.

- A *chef*, on the other hand, typically has a far greater ability to prepare a much broader range of more sophisticated dishes using much more exotic ingredients in some cases. The chef's knowledge of how to prepare those meals is not limited to predefined recipes, and in many cases, a chef will create entirely new and innovative recipes for a given situation. The best chefs are not limited to a single cuisine and are capable of combining dishes from entirely different kinds of cuisine.

I think that sums up the transformation that needs to take place—we need to develop more project managers who are "chefs" rather than "cooks."

Here are five specific benefits of developing an Agile Project Management approach.

1. **Increased focus on business outcomes:** Many people think that the primary benefit of an Agile project is just getting it done faster, but that is not always the case.

   - The primary emphasis in an Agile project is really to deliver value in the form of very successful business outcomes by taking an adaptive approach to maximize the value that is delivered.

■ That doesn't always result in the fastest delivery times. In some cases, it may require some experimentation and trial-and-error prototyping to find an optimum solution—that may or may not be the quickest way to get it done, but it should result in a better product in the end.

2. **Reduced time to market:** Time to market is, of course, an important consideration, and Agile accomplishes that in a couple of ways:

■ By reducing the startup time required for projects as a result of simplifying some of the requirements definition practices.

■ By improving the efficiency of the overall project and delivering functionality incrementally as much as possible.

■ By focusing on simplicity and eliminating non-value-added work.

3. **Higher productivity and lower costs:** Agile can also result in higher productivity and lower costs by eliminating unnecessary overhead and bottlenecks and doing work concurrently rather than sequentially.

4. **Higher quality:** A very important benefit of Agile is higher quality.

■ In a classical Waterfall project, quality is sequential and is often perceived as a separate effort that is the responsibility of the quality assurance (QA) department.

■ The developers many times develop the software and then "toss it over the wall" to be tested by QA.

■ In an Agile project, the team, as a whole (which includes QA testers), jointly owns responsibility for building quality into the design of the products that they produce—it's not someone else's responsibility.

■ The development effort is broken up into short iterations called sprints that are typically two to four weeks in length. There is an emphasis on producing a shippable product at the end of every sprint, which means that quality testing must be more integrated with development and cannot be put off indefinitely.

5. **Organizational effectiveness:** Finally, a very important benefit of Agile is a more effective organization with higher morale:

■ People at all levels are motivated and empowered to do their work and take pride in doing it well because the environment is built on solid values, including respect for people.

■ All parts of the organization work together more collaboratively in a spirit of partnership toward common goals.

SUMMARY OF KEY POINTS

1. Closing the Chasm

There has been a widespread and rapid adoption of Agile methodologies over the last 15 to 20 years:

- However, our progress in developing an integrated approach to project management that embraces Agile as well as classical plan-driven project management principles and practices has been somewhat limited.

- To make further progress in that direction, we need to get past a number of well-established stereotypes and develop a much broader vision of what project management is.

2. Comparison of Agile versus Waterfall

The typical discussion that compares Agile and Waterfall as if they were two discrete, mutually exclusive, binary choices oversimplifies what should be more accurately thought of as a range of adaptive and plan-driven approaches:

- The Agile versus Waterfall comparison has also created an impression that the approaches are competitive, and that has created some polarization.

- In fact, adaptive (value-driven) and plan-driven approaches really should be thought of as much more complementary to each other.

3. Transformation of the Project Management Profession

The project management profession is at a major turning point in its history:

- The project management profession has developed over a number of years into a well-planned and disciplined approach to how projects are managed in reaction to the need for managing very large and complex projects that evolved in the early 1950s and 1960s.

- That approach has worked well for projects that can be heavily plan-driven; however, it has serious limitations in highly uncertain and rapidly changing environments that are difficult or impractical to develop a detailed plan for.

4. Agile Project Management Benefits

Developing a more adaptive approach to project management and tailoring the approach to fit the project will generally result in a number of benefits:

- The benefits come from matching the approach to the project rather than always using a plan-driven approach for all projects.

■ These benefits are not limited only to Agile projects—even if a project manager is never involved in an Agile project at all, developing a broader and deeper knowledge of both adaptive (value-driven) and plan-driven (predictive) principles and practices is likely to significantly improve a project manager's skills for many different projects by developing a more adaptive (value-driven) approach that can be optimized for the nature of the project.

DISCUSSION TOPICS

Closing the Chasm

1. Have you observed the "chasm" between the Agile and classical plan-driven project management communities? How does it manifest itself? What is the impact? What needs to be done to "close the chasm"?

Agile versus Waterfall

1. Research the usage of the terms *Agile* and *Waterfall*. Identify and discuss how this comparison is often misleading. Explain the difference between an *adaptive (value-driven)* and a *plan-driven (predictive)* approach and how that helps to provide a more objective frame of reference.

Transformation of the Project Management Profession

1. How do you see the transformation that is going on in the project management profession? Do you think that there is a significant change, and if so, what impact will it have on the project management profession as a whole? What needs to be done to make this transformation happen?

Agile Project Management Benefits

1. What do you think are the most important benefits of developing a more adaptive/Agile (value-driven) approach? How would it affect the way you manage projects?

Balancing Agility and Control

1. How would you go about determining the appropriate balance of agility and control for a project? What factors would you consider, and why? Provide an example of a real-world project and discuss how you might do it differently based on these factors.

Agile Benefits

1. What do you think are the most important benefits of an Agile approach to some typical projects? Discuss a real-world example of how an organization might have benefited from adopting a more Agile approach.

Project Management Career Direction

1. How do you think Agile affects the career direction of project managers? What impact do you think it might have on your own career? What do you think you might have to do differently as a result of Agile?

NOTES

1. "15th Annual State of Agile Report," https://digital.ai/resource-center/analyst-reports/state-of-agile-report.
2. "15th State of Agile Report Shows Notable Rise in Agile Adoption Across the Enterprise," https://www.businesswire.com/news/home/20210713005631/en/15th-State-of-Agile-Report-Shows-Notable-Rise-in-Agile-Adoption-Across-the-Enterprise.
3. PGDCA, Department of Computer Science, D.K. College, Mirza, https://www.facebook.com/dkcomputershell/posts/633083063450334:0.
4. Winston Gonzalez, email comments to author on book review.
5. *Manifesto for Agile Software Development*, https://agilemanifesto.org/.
6. Ibid.
7. Winston Gonzalez, email comments to author on book review.
8. Project Management: Past and Present, https://opentextbc.ca/projectmanagement/chapter/chapter-1-project-management-in-industry-project-management/.
9. Ibid.
10. "Physics," http://www.conservapedia.com/Physics.
11. Ibid.
12. Winston Gonzalez, email comments to author on book review.
13. Manfred Saynish, "Mastering Complexity and Changes in Projects, Economy, and Society via Project Management Second Order (PM-2)," *Project Management Journal* 41, no. 5 (Dec. 2010).
14. Ibid.

PART 1

Fundamentals of Agile

THE FIRST STEP IN LEARNING to become an Agile Project Manager is to learn the fundamentals of Agile, which includes understanding not only the mechanics of how an Agile project based on Scrum works but also the principles behind it at a deeper level, so that you can go beyond just implementing it mechanically, by the book.

Chapter 2 – Agile History and the Agile Manifesto:

Agile is based on the values and principles expressed in the Agile Manifesto that was originally created in 2001. The values and principles in the Agile Manifesto were a reaction to many of the existing project management approaches at that time that were perceived to be overly prescriptive, bureaucratic, and not very effective in a software development environment where the level of uncertainty is typically very high.

Chapter 3 – Scrum Overview:

Scrum has become, by far, the most widely used Agile methodology in the United States and is rapidly being used in other areas of the world as well. In fact, Scrum is so widely used that it means Agile to many people, just as "Coke" is sometimes used generally for carbonated soda drinks.

- Although Scrum is most heavily used in a software development environment, it provides a framework that can easily be adapted to a wide range of projects not limited to software development.

- An understanding of how Scrum works is essential for any Agile Project Manager since it embodies all the values and principles of the Agile Manifesto.

Chapter 4 – Agile Planning, Requirements, and Product Backlog:

Probably the most important area of any Agile process, and the area that is most significantly different from traditional plan-driven projects, is the area of how planning is done and how requirements are defined and managed. An understanding of these differences is probably one of the most critical areas for an Agile Project Manager to have. This chapter provides an overview of Agile planning. It will be discussed in more detail in Chapter 15.

② Agile History and the Agile Manifesto

AGILE EARLY HISTORY

THE EARLY HISTORY OF AGILE was influenced by a number of different trends, including:

- the evolution of new concepts in manufacturing, such as just-in-time and lean manufacturing processes

- new approaches to quality management, including the Total Quality Management movement developed by Dr. W. Edwards Deming.

Those fundamental trends will be discussed in Chapter 11, "The Roots of Agile"; however, the strongest factor that has driven the Agile movement is the unique risks and challenges associated with large, complex software development projects. I can remember a time when a computer that had 16 KB (kilobytes) of memory and a 5-MB (megabyte) disk was a lot. Even before that, I can remember the old days of developing programs on punched cards and paper tape, or even toggling in some binary, assembly language code into the switch register of an early Digital Equipment Corporation (DEC) mini-computer.

In those early days, software development was very limited and primitive; however, computer technology has changed rapidly and significantly since that time. Today, my iPhone Smartphone has thousands of times more power than some of those early computers I worked with in the 1980s, and that has enabled the development of much more powerful and complex software to use that additional processing power. As software development has become much more widespread and has grown into much larger and more complex applications, it has created a number of new challenges for managing large, complex software development projects and a number of different software development methodologies have evolved over the years to meet these new challenges.

Note

While software development has been a key driving force behind Agile and is still heavily associated with Agile today, it should be clearly understood that Agile is not limited to software development and is rapidly spreading to other areas. Any area that has a high level of uncertainty that cannot be easily resolved and/or high levels of complexity is probably a good candidate for an Agile approach.

Dr. Winston Royce and the Waterfall Model (1970)

In 1970, Dr. Winston Royce published a famous paper called "Managing the Development of Large Software Systems."[1] That paper is widely associated with the Waterfall approach as we know it today and outlined an approach for breaking up a large complex software project into sequential phases to manage the effort. The model Dr. Royce proposed in his original 1970 paper is shown in Figure 2.1.

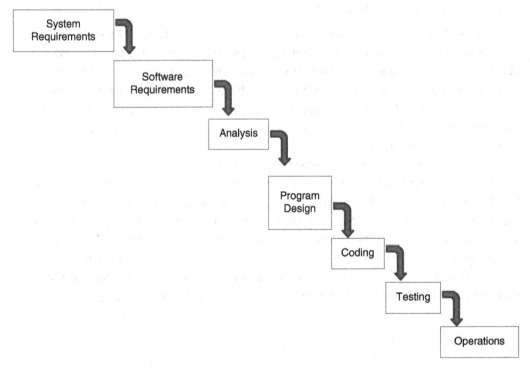

FIGURE 2.1   Original Waterfall model.

Source: Dr. Royce 1970/IEEE.

However, even Dr. Royce recognized the weaknesses in this approach at the time:

> I believe in this concept, but the implementation described above is risky and invites failure. The problem is illustrated in Figure 2.2. The testing phase which occurs at the end of the development cycle is the first event for which timing, storage, input/output transfers, etc., are experienced as distinguished from analyzed. These phenomena are not precisely analyzable. They are not the solutions to the standard partial differential equations of mathematical physics, for instance. Yet if these phenomena fail to satisfy the various external constraints, then invariably a major redesign is required. A simple octal patch or redo of some isolated code will not fix these kinds of difficulties. The required design changes are likely to be so disruptive that the software requirements upon which the design is based, and which provides the rationale for everything are violated. Either the requirements must be modified, or a substantial change in the design is required. In effect, the development process has returned to the origin and one can expect up to a 100-percent overrun in schedule and/or costs.[2]

One of the fundamental problems with the waterfall approach that Dr. Royce recognized was that the entire process was sequential and an error or omission made in one of the very early phases of the project may not be discovered until the very end of the project which might require huge amounts of rework to the work that had already been done in the early phases of the project. For that reason, more iterative approaches began to evolve that featured more incremental and evolutionary development approaches.

Figure 2.2 shows a simplified view of what a Waterfall model might look like. The project is broken up into sequential phases and each phase is terminated by a "phase gate" to review the results of that phase prior to moving on to the next phase.

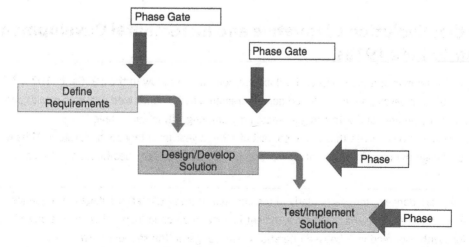

FIGURE 2.2   General phase-gate model

Early Iterative and Incremental Development Methods (Early 1970s)

Walker Royce, the son of Dr. Winston Royce and a contributor to the development of iterative and incremental development (IID) methods in the 1990s, made the following comment regarding his father's thinking:

> He was always a proponent of iterative, incremental, evolutionary development. His paper described the waterfall as the simplest description, but that it would not work for all but the most straightforward projects. The rest of his paper describes [iterative practices] within the context of the 60s/70s government contracting models (a serious set of constraints).[3]

Many of the earliest efforts to break up large projects into incremental and iterative development efforts focused primarily on breaking up the development phase into iterations.

> The next earliest reference in 1970 comes from Harlan Mills, a 1970s' software engineering thought leader who worked at the IBM FSD. In his well-known "Top-Down Programming in Large Systems" Mills promoted iterative development. In addition to his advice to begin developing from top-level control structures downward, perhaps less appreciated was the related life-cycle advice Mills gave for building the system via iterated expansions . . .
>
> Clearly, Mills suggested iterative refinement for the development phase, but he did not mention avoiding a large up-front specification step, did not specify iteration length, and did not emphasize feedback and adaptation-driven development from each iteration. He did, however, raise these points later in the decade.[4]

Further Evolution of Iterative and Incremental Development (Mid-to-Late 1970s)

Support for iterative and incremental development continued to grow in the 1970s. In 1976, Mills wrote: "Software development should be done incrementally, in stages with continuous user participation and replanning, and with design-to-cost programming within each stage."[5]

Larman and Basili write that in the context of a three-year inventory control system, Mills went on to challenge the idea and value of up-front requirements or design specifications. Mills said:

> There are dangers, too, particularly in the conduct of these [Waterfall] stages in sequence, and not in iteration—i.e., that development is done in an open loop, rather than a closed loop with user feedback between iterations. The danger in the sequence [Waterfall approach] is that the project moves from being grand to being grandiose and exceeds our human intellectual capabilities for management and control.[6]

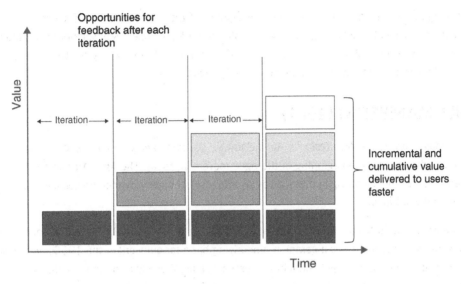

FIGURE 2.3   Incremental and iterative model.

Source: Adapted from Dr. Royce 1970.

Figure 2.3 shows a simplified view of what an incremental and iterative model might look like.[7]

Early Agile Development Methods (1980s and 1990s)

During the 1980s and early 1990s, there was a proliferation of approaches designed to further improve the methodologies for large, complex software development projects:

> Agile methods were direct spinoffs of software methods from the 1980s, namely Joint Application Design (1986), Rapid Systems Development (1987), and Rapid Application Development (1991). However, they were rooted in earlier paradigms, such as Total Quality Management (1984), New Product Development Game (1986), Agile Leadership (1989), Agile Manufacturing (1994), and Agile Organizations (1996).[8]
>
> Agile methods formally began in the 1990s with Crystal (1991), Scrum (1993), Dynamic Systems Development (1994), Synch-n-Stabilize (1995), Feature Driven Development (1996), Judo Strategy (1997), and Internet Time (1998). Other Agile methods included New Development Rhythm (1989), Adaptive Software Development (1999), Open Source Software Development (1999), Lean Development (2003), and Agile Unified Process (2005). However, the popularity of Extreme Programming (1999) was the singular event leading to the unprecedented success of Agile methods by the early 2000s.[9]

Another major influence during this same period of time was the evolution of the Rational Unified Process (RUP), which became another widely used iterative approach. (Variations later emerged from that, including the Enterprise Unified Process (EUP) in 2003 and the Agile Unified Process (AUP) in 2005.)[10]

During that same period, there was early recognition of the need for an Agile Project Management approach and in 1994, the first literature was published on Agile Project Management.[11] By the early 2000s, there was a broad and confusing proliferation of different methodologies. Ultimately, however, they evolved into a much more cohesive Agile approach.

AGILE MANIFESTO (2001)

Many people point to the *Manifesto for Agile Software Development*, or the Agile Manifesto, published in 2001, as the true beginning of the Agile movement today. The Agile Manifesto clearly condensed all of the earlier Agile methodologies into a set of clearly defined values and principles that are still valid today.

> The whole fuss started with the meeting in Feb. 2001 at Snowbird Ski Resort. 17 of us met to discuss whether there was a common, underlying basis for our work in the 1990s around what had been referred to as "light-weight processes." None of us liked the term "light-weight," feeling it was a reaction against something, instead of something to stand for.[12]

One of the most important things to recognize about the Agile Manifesto and the values and principles behind it is that they must be interpreted in the context of whatever project they're applied to. The Agile Manifesto consists of four values and a number of principles that back up those values. The value statements are not meant to be absolute statements at all, and the Agile Manifesto clearly states that; however, in spite of that, many people have made the mistake of dogmatically treating these statements as absolutes. Think about these statements in a broad sense and interpret them in the context of a particular project. If you take that approach, most of these values and principles make sense to be applied to almost any project, not just "Agile" projects.

Agile Manifesto Values

The Agile Manifesto values that were published originally in 2001 are:

> We are uncovering better ways of developing software by doing it and helping others do it. Through this work we have come to value:
>
> - Individuals and interactions over processes and tools
> - Working software over comprehensive documentation
> - Customer collaboration over contract negotiation
> - Responding to change over following a plan.
>
>   That is, while there is value in the items on the right, we value the items on the left more.[13]

Individuals and Interactions over Process and Tools

The first statement indicates a preference for "individuals and interactions over process and tools." This statement is essentially a response to "command-and-control" project management practices that had been perceived as very impersonal and insensitive to people and rigidly defined processes where the "process manages you." From a project management perspective, it calls for a softer leadership approach with an emphasis on empowering people to do their jobs as well as flexible and adaptive processes, rather than a rigid, control-oriented management approach that is highly directive. Agile processes generally depend very heavily on empowered people making intelligent decisions and the power of collaborative teamwork.

The last part of this statement regarding "process and tools" might imply that there is no place in an Agile project for process and tools, but that is certainly not the case:

- Agile uses well-defined processes like Scrum (see Chapter 3), but they are meant to be interpreted and implemented intelligently rather than followed rigidly and mechanically, and they are also designed to be adaptive rather than prescriptive.

- Tools can also play a supporting role, but the important thing is to keep the tools in the right context—they should be there to leverage and facilitate human interactions, not to replace them.

Working Software over Comprehensive Documentation

The second value statement indicates a preference for "working software over comprehensive documentation." This statement was essentially a response to typical phase-gate project management processes that called for extensive documentation deliverables at the end of each phase. There was entirely too much emphasis on producing documentation, to the extent that the documentation took on a life of its own, and there was insufficient emphasis on producing working software.

Note: Although the Agile Manifesto was written around software development projects, there's really no reason why most of the values and principles can't be extended to other non-software products/projects and/or programs if they are used intelligently in the context of whatever products, projects, and programs that they are applied to. For example, this value could read "working products or projects over comprehensive documentation."

One of the problems with documentation is that it can inhibit normal communication. The typical Waterfall project was heavily based on documentation. The project team would develop an elaborately detailed requirements specification, and the software would be tested against meeting that specification. In many Waterfall projects, the end-user didn't even see what was being developed until the final user acceptance testing at the end of the project. This approach presents several opportunities for problems:

- Many users have a difficult time defining detailed requirements upfront in a project, especially in a very uncertain and changing environment.

- Relying too heavily on documentation can lead to significant miscommunications and misunderstandings about the intent of the requirements.

It's important to note that this statement doesn't imply that there should be no documentation at all in an Agile project. The key thing to recognize is that any documentation should provide value in some way. *Documentation should never be an end in itself.*

Customer Collaboration over Contract Negotiation

The third value statement indicates a preference for "customer collaboration over contract negotiation." The typical project prior to Agile has been sometimes based on an "arm's-length" contracting approach. Project managers for many years have been measured on controlling costs and schedules. And, doing that has required some form of contract to deliver something based on a defined specification. Of course, that also requires some form of change control to limit changes in the requirements as the project progresses.

Agile recognizes that, particularly in an uncertain environment, a more collaborative approach can be much more effective. Instead of having an ironclad contract to deliver something based on some predefined requirements, it is better in some cases to create a general agreement based on some high-level requirements and work out the details as the project progresses. Naturally, that approach requires a spirit of trust and partnership between the project team and the end-customer that the team will ultimately deliver what is required within a reasonable time and budget.

Again, it is important to recognize that these values are all relative, and you have to fit this statement to the situation. At one extreme, I have applied an Agile approach to a government-contracting environment. Naturally, in that environment we had to have a contract that called for deliverables and milestones and expected costs, but even in that environment, the government agency understood the value of collaboration over having a rigidly defined, arm's-length kind of contract, and we were able to find the right balance. At the other extreme, you might have a project where the requirements are very uncertain and a much more adaptive approach is needed to work collaboratively with the customer to define the requirements as the project progresses, without much of a contract at all.

Responding to Change over Following a Plan

The final value statement indicates a preference for "responding to change over following a plan." This statement is in response to many projects that have been oriented toward controlling costs and schedules. They made it difficult for the customer to change the requirements in order to control the scope and, hence, the cost and schedule of the project.

The problem in applying that approach to environments where the requirements for the project are uncertain and difficult to specify is that it forces the user to totally define the requirements for a project upfront without even envisioning what the final result is going to look like, and that's just not a very realistic approach in many cases.

In many situations, it is more effective to recognize that, at some level, the requirements are going to change as the project progresses and design the project delivery approach around that kind of change. It's important to recognize that this is not an all-or-nothing decision—to have either completely undefined requirements or highly detailed requirements. There are a lot of alternatives between those two extremes, and you have to choose the right approach based on the nature of the project.

All of these statements are somewhat interrelated, and you have to consider them in concert to design an approach that is appropriate for a particular project. From a project management perspective, this calls for some skill. Instead of force-fitting a project to some kind of canned and well-defined methodology like Waterfall, the project manager needs to intelligently develop an approach that is well suited to the project—that applies to any approach, not just Agile.

Agile Manifesto Principles

The four value statements form the foundation of the Agile Manifesto. The principles in the Agile Manifesto expand on the values and provide more detail. Again, as with the value statements, these statements should also be considered as relative preferences and not absolutes. The principles are summarized as follows and are discussed in more detail in the following sections:

1. Our highest priority is to satisfy the customer through early and continuous delivery of valuable software.
2. Welcome changing requirements, even late in development. Agile processes harness change for the customer's competitive advantage.
3. Deliver working software frequently, from a couple of weeks to a couple of months, with a preference to the shorter timescale.
4. Business people and developers must work together daily throughout the project.
5. Build projects around motivated individuals. Give them the environment and support they need, and trust them to get the job done.
6. The most efficient and effective method of conveying information to and within a project team is face-to-face conversation.
7. Working software is the primary measure of progress.
8. Agile processes promote sustainable development. The sponsors, developers, and users should be able to maintain a constant pace indefinitely.
9. Continuous attention to technical excellence and good design enhances agility.
10. Simplicity—the art of maximizing the amount of work not done—is essential.
11. The best architectures, requirements, and designs emerge from self-organizing teams.
12. At regular intervals, the team reflects on how to become more effective, then tunes and adjusts its behavior accordingly.[14]

> 1. Our highest priority is to satisfy the customer through early and continuous delivery of valuable software.

The first principle emphasizes "early and continuous delivery of valuable software." In many classical plan-driven projects prior to Agile, the end-user customer didn't see anything until the final user acceptance test phase of the project, and by that time it was very difficult and expensive to make any changes that might be needed.

Emphasizing early delivery of the software accomplishes two major goals:

1. It provides an opportunity for the customer to see the software early in the development cycle and provide feedback and input so that corrections can be made quickly and easily.
2. Working software is a good measure of progress. It's much more accurate and effective to measure progress in terms of incremental software functionality that has actually been completed, tested, and delivered to the user's satisfaction rather that attempting to measure the percentage of completion of a very large development project that is incomplete.

It is very difficult to accurately measure progress of a large software development project as a whole without breaking it up into pieces. That can be a very subjective judgment with some amount of guesswork. Breaking up the effort into well-defined pieces that each have clearly defined criteria for being considered "done" provides a much more factual and objective way of measuring progress.

> 2. Welcome changing requirements, even late in development. Agile processes harness change for the customer's competitive advantage.

The second principle emphasizes creating an environment where change is expected and welcomed rather than rigidly controlled and limited; but, of course, that doesn't mean that the project is totally uncontrolled. There are lots of ways to manage change effectively and collaboratively based on a partnership with the customer. The important thing is that the project team and the customer should have a mutual understanding upfront of how change will be managed.

> 3. Deliver working software frequently, from a couple of weeks to a couple of months, with a preference to the shorter timescale.

The third principle emphasizes using an incremental and iterative approach to break up a project into very small increments called *sprints* or *iterations*, which are typically in the range of two to four weeks. There are a couple of reasons why this makes a lot of sense:

1. All Agile development processes, such as Scrum, are based on continuous improvement.
   - Instead of having a rigidly defined process that never changes, the team is expected to take an empirical approach to learn what works and what doesn't work as the project progresses and make adjustments as necessary.

- If the project is broken up into very short increments and learning takes place at the end of each increment, learning and continuous improvement can happen much more rapidly.
- A popular Agile mantra is, "Fail early, fail often." In other words, it's better in many cases to try something quickly and learn from it and make adjustments, rather than taking all the time that may be needed to try to design an approach that is going to work flawlessly the first time.

2. People work more productively given short time-boxes to get things done. If it is done correctly, the team develops a cadence and a tempo that are very efficient for producing defined increments of work quickly and efficiently, like a manufacturing assembly line.

> **4.** Business people and developers must work together daily throughout the project.

The fourth principle emphasizes a partnership approach between the project team and the business sponsors. This is very consistent with the Agile Manifesto value of "customer collaboration over contracts." To implement this principle, both the business sponsors and the project team need to feel joint responsibility for the successful completion of the project. This calls for a much higher level of engagement of the business sponsors than is commonly found in many classical plan-driven projects where the implementation of the project might be almost totally delegated to the project team.

The degree of engagement, of course, should be appropriate to the nature of the project and how the business sponsors are engaged might be different depending on the circumstances.

- For example, Scrum has a role called the *Product Owner* that provides the day-to-day business direction for the project, but the direction might not be limited to that and might also include higher-level business direction and strategy.
- In a large, enterprise-level project, there might be a number of other stakeholders who need to provide input and need to be engaged somehow.

Designing an approach that gets the right people engaged at the right time is very important for making the project successful.

> **5.** Build projects around motivated individuals. Give them the environment and support they need and trust them to get the job done.

The fifth principle emphasizes the importance of properly motivated individuals on a project.

- Too often in the past, some Project Managers have used high-pressure, command-and-control tactics to pressure project teams into delivering results faster.

- Many of us have been involved in "death march" projects in our careers where people are given an absolute deadline for getting something done and have to work nights and weekends if necessary to get it done.

- When you're in an environment that requires high levels of creativity and innovation, that approach just doesn't work very well.

The philosophy of Agile is based on a high level of empowerment and individual initiative by the people on the project. Instead of being told specifically what to do and how to do it and being pressured into doing it to meet deadlines, Agile teams are given general direction and are expected to figure out how to get it done most effectively and efficiently themselves. Making that kind of approach work requires a people-oriented leadership style. However, it doesn't mean that there is no need for leadership whatsoever.

"This principle is an invitation to Project Managers and leaders to take the initiative in adopting an Agile approach with an environment that promotes the psychological safety that every adaptive workplace needs to learn, innovate, and grow into a more lean-agile organization."[15] An Agile Project Manager needs to adapt his or her leadership style to fit the situation and that will typically depend on several factors including the nature of the project and the level of maturity and experience of the team.

> **6.** The most efficient and effective method of conveying information to and within a project team is face-to-face conversation.

The sixth principle emphasizes face-to-face conversation. This is another statement that you have to not take as an absolute but think of it as relative. It is not always possible with distributed teams to have face-to-face communications, but it is certainly desirable *if it is* possible.

This statement also doesn't mean that the *only* form of communication is direct, face-to-face communications.

- It is a reaction to the history of waterfall projects that heavily relied on documented requirements as a way of communication.

- There are many ways to communicate information in various forms, and you need to choose the optimum mix to fit a given situation.

- The right mix will depend on a number of factors, including the scope and complexity of the project, the distribution of the team working on the project, and the level of maturity of the team.

> **7.** Working software is the primary measure of progress.

Measuring progress on a software development project can be difficult and problematic. The classical method is to break a project into tasks and track percent completion of those tasks as a way to measure progress; however, that can be very misleading, because often the list of tasks is incomplete and the level of completion often requires some subjective judgment, which is difficult to make and often inaccurate.

Testing is another factor in this—very often in the past, the entire development process and the testing process might have been sequential.

- The result is that even though the development of the software might have seemed to be complete, you don't know how complete it really is until it has been tested and validated to be complete.

- An Agile approach emphasizes doing testing much more concurrently as the software is developed.

- There is a concept in Agile called the Definition of Done that you will hear quite often. The team should clearly define what "done" means—it generally means that the software has been tested and accepted by the user.

- In other environments, the Definition of Done might be a lot more ambiguous and subject to interpretation. If you don't have a clear Definition of Done, any estimate of percent complete is likely to be suspect.

A more accurate measure of progress is to break up a software project into chunks of functionality where each chunk of software has a clear Definition of Done and can be demonstrated to the user for feedback and acceptance.

In order to measure real progress, we must have something useful and tangible to review; something that meets the team's Definition of Done. Examples of this include a new version of the software, a working prototype incorporated in the finished product, a mockup simulating the end result, etc.[16]

> **8.** Agile processes promote sustainable development. The sponsors, developers, and
> users should be able to maintain a constant pace indefinitely.

Many of the underpinnings of Agile come from Lean Manufacturing and Total Quality Management (TQM). In a manufacturing environment, companies learned many years ago that running a manufacturing plant like a sweatshop and forcing workers to work an excessive number of hours under poor conditions often do not result in high-quality products.

A similar thing is especially true in an Agile environment, because the success of the effort is so critically dependent on the creativity and motivation of the team. In that kind of situation, it is even more important to create an environment where work is sustainable over a long period of time.

This principle "is also about improving collaboration and communication, from one sprint to the next so that the team can keep up a constant pace indefinitely or until the solution/product/software meets the Definition of Done."

Some people interpret this principle thinking that Agile development goes on and on-and-on and that a product will never get done. However, this principle makes us reflect on when will we be done with our finished product/solution; and the answer should be: when we meet the Definition of Done using a customer-centric approach.[17]

---

9. Continuous attention to technical excellence and good design enhances agility.

This next statement is an interesting one. Many people might have the image of an Agile software project team as a bunch of "cowboys" who just get together and hammer out code without much design planning and without any coding standards. That is not usually the case.

Agile recognizes the need for doing things the right way to avoid unnecessary rework later.

- However, an Agile approach should not result in overdesigning or "gold-plating" a product.

- A comment you will hear often in an Agile environment is the concept of "just barely good enough." In other words, the work should be done to a sufficient level of completeness and quality to fulfill the purpose it was intended to fill, and nothing more. Going beyond that level of "just barely good enough" is considered waste.

This principle is also about continuous improvement. An Agile approach is not static. It is meant to continuously improve as the project is in progress both in terms of continuously improving the solution as well as the process used to develop the solution.[18]

---

10. Simplicity—the art of maximizing the amount of work not done—is essential.

This tenth principle emphasizes *simplicity*. How many times have we seen projects spin out of control because the requirements become much too complex and very difficult to implement and the solution becomes overdesigned to try to satisfy every possible need you can imagine?

This is also related to the concept of "just barely good enough"—don't overdesign something; keep it as simple as possible. In some cases, it might make sense to start with something really simple, see if it fills the need, and then expand the functionality later only if necessary. An important concept in Agile is called the *minimum viable product*, which defines the minimum set of functional features a product has to have to be viable at all in the marketplace.

It's generally much more effective to take an incremental and iterative approach to start with something simple and then expand it as necessary, rather than starting with something overly complex that may be overkill for the requirement. An example would be to develop and communicate a vision statement for the project in a succinct and explicit manner which is essential for requirement to unfold and evolve over time.[19]

> **11.** The best architectures, requirements, and designs emerge from self-organizing teams.

Agile is heavily based on the idea of self-organizing teams but that needs some interpretation.

- Sometimes, developers have used the idea of "self-organizing" as an excuse for anarchy, but that is not what was intended.
- The intent is that if you have the right people on a cross-functional team and the team is empowered to collectively use all the skills on the team in a collaborative manner, it will generally deliver a better result than a single individual could deliver acting alone.

> **12.** At regular intervals, the team reflects on how to become more effective, then tunes and adjusts its behavior accordingly.

Agile is adaptive in two respects:

- **Solution Adaptation:** The design is adaptive to uncertain and changing user requirements—it can start with a high-level view of requirements and progressively elaborate requirements after the project is initiated.
- **Process Adaptation:** The process itself is adaptive rather than highly prescriptive. Agile is based heavily on continuous improvement, using short intervals to reflect on what's working and what's not working and taking quick corrective action as necessary. In Scrum, this is called a *retrospective*, and it happens at the end of each sprint.

The team is expected to continuously improve and adapt the Agile process as needed as the project progresses.

SUMMARY OF KEY POINTS

1. Agile History

Agile has evolved over a number of years from a proliferation of approaches going back to the 1980s and 1990s. The Agile Manifesto, which was published in 2001, was a key turning point that defined some principles and values that were needed to simplify and synthesize that diverse array of different methodologies.

2. Agile Manifesto Values and Principles

The Agile Manifesto values and principles provide a strong foundation for understanding Agile at a deeper level. These values and principles were not meant to be applied rigidly; they were intended to be interpreted and applied in the context of a given project and business environment.

That is the essence of an adaptive approach; rather than rigidly applying a fixed set of rules to all projects, it is much better to focus on the values and principles and use good sense to apply them intelligently to fit a given project and business environment.

Many of these values and principles are applicable to almost any project in some way. It's not a matter of having one set of values and principles for Agile projects and another set of completely different values and principles for other projects; it's a matter of applying these values and principles intelligently in the right context to fit the project.

DISCUSSION TOPICS

Agile History

1. What do you think is most significant about the role of the Agile Manifesto in the overall history of the development of Agile principles and practice up to that point?

Agile Manifesto Values

1. What are some of the trade-offs that you think might have to be made in applying the Agile Manifesto values to real-world projects? How would you go about resolving these trade-offs?
2. Provide an example of a project and discuss how you might have applied the Agile Manifesto values to that project.

Agile Manifesto Principles

1. Which two or three of the Agile Manifesto principles do you think are likely to have the biggest impact on the success or failure of a typical project, and why?
2. Are there any Agile Manifesto principles that you think would not be applicable to every project (Agile or not)? Why not?
3. Provide an example of a project and discuss how you might have applied the Agile Manifesto principles to that project.

NOTES

1. Dr. William Royce, "Managing the Development of Large Software Systems," *Proceedings, IEEE Wescon* (August 1970), pp. 1–9.
2. Ibid., p. 2.
3. Vic Basili and Craig Larman, *Iterative and Incremental Development: A Brief History* (IEEE Computer Society Digital Library, June 2003), http://www.craiglarman.com/wiki/downloads/misc/history-of-iterative-larman-and-basili-ieee-computer.pdf, p. 2.
4. Ibid., p. 3.
5. Ibid., p. 4.
6. Ibid., p. 4.
7. Winston Gonzalez, email comments to author on book review.
8. David F. Rico, "The History, Evolution and Emergence of Agile Project Management Frameworks," ProjectManagement.com (February 4, 2013), http://davidfrico.com/rico-apm-frame.pdf, p. 1.
9. Ibid., p. 1.
10. Scott W. Ambler, "History of the Unified Process," Enterprise United Process (2013), http://www.enterpriseunifiedprocess.com/essays/history.html.
11. Jim Highsmith, *Agile Project Management* (Reading, MA: Addison-Wesley, 2010).
12. Blogroll, "10 Years of the Agile Manifesto—It started in 2001 with the Manifesto," posted by Alistair on January 2, 2010, http://10yearsagile.org/it-started-in-2001-with-the-manifesto.
13. http://agilemanifesto.org/.
14. http://agilemanifesto.org/principles.html.
15. Winston Gonzalez, email comments to author on book review.
16. Ibid.
17. Ibid.
18. Ibid.
19. Ibid.

③ Scrum Overview

SCRUM HAS BECOME RAPIDLY ACCEPTED as the most widely used agile approach:

- It provides a good general foundation that can be adapted to fit a very broad range of projects. Scrum is also not limited to software development, but that is where it is most widely used at the current time.

- For that reason, much of the discussion in this book will be focused on software development, but it should be understood that Agile and Scrum are not limited to software development. For example, I used a Scrum approach to plan and organize the writing of this book.

It should be noted that Scrum, in itself, is considered by some to be more of a product development framework rather than a project framework because it does not explicitly include any project planning/initiation/startup activities or project closeout phases/activities that would normally be required when using Scrum in a project context. Although these activities are not explicitly defined by Scrum, they can easily be added to extend Scrum for project work.

Scrum, is, by definition, an *empirical process* as opposed to a "defined and predictive process." The following is how these two types of processes are different:

1. **Empirical process:** The empirical process control model was defined to exercise or control the process via following some frequent adaptations as well as frequent inspections.

   - It works best in situations with high levels of uncertainty where it is difficult, if not impossible, to clearly define the solution in advance and an experimental, trial-and-error approach is needed to converge on an acceptable solution.

   - The term *empirical process* control model is based on the word *empirical*, which means the information is acquired by the means of experimentation and observation.[1]

2. **Defined and predictive process:** "The defined process control model can be thought of as a theoretical approach. When a well-defined set of inputs is given, it is obvious that the same outcomes will be generated every time the program executes. With the well-understood technologies and stable requirements, one can very well predict a whole software project."[2]

In other words, a defined and predictive process model is appropriate if both the requirements for the project and the solution to those requirements can be accurately predicted in advance; however, that is typically not the case in a software development project. That is why an empirical process like Scrum is typically so much more successful in any environment such as software development with high levels of uncertainty.

Scrum is adaptive in two ways:

1. **Solution adaptation:** Scrum is adaptive in the sense of progressively defining the solution. You can start a Scrum project with only a very fuzzy idea of the requirements and the solution and progressively elaborate the requirements and the solution as the project proceeds to ultimately converge on a solution that meets the business need.
2. **Process adaptation:** The process itself behind Scrum is also adaptive. A Scrum project is broken up into short, fixed-length sprints. At the end of each sprint as the project proceeds, the project team pauses to determine if the process is working or needs to be further adapted to fit the nature of the problem.

Ken Schwaber, one of the original developers of Scrum, provides this overview of Scrum:

> I offer you Scrum, a most perplexing and paradoxical process for managing complex projects. On one hand, Scrum is disarmingly simple. The process, its practices, its artifacts, and its rules are few, straightforward, and easy to learn . . . On the other hand, Scrum's simplicity can be deceptive. Scrum is not a prescriptive process; it doesn't describe what to do in every circumstance. Scrum is used for complex work in which it is impossible to predict everything that will occur. Accordingly, Scrum simply offers a framework and a set of practices that keep everything visible. This allows Scrum's practitioners to know exactly what's going on and to make on-the-spot adjustments to keep the project moving toward desired goals.
>
> Common sense is a combination of experience, training, humility, wit, and intelligence. People employing Scrum apply common sense every time they find the work is veering off the path leading to the desired results. Yet most of us are so used to prescriptive processes—those that say 'do this, then do that, and then do this'—that we have learned to disregard our common sense and instead await instructions.[3]

SCRUM FRAMEWORK

Scrum is generally called a *framework* rather than a *methodology* because it is meant to provide a framework for organizing the work rather than a more specific, well-defined, and prescriptive methodology on how the work should be done. It's important to keep that in mind—some people get consumed with implementing Scrum "mechanically by the book" and lose sight of the fact that it is only a framework, and it may take a considerable amount of skill to adapt it to particular situation. A high-level overview of the Scrum framework is shown in Figure 3.1.

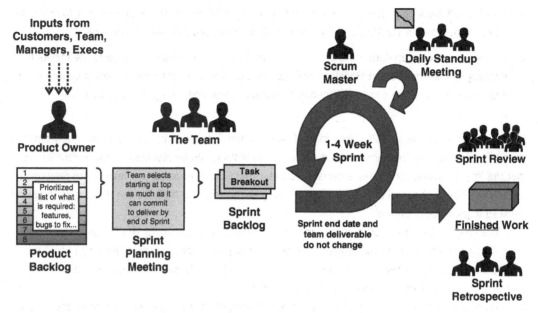

FIGURE 3.1   Scrum framework.

Source: Courtesy of Rally.

Each of the elements of the framework is discussed in more detail in the following sections.

Sprints

The heart of the Scrum process is the *sprint*. A "sprint is typically a fixed-length time-box and is generally from two to four weeks in duration and is where the development work is actually done.

■ Instead of a traditional *schedule-boxing* approach in which the length of a project or an iteration is expanded as needed to be long enough for whatever is being developed, the length of a sprint is fixed and the work to be done is broken up into small increments where no single increment is so big that it cannot be accomplished in a single sprint.

■ Instead of expanding the length of the sprint to fit the work to be done, the challenge is to see how many of these small increments of work can be fit into a single fixed-length sprint.

Product Backlog

The Product Backlog is a queue of the work to be done organized in the form of small increments of work typically in the form of *user stories*. (User stories will be discussed in more detail later; however, they are a very streamlined way of defining a very small and individual user requirement.)

■ A good practice with user stories is to describe the user requirement in very concise terms and also briefly define what the user expects to accomplish with the user story as an acceptance criteria.

- The Product Backlog is dynamic and is continuously prioritized by the product owner to order the items in the Product Backlog to deliver the most business value as early as possible.

- As new ideas come up during the course of the project, they will also be added to the Product Backlog so that the Product Backlog continuously expands and contracts as work is completed and new work is added and is continuously reviewed, approved, and prioritized by the product owner.

The Product Backlog is also used to hold all work that is to be completed, including any defects that are deferred from the current sprint. Normally, defects would be resolved in the current sprint, but the Product Owner can make a decision to defer that work if necessary. However, building up an excessive amount of defects in the Product Backlog is referred to a *technical debt* and should not be allowed to grow out of control.

The Product Owner in Scrum is responsible for defining and managing the Product Backlog on an ongoing basis throughout the project; however, in many situations he/she is assisted in that role by the Scrum Master. If the project warrants it, in some cases, a Business Analyst may also assist the Product Owner with managing the Product Backlog.

There are different ways that the Product Backlog might be managed, depending on the nature of the project:

- In a very adaptive project approach with uncertain requirements, the Product Backlog might be limited to only looking two to three sprints into the future. However, that approach would not provide much of an ability to plan the schedule and costs of a project.

- If there is a need for longer-term planning, naturally, the Product Backlog would need to be defined further out in time; however, it could be limited to a high-level definition only for items that are very far out in time. The Product Backlog will be discussed in more detail in a later section.

A very important step in the Scrum process that is not explicitly shown in Figure 3.1 is Product Backlog Grooming which is also called "Product Backlog Refinement." In Product Backlog Grooming (refinement), the team (or part of the team including the Product Owner) meets regularly to "groom the Product Backlog," in a formal or informal meeting which can lead to any of the following:[4]

- removing user stories that no longer appear relevant

- creating new user stories in response to newly discovered needs

- reassessing the relative priority of stories

- assigning estimates to stories which have yet to receive one

- correcting estimates in light of newly discovered information

- splitting user stories that are high priority but too coarse-grained to fit in an upcoming iteration.

The Product Owner is responsible for Product Backlog Grooming (Refinement); the Scrum Master might facilitate it; expert "leads" (SMEs, developers, QA) sometimes need to opine; and stakeholders need to contribute.

The review and refinement (especially business prioritization and sizing) of the Product Backlog are typically done in parallel *as the team is building* during a sprint. If Product Backlog Grooming (Refinement) is not done, the next Sprint Planning will not have a good Product Backlog ready for the product owner and team to agree on what the next sprint needs to include. Instead, they may have to spend a day or two doing the evaluation, which could delay the start of the next sprint.

Scrum Meetings

Sprint Planning

Sprint planning plays a very important role in Scrum. The Sprint Planning meeting takes place prior to the beginning of every sprint. It has two major goals and is typically broken into two parts to align with those two goals:

1. The goal of the first part is for the Product Owner and the team to negotiate what stories will be taken into the sprint.

   - The Product Owner makes any necessary final revisions in the priority ranking of the potential stories to be taken into the sprint.
   - The team decides how many of those stories can be completed in the sprint based on the level of effort required for each story and the estimated team capacity or velocity.

Note

Velocity is the measure of the throughput of an Agile team per iteration. Since a user story, or story, represents something of value to the customer, velocity is actually the rate at which a team delivers value. More specifically, it is the number of estimated units (typically in story points) a team delivers in an iteration of a given length and in accordance with a given definition of "done."[5]

This part of the planning meeting is essentially a negotiation between the product owner and the team.

2. The goal of the second part is for the team to define the tasks that will be needed to implement those stories and to plan how those stories and tasks will be allocated among the members of the team. The Product Owner typically does not participate in this portion of the Sprint Planning meeting.

The Product Backlog should be groomed and prioritized by the Product Owner prior to the Sprint Planning session; however, final adjustments in the priority ranking are likely to take place in the Sprint Planning meeting as the team and the Product Owner negotiate and plan what can be taken into the sprint and resolve any trade-offs.

The Sprint Planning meeting is typically time-boxed to four hours. Most of that time is typically spent in the second part of task-level planning by the team. During the Sprint Planning meeting, the Product Owner and the team should agree on an overall goal to be accomplished in the sprint in addition to negotiating the stories that will be taken into the sprint. Once that is done, the team typically does the task-level planning portion of the meeting without the product owner.

Daily Scrum

The Daily Scrum meeting is basically a check-in for everyone on the team to coordinate what's going on, monitor progress, and to identify any obstacles that may be inhibiting progress. Note that at one time this meeting was called a "Daily Standup" because at one time, it was customary for the team to stand during the whole meeting to keep the meeting short. It is typically limited to a fixed duration of 15 minutes or less. Traditionally, during the Daily Scrum, each person on the team typically answers three questions:

1. What did you accomplish yesterday?
2. What are you going to accomplish today?
3. What obstacles are in your way?

However, some teams have moved away from this format and may not use these same questions. The meeting is often facilitated by the Scrum Master and it is often done in front of a Scrum board that indicates the progress of the tasks in the current sprint. Figure 3.2 shows an example of a Scrum board. Please keep in mind that there are different ways to show the stages in a Scrum process and this is only an example. Alternatively, for many teams (especially distributed teams), an online Agile Project Management tool is typically used in lieu of a physical Scrum board.

Usually, everyone on the team stands up during these meetings as an incentive to keep the meeting short and focused (that's why it's called a *Standup*). However, it is common to deviate from those rules with distributed teams. With distributed teams, the Daily Scrum may be one of only a few opportunities for communication among the team, so it might be necessary to go beyond the three basic questions, but it is a good idea to still limit the length of the meeting and defer lengthy topics to other meetings.

Sprint Review

The Sprint Review is where the team presents the finished work to the Product Owner for his/her final review and approval. It is essentially a brief User Acceptance Test at the end of each sprint.

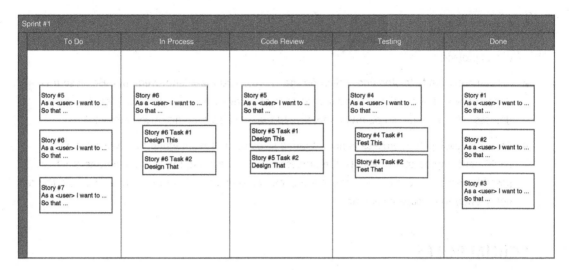

FIGURE 3.2   Example Scrum board

- It should be noted that the Sprint Review should not be the very first time that the Product Owner has seen the results of the software that the team is developing.

- The team should demo and preview the results of their work to the Product Owner as it is being developed during the sprint to get feedback and inputs.

- The Sprint Review is essentially a formal review of all the items in the sprint.

   All defects in the product should typically be resolved prior to the Sprint Review unless the Product Owner has specifically approved deferring the resolution to a later sprint.

- If the Product Owner requests significant changes or enhancements in the stories, those will typically be added to the Product Backlog and deferred until a future sprint.

- Changes to stories that are in progress should not be allowed in the middle of a sprint—changes while the sprint is in progress should be limited to clarification and refinement of user stories only.

   Although the Product Owner has primary responsibility for approving the results of the Sprint Review and should represent the interests of all business users, it is a very good practice to include business users and other stakeholders who might have input into the Sprint Review in these meetings.

Sprint Retrospective

The Sprint Retrospective is an opportunity to review and discuss lessons learned at the end of the sprint where the team looks back, reflects on what went well and what didn't go well, and identifies opportunities for process improvement in the next sprint.

■ This is a very important element of the Scrum process. A Sprint Retrospective is to a sprint in an Agile project as a postmortem (or lessons learned meeting) is to a project in a classical plan-driven project.

■ It is an opportunity to stop and reflect on the process and make improvements. Because it is done at the end of each sprint and sprints take place every two to four weeks, learning and process improvement can be much more rapid than a classical plan-driven project.

■ An Agile development process is not intended to be rigid; it is intended to rely heavily on continuous improvement to adapt the process to the project and the environment as much as possible. Because sprints are short, learning can be very fast so that mistakes and problems are not allowed to propagate very far.

SCRUM ROLES

Scrum is based on some very well-defined roles that are discussed in this section.

Product Owner Role

The Product Owner's role in Scrum is defined as follows:

> The Product Owner is responsible for maximizing the value of the product and the work of the Development Team. How this is done will vary widely across organizations, Scrum Teams, and individuals.
>
> The Product Owner is the sole person responsible for managing the Product Backlog. Product Backlog management includes:
>
> ■ clearly expressing the Product Backlog items
>
> ■ ordering the items in the Product Backlog to achieve goals and missions
>
> ■ optimizing the value of the work the Development Team performs
>
> ■ ensuring that the Product Backlog is visible, transparent, and clear to all, and shows what the Scrum Team will work on next
>
> ■ ensuring the Development Team understands items in the Product Backlog to the level needed.
>
> "The Product Owner may do the above work or have the Development Team do it. However, the Product Owner remains accountable."[6]

The Product Owner represents the Business Sponsor, is a decision-maker for the project, provides direction to the team on what will be built, and prioritizes the work to be done. On a typical Scrum project, there is only one Product Owner; however, on large enterprise-level projects, Product Owners might not act alone. They might gather inputs from multiple stakeholders and, in some cases, need to coordinate with Product Owners of related efforts.

Relationship to Project Management Role

The Product Owner role has many elements of functions that might be considered project management in providing direction to the project, but it goes beyond a typical project management role and includes the domain knowledge to provide the business direction to a project as well. This role is actually very similar to a Product Manager in a product development company. The Product Owner has some responsibilities that might be similar to a project manager, but it is a very different role. In a classical plan-driven project, a project manager doesn't define the requirements of what should be developed and that is what a Product Owner does.

The Product Owner role has some of the attributes of a Business Analyst in collecting and defining requirements, but the role goes well beyond the role that a Business Analyst would be expected to play in being a decision-maker on the product features and capabilities.

Scrum Master Role

The Scrum Master's role in Scrum is defined as follows:[7]

> The Scrum Master is responsible for ensuring Scrum is understood and enacted. Scrum Masters do this by ensuring that the Scrum Team adheres to Scrum theory, practices, and rules.
>
> The Scrum Master is a servant-leader for the Scrum Team. The Scrum Master helps those outside the Scrum Team understand which of their interactions with Scrum Team are helpful and which aren't. The Scrum Master helps everyone change these interactions to maximize the value created by the Scrum Team.
>
> The Scrum Master serves the Product Owner in several ways, including:
>
> - finding techniques for effective Product Backlog Management
>
> - helping the Scrum Team understand the need for clear and concise Product Backlog items
>
> - understanding product planning in an empirical environment
>
> - ensuring the Product Owner knows how to arrange the Product Backlog to maximize value

- understanding and practicing agility

- facilitating Scrum events as requested and needed.

The Scrum Master serves the Development Team in several ways, including:

- coaching the Development Team in self-organization and cross-functionality, particularly in organizational environments in which Scrum is not yet fully adopted and understood

- helping the Development Team to create high-value products

- removing impediments to the Development Team's progress

- facilitating Scrum events as requested or needed.

The Scrum Master serves the organization in several ways, including:

- leading and coaching the organization in its Scrum adoption

- planning Scrum implementation within the organization

- helping employees and stakeholders understand and enact Scrum and empirical product development

- causing change that increases the productivity of the Scrum Team

- working with other Scrum Masters to increase the effectiveness of the application of Scrum in the organization.

The Scrum Master is what is known in Agile as a *servant-leader*:

- He/she has the responsibility to facilitate the team. He/she is not expected to provide a significant amount of direction to the team as a *project manager* would.

- The team is supposed to be self-organizing and empowered so the role of the Scrum Master in providing direction is limited. However, in the real world, many teams are not at a level of maturity to act effectively as a self-organizing and empowered team, and the Scrum Master might have to be somewhat of an Agile Coach to help the team reach that level of maturity.

Relationship to Project Management Role

Some of the responsibilities of the Scrum Master might be considered similar to a project manager. For example, the Scrum Master is expected to track and remove obstacles that are inhibiting the performance of the team. However, it is also a very different role—the Scrum Master is expected only to facilitate the team without providing much direction. There are a couple of different views on this:

- The "idealistic" view is that the team is totally empowered and self-organizing, and very little or no direction of any kind is required.

- A more pragmatic view is that not all teams are at that level of maturity, and some level of leadership is needed. The Scrum Master has to really "fill the cracks" to fit the situation. Examples of cracks to be filled might include:[8]

  - novice team members who need further training and coaching
  - conflict resolution within the team
  - subject matter expertise in certain areas
  - working with business people and management.

The right solution, in my experience, has been to take an adaptive approach to provide whatever leadership is needed to help the team be successful; however, a very important key point is that this is not a classical plan-driven project management role. At the team level, there is no defined role for a project manager in a Scrum project.

Team Role

The role of the team in an Agile project is defined as follows:[9]

The Development Team consists of professionals who do the work of delivering a potentially releasable increment of "Done" product at the end of each Sprint. Only members of the Development Team create the increment.

Development Teams are structured and empowered by the organization to organize and manage their own work. The resulting synergy optimizes the Development Team's overall efficiency and effectiveness.

Development Teams have the following characteristics:

- They are self-organizing. No one (not even the Scrum Master) tells the Development Team how to turn Product Backlog into increments of potentially releasable functionality.

- Development Teams are cross-functional, with all of the skills as a team necessary to create a product increment.

- Scrum recognizes no titles for Development Team members other than Developer, regardless of the work being performed by the person; there are no exceptions to this rule.

- Scrum recognizes no sub-teams in the Development Team, regardless of particular domains that need to be addressed like testing or business analysis; there are no exceptions to this rule.

- Individual Development Team members may have specialized skills and areas of focus, but accountability belongs to the Development Team as a whole.

The team is expected to act as a single entity where all the members of the team act collaboratively and cohesively as a single unit. A team may consist of people with different specialties. For example, there could be people who specialize in development on the team and others who specialize in testing. It might also include business analysts within the team.

There are also different views regarding how a team is made up:

- The idealistic view is that everyone on the team is equally capable of performing any task required by the team.

- The more pragmatic view is that, in the real world, some specialization by skill within the team is beneficial.

  - Having different skills within the team is not inconsistent with having good teamwork, in my opinion. In a sports team, you have people with very different skills and abilities who play different positions on the team that are well aligned with their unique skills and abilities.

  - Those are very different skills, but they can still work together as a well-coordinated, cohesive team. In fact, an argument can easily be made that the team should be cross-functional and having different perspectives within the team is more effective in many circumstances than having a team of people who all think absolutely alike in all situations.

Relationship to Project Management Role

The team also has some responsibilities that might be considered similar to a project manager in a classical plan-driven project. Each member of the team is expected to plan their activities and be accountable for delivering the results that they committed to. In a sense, Agile distributes responsibility downward into the team.

SCRUM VALUES

Having a consistent set of values among everyone who participates in a Scrum team plays a critical role in helping to develop a collaborative, cross-functional approach. When people act from common values, it helps to unify everyone on the team, reduces potential conflict, and provides a solid foundation for guiding the team. The following is a summary of the Scrum values.

Commitment and Focus

The first Scrum value is commitment and focus. As stated in the Scrum Alliance Code of Ethics:

> Commitment is our willingness to dedicate ourselves to a goal and to do our best to meet that goal. Focus means that we concentrate on and are answerable for doing the things that we have committed ourselves to do, rather than allowing ourselves to become distracted or diverted.
>
>   As practitioners in the global Scrum community:
>
> - We take responsibility for and fulfill the commitments that we undertake—we do what we say we will do.
>
> - We make decisions and take actions based on the best interests of society, public safety, and the environment.
>
> - When we make errors or omissions, we take ownership and make corrections promptly. When we discover errors or omissions caused by others, we promptly communicate them to the appropriate individual or body. We accept accountability for any issues resulting from our errors or omissions and any resulting consequences.
>
> - We protect proprietary or confidential information that has been entrusted to us.
>
> - We proactively and fully disclose any real or potential conflicts of interest to the appropriate parties.[10]

Commitment is very important:

- The team needs to agree on a goal and the items that will be accomplished for each sprint during the Sprint Planning meeting and the team should hold itself accountable for successfully completing those items and goals as the sprint progresses.

- The idea of team accountability is a very important notion in Scrum. The whole idea of a self-organizing team relieves a lot of burden on the need for someone to manage the efforts of the team. If it works properly, there is no need for a project manager to provide direction to the team and track progress at the team level.

The idea of focus is also very important—once the items to be completed during the sprint have been agreed to, they should not be radically changed in the middle of the sprint. It is understood that details associated with items will be elaborated during the sprint; however, nothing should be allowed that changes the focus of the team during the sprint.

Openness

The second Scrum value is openness. Being able to be open with others is a sign of emotional maturity and it is essential for developing high performance teams that work collaboratively with each other. We've also seen immature people and teams who have hidden agendas and political goals that might supersede and conflict with the goals of the team.

Once people have worked on Scrum teams and have learned to practice these values, it becomes much easier to assemble a team who has shared values, but it can be difficult to get an immature team with no prior experience in Scrum to this level. The *Scrum Alliance Code of Ethics* says that, as practitioners in the global community:

- We earnestly seek to understand the truth.

- We strive to create an environment in which others feel safe to tell the truth.

- We are truthful in our communications and in our conduct.

- We demonstrate transparency in our decision-making process.

- We provide accurate information in a highly visible and timely manner.

- We make commitments and promises, implied or explicit, in good faith.

- We do not engage in or condone behavior that is designed to deceive others.[11]

Teams typically go through stages of maturity in developing openness with each other. An excellent model of team dynamics is the *Tuckman model* that describes several different stages of team dynamics:[12]

1. **Forming**
   - No prior experience working with others on the team; team members may be somewhat reserved, formal, and polite with each other.
   - Some members of the team may be apprehensive and unsure of themselves with the rest of the team.
   - There is uncertainty about how the team will evolve.

2. **Storming**
   - People on the team begin to openly challenge each other.
   - Some conflict occurs and is essential to get past before the team can progress.

3. **Norming**
   - The team agrees to work together around some defined rules.

4. Performing

- The team reaches the highest level of performance and works naturally and collaboratively and is able to dynamically adjust to new situations.

The theory is that teams need to go through these stages in order to make progress, but it isn't necessarily totally sequential.

- Sometimes teams will move back and forth between stages in order to make progress.

- For example, *storming* and conflict should be viewed as a sign of progress because it is essential for teams to go through that stage before making real progress toward the higher level of "performing" as a truly cross-functional and collaborative team based on openness.

Respect

Showing respect is a very important Scrum value. It is well known that high-performance teams may be composed of people with different opinions. If everyone on the team was a "yes man" and went along with everything others on the team said automatically, the team would probably not be very high-performing. There is a lot of value in hearing different points of view and reaching consensus, and that calls for respect. The *Scrum Alliance Code of Ethics* puts it this way:

> Respect means that we show a high regard for ourselves, others, and the resources entrusted to us. Resources entrusted to us may include people, money, reputation, the safety of others, and natural or environmental resources.
>
> An environment of respect engenders trust, confidence, and performance excellence by fostering mutual cooperation and collaboration—an environment where diverse perspectives and views are encouraged and valued.
>
> As practitioners in the global Scrum community:
>
> - We respect the rights and beliefs of others.
>
> - We listen to others' points of view, seeking to understand them.
>
> - We approach directly those persons with whom we have a conflict or disagreement.
>
> - We conduct ourselves in a professional manner, even when it is not reciprocated.
>
> - We negotiate in good faith.
>
> - We do not exercise the power of our expertise or position to influence inappropriately the decisions or actions of others in order to benefit personally at their expense.
>
> - We do not discriminate against others based on, but not limited to, gender, race, age, religion, disability, nationality, or sexual orientation.
>
> - We do not engage in any illegal behavior.[13]

Courage

The final Scrum value is courage. In my experience, this value becomes important in several different ways:

1. Scrum is difficult and challenging, and doing it successfully requires courage to face and overcome many obstacles.
2. It requires facing up to reality openly and honestly.
3. You need courage to overcome your own personal inhibitions to interact with others. The *Scrum Alliance Code of Ethics* puts it this way:

> Courage means that we have the daring to do the best that we can and the endurance not to give up. We have the determination and resolution to take ownership of the decisions we make or fail to make, the actions we take or fail to take, and the consequences that result.
>
>   As practitioners in the global Scrum community:
>
> - We share bad news even when it may be poorly received.
>
> - We avoid burying information or shifting blame to others when outcomes are negative.
>
> - We avoid taking credit for the achievements of others when outcomes are positive.
>
> - We would rather say, "No," than make false promises.
>
> - We accept the possibility of failure but also know that we can learn from failure and apply those learnings to our next attempt.
>
> - We acknowledge that change is inevitable, that change leads to growth, and that growth guides us toward improvement.
>
> - We admit when we need help and we ask for help.[14]

GENERAL SCRUM/AGILE PRINCIPLES

It's important not to get lost in the mechanics of Scrum and to understand the principles behind it in order to apply the methodology intelligently in a variety of different project and business environments.

> People tend to interpret Scrum within the context of their current project management methodologies. They apply Scrum rules and practices without fully understanding the underlying principles of self-organization, emergence, and visibility and the inspection/adaptation cycle. They don't understand that Scrum involves a paradigm shift from control to empowerment, from contracts to collaboration, and from documentation to code.[15]

Kenneth Rubin has provided a very nice summary of the important principles behind Scrum in his book, *Essential Scrum: A Practical Guide to the Most Popular Agile Process*[16] that are further discussed here. These principles are really not unique to Scrum; they are general enough to apply, to some extent, to almost any project management approach (either plan-driven or adaptive). They just have to be used intelligently in the right proportions based on the nature of the project.

Variability and Uncertainty

An understanding of variability and uncertainty and how to manage them effectively is probably one of the most critical requirements for any Agile Project Manager to master. Kenneth Rubin identifies several important principles associated with managing variability and uncertainty:[17]

- **Embrace helpful variability:** Plan-driven processes are built around controlling and limiting change.

  - In a plan-driven process, any change is considered an exception to the norm and must be controlled in order to manage the scope, costs, and schedule of the project.
  - That can lead to a very inflexible process where you may wind up meeting cost and schedule goals but fail to provide the desired business value because the process wasn't adaptive enough to meet customer needs.
  - This is probably the primary reason Agile has been so successful in a very dynamic and rapidly changing business environment.

      An Agile process is based on easily adapting to customer needs and embracing change; however, change in an Agile process is not totally unrestricted and a sensible approach is still needed to manage changes.

- **Employ iterative and incremental development:** Agile is heavily based on an incremental development; instead of trying to build the entire product all at once, the product is broken up into smaller pieces and built incrementally with an opportunity for feedback and learning after each increment is completed.

- **Leverage variability through inspection, adaptation, and transparency:** Scrum and Agile are heavily based on inspection and adaptation.

  - Not only is the product being built continuously and inspected and adapted as needed to maximize the value it provides to the user, but the process itself used to build the product is also continuously inspected to determine if the process is working optimally and any adjustments are made to the process as necessary.
  - Transparency about both the product and how the process works is essential to accomplish that.

- **Reduce all forms of uncertainty simultaneously:** In any project there are a variety of kinds of uncertainty:

    - End uncertainty deals with uncertainty associated with the features of the final product.
    - Means uncertainty surrounds the process and technologies used to develop the product.
    - Customer uncertainty relates to who the ultimate customer of the product will be and what their needs are.

        A plan-driven project attempts to remove all uncertainty about the product and how it will be built upfront, and that's just not realistic in many situations. An Agile project is built around recognizing, managing, and reducing uncertainty as the project progresses.

Prediction and Adaptation

Many people have tried to implement Scrum dogmatically and rigidly "by the book," when it was intended to be adapted to the nature of the project. An important set of principles behind Scrum identified by Kenneth Rubin is related to prediction and adaptation. The following are the key elements of this principle that he has identified:[18]

- **Keep options open:** Plan-driven approaches frequently try to make all decisions about the requirements upfront to resolve any uncertainty about the project before it starts.

    - Scrum and Agile are based on delaying decisions until "the last responsible moment." That is, in general, you should delay making decisions as long as they can be delayed without impacting the project.
    - The reasoning behind that is that if you delay making decisions, you will typically have better information for making those decisions and ultimately make better decisions.
    - If you try to make decisions too far in advance it will frequently require speculation and many times that speculation will be wrong and the effort may be wasted because that decision will only need to be re-planned later when better information is available.

- **Accept that you can't get it right upfront:** An important principle in Scrum and Agile is to accept that some amount of trial-and-error experimentation may be necessary to find the best solution. This requires a mature attitude to recognize that we "don't know what we don't know."

- **Favor an adaptive, exploratory approach:** There is an economic decision associated with weighing the cost of experimenting and adapting to the results of the experimentation versus the cost of designing the product upfront and the risk of major rework and redesign if it is wrong.

    - In the 1990s, the cost of making changes was significant, which is one of the reasons that led to developing heavily plan-driven approaches.
    - However, tools and technologies have improved greatly since that time, and the cost of exploratory development has come down significantly.

- **Embrace change in an economically sensible way:** Agile and Scrum provide sensible ways to manage change to minimize the economic impact to the project of those changes.

  - With classical plan-driven approaches, the cost of changes can rise significantly later in the project because the impact can be significant.
  - By using an incremental approach to development in an Agile project, the impact of changes can be limited and the costs of making changes can be relatively flat throughout the project.

- **Balance predictive upfront work with adaptive just-in-time work:** In any project, there is always a certain amount of information that is known or can easily be known upfront and some that is much more difficult to determine upfront.

  - It would be foolish to ignore what is already known.
  - A sensible approach is to balance a predictive upfront approach with an adaptive just-in-time approach to resolve uncertainties based on the nature of the project.

Validated Learning

A third area of principles that Kenneth Rubin has identified is related to "Validated Learning."[19] As we've mentioned, Agile is based heavily on an empirical and experimental approach to try things and see what works. To make that approach work, it is important to recognize that we don't have all the answers and we might have to make some assumptions, but it is important to recognize that those assumptions are only assumptions and need to be validated.

- **Validate important assumptions fast:** If there are specific assumptions or decisions that might have a significant impact on a number of areas of the project that might require significant rework if they're not made correctly, it makes good sense to identify those areas and resolve them as quickly as possible to minimize that impact.

- **Leverage multiple concurrent learning loops:** There are multiple concurrent learning loops in a Scrum project that operate at different levels, and it is best to take advantage of all of these different levels concurrently.

  - For example, there is a level of learning that takes place in the daily Scrum meetings and there is also a level of learning that takes place at the end of each sprint in the Sprint Review and Sprint Retrospective.
  - We should take advantage of all these sources of learning.

- **Organize workflow for fast feedback:** We should organize the workflow in the project so that we get feedback as rapidly as possible on the items where feedback is most essential to resolve uncertainties and validate assumptions.

Work in Progress

The next group of principles identified by Kenneth Rubin is related to work in progress, or WIP. The important elements of the principles that he has identified in this area are:[20]

- **Use economically sensible batch sizes:** Classical plan-driven development processes have been based on the principle of economy of scale that comes from a manufacturing environment.

  - In a manufacturing environment, it may make sense to perform a repetitive manufacturing operation on large batch sizes to reduce costs, but there are very different economies of scale associated with a product development process and small batch sizes tend to be more efficient in that environment for a number of reasons:
  - Work can be distributed more evenly and flow through the process is maximized where large batch sizes can cause serious bottlenecks and reduce overall flow.
  - Faster cycle time means feedback is quicker, which results in faster learning and adaptation. Risk is also reduced.

- **Recognize inventory and manage it for good flow:** Manufacturing plants are very much aware of the cost and impact of carrying a large amount of inventory.

  - In a software development process, "inventory" is not as visible, and the impact of carrying inventory is not as well understood.
  - Inventory in a product development process consists of a variety of different kinds of items of work in process including requirements, code, and so on.
  - There is always a cost and a certain amount of overhead associated with managing those items of inventory, so it is best to keep work in process inventory as small as possible to make the overall process more efficient.

- **Focus on idle work, not idle workers:** There are two kinds of waste that are important to recognize in a project:

  - One is *idle worker waste* where workers are not being fully utilized at 100% of their capacity and some component of their capacity is not being used and wasted that might be better utilized.
  - The other is *idle work waste*, which is when work sits idle until people are available to work on it.

    Both of these forms of waste are important, but classical plan-driven projects are typically very heavily focused on idle worker waste and attempt to make sure that everyone in the project is working at 100% of capacity. An Agile project recognizes that both of these forms of waste are important and idle work waste is at least equally important if you want to optimize the flow of work through the process.

- **Consider cost of delay:** A related consideration is the cost of delay. If a project is delayed waiting for resources to be available, the impact on the cost of the project might be significant because everyone on the project might be delayed for some period of time. The example that Kenneth Rubin gives is that delaying the completion of documentation until the end of the project might delay the entire project by two to three months, and there could be a significant cost in that delay.

Progress

Another set of principles identified by Kenneth Rubin is related to progress. He has identified the elements of these principles as follows:[21]

- **Adapt to real-time information and replan:** Classical plan-driven projects are based on conformance to plan, which overlooks the fact that the plan could be wrong, or the plan might need to be adapted to changing conditions as the project progresses. An Agile/Scrum project is much more adaptive and is based on the assumption that the plan will be continuously revised based on learning and information that emerges throughout the process.

- **Measure progress by validating working assets:** Classical plan-driven projects often measure progress by estimating percent complete of the tasks required to complete the project. That is typically based on a very subjective judgment that can be very inaccurate.

    A classical plan-driven project also has some major risks to progress that aren't fully considered in evaluating progress. For example, a task might be complete but if it isn't tested and accepted by the customer, is that really complete? A much better way to measure progress is measuring completed working assets that have been validated and accepted by the customer as they are being built.

- **Focus on value-centric delivery:** Classical plan-driven projects typically perform final integration, testing, and delivery of all features at the end of the project.

    - With this approach, there is a risk that the project will run out of time and money before delivering all of the important value to the customer.
    - Also, we all know that there are many classical plan-driven projects that are bloated and/or "gold-plated" where a large percentage of the features are not really essential. This often results from the customer's belief that if they don't get some item that they might want into the requirements upfront, they may never get it at all.

        Agile works by prioritizing the features to be delivered based on value and developing and by delivering those features incrementally, we can avoid the "all-or-nothing" approach that typically occurs in classical plan-driven projects and at least deliver the most important value to the customer quickly.

        There have been many situations where after some percent of the value is delivered, the customer believes that their essential needs are satisfied, there may be diminishing returns associated with completing the rest of the functionality, and there may be no need to complete the remainder of the project. This can result in a considerable savings in development costs.

Performance

The final set of principles identified by Kenneth Rubin is related to performance. He has identified the elements of these principles as follows:[22]

Go Fast But Never Hurry

An Agile project is based on being nimble, adaptive, and speedy; however, rushing too quickly to get things done may not be the best approach and may have undesirable side-effects such as burning out the people who work on the project and perhaps even compromising the quality of the product.

Build in Quality

In a classical plan-driven project, testing might typically be done in a later phase so the approach to quality might be somewhat reactive and focused on finding and fixing defects well after the development has been completed. There are a number of very significant potential problems associated with that approach:

■ Deferring finding problems until very late in the project, where it might be much more difficult to resolve the problems, can make detecting and resolving individual defects much more difficult.

■ When problems are allowed to accumulate without being resolved, there is a compounding effect that can make it very difficult to isolate and resolve individual problems. It's like trying to untangle a gigantic ball of twine with lots of knots in it.

As an example, I worked at a large electronics company in the early 1990s that spent over $150 million developing a large, complex communications switching system. Much of the testing to integrate all of the components was deferred until the very end, and because it was such a complex system, it became very difficult to isolate individual problems. The whole project was canceled, and the company had to default on delivery commitments to several very large and important beta test customers.

■ There is also a large risk that problems won't be found at all if testing is not directly associated with incremental functionality as it is developed, and testing is delegated to an independent group that doesn't have primary responsibility for developing the product. In that situation, the people doing the testing may not be fully aware of the functionality that has been developed; and, as a result, the testing may be incomplete.

Agile is heavily based on the principles in TQM, which emphasizes a more proactive process to eliminate defects at the source rather than finding and fixing them later. Manufacturing environments learned that lesson a long time ago.

- Years ago, there used to be a lot of quality control inspectors at the end of an assembly line that would inspect products and reject any that were defective.

- That was a costly process, not only in terms of products that had to be discarded or reworked but also in the cost of the many inspectors it took to find those defects.

- Also, any inspection approach like that is based on sampling, and it would be prohibitively expensive to inspect 100 percent of the product, so some defects are bound to get through undetected.

Going upstream in the process and designing the process to be inherently reliable in producing products that were free of defects, companies were able to significantly reduce the cost of QC inspectors and produce much more reliable products. Also, putting the responsibility on the people doing the development of the product to produce quality products that are free of defects rather than relying on an independent group to try to find defects later in the process has a very significant impact.

Employ Minimally Sufficient Ceremony

Classical plan-driven projects tend to be high ceremony, document-centric, process-heavy approaches. Scrum and Agile use a much leaner process where ceremony, documentation, and process are limited as much as possible to only items that contribute value to the project.

- A good example is documentation. There is a common misconception that an Agile project does not produce any documentation. That is not the case, but documentation should not be an end-in-itself. We shouldn't produce documentation for the sake of documentation.

- If a document serves an important purpose and provides value in some way, it should be included in the project, but there are many ways to simplify documentation to make it satisfy the purpose it was intended for more efficiently. For example, if a document is required in some instances (government contracts, audit purposes, etc.), it may be required but can still be minimal.

Here's an example of a streamlined approach to documentation:

- Instead of writing detailed functional specs to define a set of features, some simple user stories can be created.

- Some very succinct acceptance tests can be defined in conjunction with the user stories to better define the requirements that story must satisfy.

- Those very simple user stories and very succinct acceptance criteria can eliminate the need for highly detailed functional specifications and, in most situations, provide a more effective solution.

SUMMARY OF KEY POINTS

1. Scrum Overview

Scrum is based on an empirical process control model; it is adaptive at two levels (1) the definition of the solution is adaptive, and (2) the process itself is adaptive.

2. Scrum Framework

Many people think of Agile as being very loosely defined, with a very minimally defined process—that is not really accurate. The Scrum process is actually very well-defined and requires a good deal of skill and discipline. It is a different kind of discipline in that the process itself is intended to be very dynamic and adaptive. It is generally considered to be a "framework" rather than a methodology because it is not prescriptive.

3. General Scrum/Agile Principles

It's very important to not get lost in the mechanics of how to do Scrum—having a defined methodology is important, but it is intended as a basis for ongoing continuous improvement. Both the process and the project deliverables are expected to evolve with the project based on rapid learning at the end of each sprint.

It's also very important to understand the principles behind Scrum in order to do whatever adaptation may be necessary to apply it to unique projects and business environments.

4. Scrum Values

Values in Scrum are important. Having a consistent set of values among everyone who participates in a Scrum team plays a critical role in helping to develop a collaborative, cross-functional approach. When people act from a common sense of values, it helps to unify everyone on the team, reduces potential conflict, and provides a solid foundation for guiding the team.

DISCUSSION TOPICS

Scrum Overview

1. Explain the difference between an empirical process control model and a defined and predictive process control model. When would each make sense?

Scrum Roles

1. How do you think the overall need for project management changes in a Scrum project, and how would the functions of a traditional project manager be absorbed into the Scrum roles on a typical Scrum team?

2. What are some of the most important differences between Scrum roles and traditional project roles, and why is the shift in roles important in an Agile environment?

Scrum Methodology

1. What is the role of the Product Backlog in a Scrum project? How does it differ from the requirements definition practices in a classical plan-driven project?

2. When is the Sprint Planning meeting held? What are the major parts of a Sprint Planning meeting? What is the output of the Sprint Planning meeting?

3. What is the purpose of the Sprint Review meeting, and when is it held? What is the desired output of the Sprint Review meeting?

4. What is the objective of the Sprint Retrospective meeting, and how is it different from a Sprint Review meeting?

5. Who attends the Daily Standup meeting, and what is its purpose?

General Scrum/Agile Principles

1. Which general Scrum/Agile principle do you think might have had the greatest impact on projects you have been involved in, and why?

2. Discuss a project and how you might have used the general Scrum/Agile principles to do the project differently. Why?

3. Which general Scrum/Agile principles might be the most difficult to implement, and why?

Scrum Values

1. Why are values important? How does having a consistent set of values affect the performance of a Scrum team?

2. Which Scrum value is likely to have the most impact on team performance? Why?

3. If a team is having conflict, what might that indicate? What action would be appropriate to resolve the conflict?

NOTES

1. Blogspot, "Explain Empirical versus Defined and Predictive Process?" April 5, 2012, http://productdevelop.blogspot.com/2012/04/explain-empirical-vs-defined.html.
2. Ibid.
3. Ken Schwaber, *Agile Project Management with Scrum* (Redmond, WA: Microsoft Press, 2004), p. 1.
4. "Backlog Grooming," Agile Alliance, http://guide.agilealliance.org/guide/backlog-grooming.html.
5. "Agile Sherpa–Velocity," VersionOne, http://www.agilesherpa.org/agile_coach/metrics/velocity/.
6. Ken Schwaber and Jeff Sutherland, "Scrum Guide," Scrum.org (July 2013), https://www.scrum.org/scrum-guide.

7. Ibid.

8. Monica Kay, email comments to author on book review.

9. Ibid.

10. *Scrum Alliance Code of Ethics*, http://www.google.com/url?sa=t&rct=j&q=&esrc=s&source =web&cd=1&ved=0CB0QFjAA&url=http%3A%2F%2Fmembers.scrumalliance.org% 2Fresource_download%2F1687&ei=tJTmU5OBL8-dygSO_YLADQ&usg=AFQjCNFUdwoEFFD_ gqPoN4g2j7hzANCqdw&sig2=GdjGuSvPhlf_tYCLOsmXGg&bvm=bv.72676100,d.aWw.

11. Ibid., p. 2.

12. Mind Tools, "Forming, Storming, Norming, and Performing," http://www.mindtools.com/pages/ article/newLDR_86.htm.

13. *Scrum Alliance Code of Ethics*, op. cit., p. 2.

14. Ibid., p. 3.

15. Schwaber, *Agile Project Management with Scrum*, op. cit., p. 26.

16. Kenneth Rubin, *Essential Scrum: A Practical Guide to the Most Popular Agile Process* (Upper Saddle River, NJ: Pearson Education, 2013).

17. Ibid., p. 32.

18. Ibid., pp. 37–39.

19. Ibid., pp. 44–46.

20. Ibid., pp. 48–52.

21. Ibid., pp. 54–55.

22. Ibid., pp. 57–58.

④ Agile Planning, Requirements, and Product Backlog

AGILE PLANNING PRACTICES

PROBABLY THE BIGGEST DIFFERENCE between Agile and more classical plan-driven project management approaches is in the area of planning. There's a misconception that Agile projects do not require planning. That is not the case—they require just as much or more planning; it is just done very differently. This chapter is just a high-level overview of Agile planning. More detail will be provided in a later chapter on Adaptive Planning.

Classical plan-driven project management approaches typically attempt to do more of the planning prior to the start of the project, while Agile Project Management approaches typically defers planning decisions to "the last responsible moment." By *the last responsible moment*, we mean the latest point in time that a decision can be made without impacting the outcome of the overall project.

That is what is called rolling-wave planning

- You typically start with a high-level plan that is sufficient for defining at least the vision, scope, and objectives of the project to whatever level of detail is needed at that point to support whatever level of planning and estimation is required for the project.

- The details of the plan and the requirements are further elaborated as the project progresses. Of course, if the details of the project can be known with some level of certainty prior to the start of the project, there is absolutely nothing wrong with including those details in the plan.

The thinking behind that strategy is that attempting to plan too far in advance naturally involves some amount of guesswork and speculation.

■ Quite often, that speculation is wrong and will result in wasted effort in replanning later, and it might also require reworking any work that has been done based on erroneous assumptions.

■ Deferring planning decisions as long as possible should result in better decisions because more information will be available at that point in time to make those decisions with less speculation.

Rolling-wave planning is not really new to Agile. The general concept has been around for a long time and has been in PMBOK® for some time. However, in actual practice, the plan-driven or waterfall approach typically requires doing more of the planning up-front. Prior to Agile, iterative approaches such as the Rational Unified Process (RUP) have recognized the need for taking a more incremental approach to planning and building a solution. Iterative approaches also deferred detailed planning of each iteration until when that level of detail is needed.

Planning Strategies

Planning is very much related to management of uncertainty in a project, and the planning approach generally attempts to remove uncertainty by more clearly defining the goals and requirements for the project. A plan-driven or Waterfall project attempts to reduce the level of uncertainty associated with the project to a very low level before the project starts, as shown in Figure 4.1. That might be a reasonable approach with some projects, but it's not reasonable to force that kind of approach on a project that has very high levels of uncertainty that can't easily be reduced.

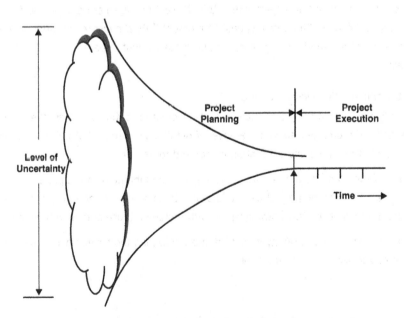

FIGURE 4.1   Typical plan-driven or Waterfall planning approach

Attempting to do that on a project with very high levels of uncertainty forces you to make a number of assumptions about the requirements for the project based on very sketchy and incomplete information, and the quality of those assumptions may be very questionable.

- Sometimes, project managers make assumptions like that, lock them into a requirements document to define the project, and then make it difficult to change those assumptions through a formal method of change control later.

- That creates an illusion of control:

    - On the surface, it looks like the project is very well controlled, but a project can only be as well controlled as the requirements are certain.
    - As a result, you might start the project with what appears to be a very detailed plan that is designed to control the project, and after some large number of change requests later, discover that it wasn't so well controlled after all.

- An Agile approach recognizes that and doesn't attempt to resolve all of the uncertainty associated with a project prior to the start of the project.

A key point I want to make is that the level of planning in an Agile project can be very different, depending on the nature of the project. It's a difference between taking a highly adaptive approach and a totally plan-driven approach. The approach should be adapted to fit the nature of the project. There are a number of factors that influence which approach is most appropriate for a given project, but probably the biggest factor is the level of uncertainty in the project.

For example, if you were building a bridge across a river, it would be ridiculous to say something like, "We're going to build the first span, see how that comes out, and then we'll decide how to build the remaining spans." A typical bridge-building project would normally be very plan-driven and all the requirements, as well as the design of the bridge, would typically be well defined to a fair amount of detail prior to beginning to build the bridge.

On the other hand, what if there was some uncertainty in the project? Suppose that the river called for building a new kind of bridge that had never been built before, or there was some uncertainty about how stable the foundation for the bridge is? You might want to prototype and test the new design on a smaller scale somehow to remove some of the uncertainty and risk before committing to the full-scale development.

Key things to remember

- You should always fit the methodology to the project based on the characteristics of the project and other factors such as the training and capabilities of the project team.

- One of the most important factors that influences the selection of an approach is the level of uncertainty (ambiguity) and complexity (difficulty) in the project.

- It is not a black-and-white decision to be totally plan-driven or totally adaptive. There are plenty of alternatives between those two extremes.

Figure 4.2 shows the general way that an Agile rolling-wave planning approach progressively reduces the uncertainty in a project as the project progresses rather than attempting to remove all the uncertainty up-front. Breaking up the project into releases and iterations means that a lot of the detailed planning can be deferred until that release or iteration is ready for development (as I've mentioned, the approach need not be this sophisticated).

Capacity-Based Planning

A significant difference between the way planning is done in a classical plan-driven approach and an adaptive (value-driven) approach is associated with the way the work to be done is planned and managed:[1]

■ A classical plan-driven project management approach tends to identify a pre-determined structure consisting of the phases, activities, and/or tasks for a project that is expected to be needed to complete the work for the project. In the planning approach, there is a lot of emphasis on

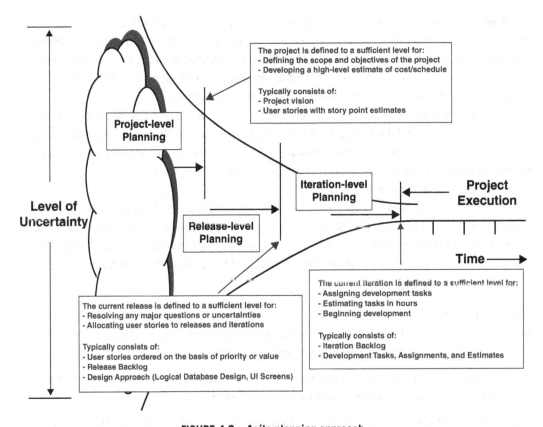

FIGURE 4.2 Agile planning approach

defining and managing the scope and structure of the project in the form of Work Breakdown Structures, Gantt charts, Pert charts, etc. In this approach:

- The resources are assigned to the project and they may not be fixed.
- Resources to do the work are allocated based on the estimated work to be done and the expected availability of resources to do the work.
- The schedule for completing the work is determined based on the expected duration of each of those tasks, given the expected resources available to do the work.
- The overall schedule is expanded or contracted as necessary based on the total time required to complete all the work to be done.

- In an adaptive (Agile) approach, there is less emphasis on defining and managing the structure of the work to be done and more emphasis is put on managing and optimizing the flow of work through the project. That is necessary because the structure of the work may not be known and may change throughout the project; and due to the typical level of uncertainty in the project, a more flexible and adaptive approach is needed. In this approach:

- The teams typically have a fixed number of resources assigned to the project, a fixed number of resources assigned to each team, and the duration of each sprint is also typically fixed.
- Instead of adjusting the number of resources based on the work to be done and adjusting the duration of the sprints to fit the work, the opposite is done.
- The planning becomes "capacity-based" and is focused on how much work can be done in each fixed-length sprint with a fixed number of resources on the teams. The capacity or the team's velocity is determined by the average number of work units that a team can complete in each sprint. As the team completes work from one sprint to the next, the team establishes its velocity or speed to complete the work which may change over the course of the project.
- The overall duration can be determined by the number of sprints that are needed to complete the overall work to be done.

Naturally, the adaptive planning process is much more fluid because the work to be done is not typically well known and is expected to change and be further defined as the project is in progress. After all, an estimate is not much more than just that—an estimate.

Spikes

Many times, there is a lot of uncertainty associated with requirements and/or what the most likely solution to those requirements should be. It's a good technique to isolate and resolve this uncertainty separately from the development effort. Letting too much uncertainty get into the development process can seriously undermine the efficiency of the team.

- For that reason, a technique called a "Spike" is often used in an Agile project to resolve uncertainty. A "Spike" is a special kind of iteration that is used to do research, possibly prototype a solution, and/or evaluate alternative approaches for resolving uncertainty associated with developing a solution.

- Isolating and time-boxing the amount of time dedicated to resolving this kind of uncertainty can help prevent it from significantly slowing down the mainstream development effort.

Note that a Spike may not require a full sprint and the team may elect to limit the time devoted to a Spike.

Progressive Elaboration

The concept of rolling-wave planning is very much related to the concept of *progressive elaboration*, which is an important concept in planning a strategy for the definition of requirements in an Agile project. However, it is not really new to Agile. Progressive elaboration has been a classical plan-driven project management practice for a long time, but Agile emphasizes its use much more heavily.

The key point is that you shouldn't get too bogged down in planning requirements too far in advance because it involves too much speculation and may lead to unnecessary rework if that speculation is wrong. Requirements should only be defined to the extent needed to support whatever decisions or action is required at that particular point in time. For example:

- At the project level, there may be a need to understand the overall scope of the project to evaluate the resources, costs, and schedule required; however, that can typically be done on the basis of a very high-level understanding of the requirements without details.

- At the release level, there may be a need for a little more accuracy in the estimated time and effort to complete a release. That might call for understanding the requirements in a little more detail, but that also should not require a significant amount of detail.

Finally, at the sprint level, there are typically two needs that should be satisfied prior to the start of a sprint:

- The team needs to have a reasonable estimate of the level of effort associated with the requirements to determine if they can fit into the capacity for a sprint.

- The team should not take any requirements into a sprint that have major uncertainties or issues associated with them that might block or delay the development effort.

However, even at the sprint level, many details of how the requirements will be implemented can be worked out while the sprint is in progress without any significant impact on the sprint-level estimates.

Value-Based Functional Decomposition

An important benefit of an Agile approach is increased focus on business value. A good approach is to start with a vision statement that clearly defines the business value that the solution will provide.

A vision statement should be short and succinct. This is an example of a format that can be used for a vision statement to keep it simple:

- For (target customer)

- Who (statement of the need or opportunity)

- The (product name) is a (product category)

- That (key benefit, compelling reason to buy)

- Unlike (primary competitive alternative)

- Our product (statement of primary differentiation)

Once a high-level vision statement has been defined, it is a good technique to use functional decomposition to break down that vision statement into the functionality that will be needed to achieve that overall vision.

- Functional decomposition becomes particularly important on large projects where there could be hundreds of User Stories.

- It provides a hierarchical approach for organizing requirements, which can be an essential technique for prioritizing requirements.

- Without an effective approach for functional decomposition that aligns with an overall value statement, it is very easy to "get lost in the weeds" of individual requirements or User Stories and lose sight of the big picture of the value you're trying to deliver.

In an Agile environment, functional decomposition normally consists of starting with a vision and breaking down that vision into epics and user stories that are necessary to accomplish that vision. Epics and user stories are normally expressed in terms of the business needs that they are expected to satisfy and are described in more detail later in this chapter.

More discussion on Agile functional decomposition is included in Chapter 15.

AGILE REQUIREMENTS PRACTICES

The Role of a Business Analyst in an Agile Project

A Business Analyst (BA) many times plays a key role in a classical plan-driven project in working with the users to analyze and document requirements prior to development. However, since Agile emphasizes direct face-to-face communications between the project team and the users as much as possible with less emphasis on requirements that are documented in detail, there is less of a need for a Business Analyst and the role, if any, is very different.

In an ideal Agile environment, there is no defined role for a Business Analyst; the Product Owner is directly responsible for defining and communicating the requirements to the developers in the form of user stories. However, in the real world, that is not always possible:

- For major projects, the workload on the Product Owner could be very large.

- For complex projects, additional analysis work may be needed to further analyze requirements and to evaluate alternative approaches.

The important points are that:

- The Business Analyst should not become too much of an intermediary, isolating the project team from the Product Owner and the business users and inhibiting communications.

- The Business Analyst should play a supporting role to enable and facilitate more effective communications with the project team.

- The Business Analyst should generally provide a higher level of value-added beyond simply developing requirements documents.

Frequently, the Business Analyst will facilitate discussions between the developers, the Product Owner, and the business users to perform Product Backlog Grooming (Refinement) and to further elaborate and define requirements as the project progresses.

In summary, if a Business Analyst is used in an Agile project, the primary functions are to support the Product Owner by doing the following:

- **Analyze a broadly defined area and use functional decomposition to define high-level Epics and stories to create a well-organized, value-driven framework to provide the required business value in the Product Backlog:** If the stories follow a logical functional hierarchy, it provides a mechanism for better understanding the relationship of the stories and Epics to each other and for satisfying overall business goals.

- **Write individual stories that are very clear and concise and are easy to understand and implement by the project team:** Writing effective User Stories is a skill that is often taken for granted. What is often overlooked in good stories is the "why" or the "so that" clause that expresses the business value the story is intended to provide. A good Business Analyst can provide that perspective that is difficult for a developer to provide.

- **Identify related user stories and epics, grouping them into a logical structure as necessary and documenting the interrelationship and associated business process flows as necessary:** The interrelationship of User Stories and Epics should be well-understood and the implementation of stories across different functional areas may require some planning and coordination so that they are consistently implemented. This overall framework can provide a mechanism for easily identifying any inconsistencies and/or missing functionality.

- **Integrate the needs of various stakeholders to produce an overall solution:** This is more crucial on large projects or programs, where there might be a need to integrate the needs of related projects as well as the needs of a number of different stakeholders.

"Just Barely Good Enough"

A common problem in many classical plan-driven projects is that requirements can become bloated. One of the main reasons that happens is that users feel that if they don't ask for everything that they could possibly need and get it into the requirements, they might not get it at all.

Also, we all know that in many situations in the past, project teams have gone beyond the basic requirements and "gold-plated" a solution. Excellence was perceived as going above and beyond the user requirements to develop a more complete solution but going far beyond the user needs to develop a solution isn't necessarily a good thing and can significantly increase the scope and complexity of the development effort.

It may seem to be counter-intuitive to classical plan-driven project management thinking, but a very important principle in Agile is simplicity and limiting a solution to what is "just barely good enough" to solve the problem.

- There is an optimum point, and beyond that point, adding additional features has diminishing value.

- It requires close collaboration with the users to find where that optimum point is.

- A good technique is to start with the simplest and most basic solution possible and then add to it incrementally and iteratively only to the extent that it adds value to the user.

An important element in making this work is a spirit of trust and partnership between the users and the project team. The users have to trust that if they don't ask for something specific and it is discovered to be important as the project progresses, that it can easily be added at that point if necessary.

Differentiating Wants from Needs and the "Five Whys"

Another very good technique in developing requirements for Agile projects is to differentiate "wants" from "needs."

- "Wants" tend to be associated with a solution that a client envisions.

- "Needs" tend to be associated with the problem.

It is typical in many instances for a user to tell you what they think they *want* for a solution before the problem the solution is intended to solve is clearly defined. You should never lose sight of the problem to be solved before you go too far into implementing a solution. It is also a good practice to dig a little deeper into the problem to be solved to get to the root cause of the problem to be solved.

The "Five Whys" method is a good approach for digging into a customer need to get to the root of the problem. The idea is that by progressively asking "why" over and over again, you eventually get to the root cause of the problem. The following shows an example of how the "Five Whys" technique can be used to get to the root of a problem:[2]

1. *Why is our client, Hinson Corp., unhappy?* Because we did not deliver our services when we said we would.
2. *Why were we unable to meet the agreed-upon timeline or schedule for delivery?* The job took much longer than we thought it would.
3. *Why did it take so much longer?* Because we underestimated the complexity of the job.
4. *Why did we underestimate the complexity of the job?* Because we made a quick estimate of the time needed to complete it and did not list the individual stages needed to complete the project.
5. *Why didn't we do this?* Because we were running behind on other projects. We clearly need to review our time estimation and specification procedures.

MoSCoW Technique

Another very simple technique for prioritizing requirements is called MoSCoW, which is defined as follows:[3]

- *Must have*—Requirements labeled as "*MUST*" have to be included in the current delivery time box in order for it to be a success. If even one "*MUST*" requirement is not included, the project delivery should be considered a failure.

- *Should have*—"*SHOULD*" requirements are also critical to the success of the project but are not necessary for delivery in the current delivery time box.

- *Could have*—Requirements labeled as "*COULD*" are less critical and often seen as *nice to have*.

- *Won't have*—"*WON'T*" requirements are either the least-critical, lowest-payback items, or not appropriate at that time.

An example of how this approach has been used successfully is the case study of General Dynamics, UK, which is discussed in Chapter 29. In that situation, General Dynamics, UK, was faced with the challenge of managing a large fixed-price government contract, and the only way to do it successfully was to work with the government customer to prioritize the requirements and eliminate some of the requirements that weren't absolutely essential to the solution using the MoSCoW method.

Note that MoSCow is only an example of a simple method of prioritizing requirements. There are many other alternative methods for prioritizing requirements. Some of those are discussed in more detail in Chapter 13.

USER PERSONAS AND USER STORIES

User Personas

A *User Persona* is used in Agile for personalizing user requirements; however, it really is a general process that can be used in any kind of development effort.

■ The idea behind a User Persona is that many traditional requirements management practices are impersonal; they result in a large documentation with lots of specific requirements that the system must meet, but it is not clear who those requirements are designed to satisfy.

■ Agile puts a strong emphasis on providing value to the user and also has a strong emphasis on engaging the user directly in the project to provide feedback and inputs as the project progresses.

■ Organizing the project requirements around specific users and their needs puts more focus on really understanding the value that the project provides and *who* the recipient of that value is.

To facilitate that process, it is useful to identify User Personas as specifically as possible so that the project team can target their development efforts at the needs of those specific users. A User Persona could be a specific category of user, or it could even be a specific user, but in either case, it is useful to model that user's personality and specific interests as a hypothetical User Persona. A User Persona helps the team visualize that user and focus its efforts on the user's needs.

The following is an example of a User Persona:[4]

■ **User:** Fred Fish, Director of Food Services: "Get me out of the office and into the kitchen."

■ **Background:** Fred is Director of Food Services for Boise Controls, a mid-sized manufacturer of electronic devices used in home security systems. He uses a computer, but he's a chef by trade and not so *computer-savvy*.

■ **Key Goals:** As a manager, Fred doesn't get his hands (literally) dirty the way he used to. He stops in at all the Boise Controls sites and sticks his fingers into things once in a while to stay in touch with cooks and cooking. He wants to learn computer tools but not at the expense of managing his kitchens. A computer is just another tool for getting his administrative tasks done.

■ **A Usage Scenario:** At the start of every quarter, he meets with the head chefs and plans out the next quarter's menus. That's one of his favorite things, because each chef gets to demonstrate a new meal. The chefs spend time in the kitchen exploring each new dish. When they're done, he sends the food to his staff and his manager.

He's not a computer whiz. On a good day, he can drag in some clip art and do some formatting with fonts. Once in a while, he'll format menus with the new editor on his MacBook Pro.

User Stories

User Stories are a succinct way of defining requirements in Agile. In a strict sense, they don't really define a requirement in the same level of detail that a plan-driven project requirement might be described. They're really primarily a short-hand way of defining a business need. Telling User Stories is a way of simplifying the definition of the requirements in a language that can easily be understood by both developers and users. It breaks the requirements into small chunks of functionality that can be built incrementally. User Stories follow the general format shown below:

As a <role> I want <to be able to do something> so that <benefit>

A User Story consists of some standard parts:

- The first part identifies *who* the actor (or role) is that needs to do something. This should be as specific as possible to clearly identify who the story needs to satisfy.

- The next part identifies succinctly *what* the user (role) needs to do so that the end result is very clear.

- The third part is the "so that" clause. This clause provides some insight into *why* this functionality is needed. Knowing this additional insight should help the developers get a deeper understanding of the user's need. (What business value does it provide?)

   Here are two examples:

   As a student I want to purchase my monthly parking passes online so that I can save time.

   As a student, I want to be able to pay for my parking passes via a credit card to avoid using cash.

User Story Advantages

There are several key advantages of a User Story:

- It provides a standardized and concise way of defining a requirement in simple language that is easy for everyone to understand (both developers and users).

- It encourages defining requirements in small, bite-sized chunks where the functionality to be developed is clearly defined and can be completed within one sprint.

   An important aspect of User Stories is that they are written in functional terms—they define an expected result and don't tell the developer *how* to design the software to achieve that result.

- User Stories are intentionally designed to be brief and don't typically provide a detailed description of what is required. They are intended to be a "placeholder for a conversation."

- That detail should take place through face-to-face communications; however, many times acceptance criteria are included in User Stories to more clearly define the expected result.

Characteristics of Good User Stories

There's a mnemonic that is well-known in Agile called "INVEST" that is a useful way of defining the essential characteristics of good User Stories.[5]

- **Independent:** Stories should be as "independent" as possible so that they can be worked on in any order. That will simplify the flow of stories through development and avoid bottlenecks that can be caused by having too many dependencies among stories.

- **Negotiable:** Stories should be "negotiable"—a story is a placeholder for conversation, and some dialog is expected to take place to explore trade-offs associated with developing the story as efficiently and as effectively as possible.

- **Valuable:** Stories should be "valuable"—the value that a story is intended to produce should be clearly defined so that the product owner can make an objective evaluation of the level of effort required versus the value to be gained from the story.

- **Estimable:** "Estimable" is the next important characteristic. A story needs to be sufficiently defined so that the team can develop a high-level estimate of the effort required for the story in story points.

- **Small:** Stories should be relatively "small" so that functionality can be developed and tested as incrementally as possible. Breaking the work into small chunks allows it to flow much more smoothly and allows the work to be distributed more evenly among the team while large efforts can easily lead to bottlenecks and inefficiencies in distributing work among the team.

- **Testable:** Finally, stories should be "testable" to determine if they have successfully fulfilled the value proposition that they are intended to fulfill.

It's good practice with Agile stories to write acceptance test criteria, along with the stories to define the tests that they need to fulfill. The test criteria essentially take the place of more detailed specifications for what the story must do.

Job Stories

A derivative of User Stories that some people prefer is called a "Job Story." "Some Agile practitioners feel that User Stories don't always convey what the user wants to get accomplished at any given circumstance. So, Job Stories are sort of alternative to a User Story, where a Job Story focuses on a triggering event that would lead to the action desired by the user. For this reason, Job Stories are written in the following format, describing the situation, motivation, and expected outcome."[6]

- When <something occurs>,

- I want <my product/solution to do something>,

- So that I can <accomplish what I want>.

"Some Developers feel that Job Stories give them more specifics when developing a software product that depends on event-driven logic and that is when Job Stories come in handy."[7]

Epics

An Epic is basically a very large User Story. An Epic serves the purpose of associating related individual User Stories with a higher-level purpose that they are collectively intended to fulfill, but an Epic is normally too large for the project team to work on directly without breaking it down into individual User Stories.

An Example of an Epic

It's a useful technique on large, complex projects for organizing User Stories into some kind of structure so that the interrelationship of User Stories is well understood. The following shows an example of a large Epic and how it can be broken down into smaller stories.

> **Epic**
>
> As an electronic banking customer, I want to be able to easily make an online deposit of a check into my bank account through my smart phone so that I can save the time required to send a check for deposit through the mail and I can have the money immediately credited to my checking account as soon as the deposit is completed electronically.

Stories

- As an electronic banking customer, I want to be able to scan an image of the front and back of a check into the smart phone so that it can be deposited electronically.

- As an electronic banking customer, I want to be able to enter the deposit information associated with an electronic deposit so that the correct amount will be deposited into the correct bank account when the electronic deposit is processed.

- As an electronic banking customer, I want to be able to electronically submit a scanned check and deposit information to the bank for deposit so that I can save the time associated with sending deposits by mail.

- As an electronic banking customer, I want to be able to receive confirmation of a completed electronic deposit so that I will know that the deposit was successfully processed.

Why Epics and User Stories Make Sense

Epics and User Stories are very efficient ways to describe the business need that a product or project solution is intended to fulfill. It's important to note that they are both intended to be short and succinct and focus on the business need, not how that business need is implemented in terms of a product or project solution.

That's a major difference from classical plan-driven projects where you might typically see a hierarchy of detailed specifications about how a product or project solution is implemented. That difference streamlines the whole project management and development effort because it is relatively quick and easy way to describe a product or project solution and it leaves the details of how it is implemented to the development team.

More discussion on Epics is included in Chapter 15.

PRODUCT BACKLOG

What Is a Product Backlog?

The Product Backlog was previously discussed in Chapter 3. It is a repository for Product Backlog Items (PBIs) and typically consists of Epics, User Stories, and other work to be done. It is dynamic and is continuously being refined and reprioritized as the project is in progress. It is essentially a queue of work to be done and the User Stories in the Product Backlog are continuously groomed and prioritized over the course of a project. There are different ways to implement a Product Backlog and there are different schools of thought on how it should be done.

- At one extreme, there is a view that the project should be totally adaptive, and you shouldn't try to define the Product Backlog any more than two to three sprints into the future.

- That approach might work fine for some projects, but a problem with that approach is that it doesn't provide a basis for planning the overall scope, costs, and schedule of a project.

- If there are some expectations about the cost and schedule of the project, you may have to define the Product Backlog more completely, at least at a high level to evaluate the scope of the project to support those estimates. However, the ability to do that is obviously related to the level of uncertainty in the project.

You should develop an approach for planning the requirements and managing the scope of the project that is appropriate for the project. I would say that the most significant factors in selecting an approach are:

1. The level of uncertainty in the project and the difficulty of defining the requirements for the project upfront.
2. The need to estimate some kind of schedule and costs for the project.

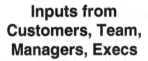

Product Owner

```
1    List of
2    requirements
3    prioritized by
4    business value
5    (highest value
6    at top or list)
7
8
```

**Product
Backlog**

FIGURE 4.3   Product Backlog flow.

Source: Courtesy of Rally.

Figure 4.3 shows the flow of information into the Product Backlog. Product Owners are responsible for defining and maintaining the Product Backlog; however, they might be assisted in that effort by the Scrum Master or others.

Product Backlog Grooming (Refinement)

There is a very general approach for thinking about how to organize the Product Backlog. Think of it as an "iceberg" in which the items in the iceberg gradually make their way to the top as they become ready for development, as shown in Figure 4.4.

The principles of rolling-wave planning are used, and items start out at the bottom of the backlog as being very roughly defined and the stories are progressively "groomed" as they move closer to development.

- Grooming (Refinement) is an important part of any Agile project, but it is often overlooked and not planned for as much as it should be.

- Product Backlog Grooming (Refinement) should be done continuously throughout the project in parallel with the development activities.

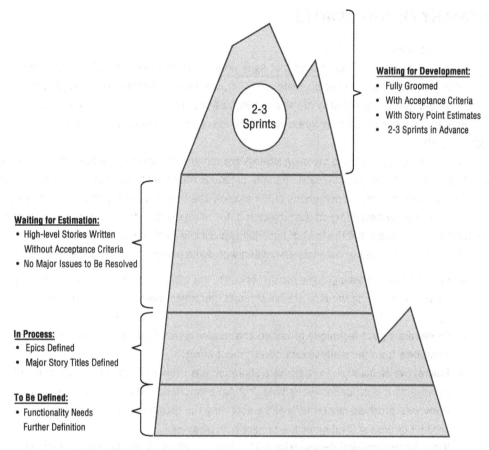

Waiting for Development:
- Fully Groomed
- With Acceptance Criteria
- With Story Point Estimates
- 2-3 Sprints in Advance

2-3 Sprints

Waiting for Estimation:
- High-level Stories Written Without Acceptance Criteria
- No Major Issues to Be Resolved

In Process:
- Epics Defined
- Major Story Titles Defined

To Be Defined:
- Functionality Needs Further Definition

FIGURE 4.4   Product Backlog Grooming (Refinement) flow

- A good technique is for the project team to allocate some time in each sprint so that the queue of stories to be developed never runs empty.

Agile is based heavily on "just-in-time" planning—obviously, you never want a project team waiting around for stories to be developed so you have to manage the grooming (refinement) process to get stories ready for development as needed; but, on the other hand, you don't want to work any farther in advance than necessary, because some of that effort might only need to be repeated once the stories are closer to being ready for development.

It is perfectly reasonable to develop a high-level backlog far enough out in time to develop an estimate of the scope, costs, and schedule of the project.

- It is a judgment call of how far it is reasonable to try to plan into the future.

- Some will say that it's a waste of time to plan any more than two to three sprints in advance, but if you have a need to estimate the costs and schedule of a project, you have to define the Product Backlog further out in time at a sufficient level to support those estimates and then elaborate the details later.

SUMMARY OF KEY POINTS

1. Agile Planning Practices

There is a common misconception that Agile projects are not planned at all. That is not the case, it is just a different kind of planning, and the planning is spread out over the project rather than attempting to do the majority of the planning upfront prior to the start of the project.

Rolling-wave planning and progressive elaboration of requirements are both used to plan Agile projects.

The planning approach and planning strategy are very much related to the level of uncertainty in a project and the planning approach attempts to reduce the level of uncertainty to an acceptable level based on the nature and complexity of the project. The level of planning in an Agile project can vary significantly depending on the project and it is always best to fit the planning approach to the nature of the project and the level of uncertainty associated with the project.

There are several useful concepts associated with Agile planning:

- **Capacity-based planning:** Agile uses a very different planning and management approach based on managing the flow of work through the project based on some fixed capacity limits.
- **Spikes** are a good technique to isolate and resolve a significant area of uncertainty separately from the main-stream development effort.
- **Progressive elaboration:** Progressive elaboration has been a classical plan-driven project management practice for a long time, but Agile emphasizes its use much more heavily. It involves progressively refining and elaborating the detailed requirements for the product or project solution as the project is in progress.
- **Value-based functional decomposition:** This is a very efficient way to plan a project by starting with a short and succinct vision statement and then breaking down that vision statement into epics and user stories that will be needed to provide the overall functionality to satisfy the business need.

2. Agile Requirements Practices

Prioritization of requirements is extremely important in an Agile project and requires a spirit of trust and collaboration between the business users and the project team to work together to do the following:

- Deliver the highest-value items first.
- Preserve simplicity of the solution.
- Avoid going beyond what is needed to satisfy the fundamental business need.

There are several useful topics associated with Agile requirements practices:

- The role of an Agile Business Analyst:
  - There is no defined role for a Business Analyst in an Agile project as there is in a plan-driven or Waterfall project and the Product Owner is ultimately responsible for

defining and prioritizing the project requirements; however, the Product Owner might be assisted in that role by a Business Analyst.

- If a Business Analyst is engaged in an Agile project, the Business Analyst should not become too much of an intermediary and isolate the project team from the Product Owner and users and inhibit communications.

- The Business Analyst should play a supporting role and facilitate more effective understanding of business requirements between the business users *and the project team.*

- "Just barely good enough":

    - In a classical plan-driven project, users often ask for everything you could possibly imagine in a product or project solution because if they don't get it into the requirements, they may not get it at all. That can lead to products and project solutions that have "bloated" functionality that no one really uses.

    - A better approach is to start with something that is "just barely good enough" to satisfy the fundamental business need and only expand it with additional features and capabilities that add value.

- Differentiating wants from needs:

    - It is typical in many instances for a user to tell you what they think they want for a solution before the problem the solution is intended to solve is clearly defined.

    - You should never lose sight of the problem to be solved before you go too far into implementing a solution. It is also a good practice to dig a little deeper into the problem to be solved to get to the root cause of the problem to be solved. The "Five Whys" technique is a good tool for doing that.

- MoSCoW technique:

    - The MoSCoW technique is a simple method of prioritizing requirements. There are many other alternative methods for prioritizing requirements. Some of those are discussed in more detail in Chapter 13.

3. User Personas and User Stories

- User Personas:

    - A User Persona is used in Agile for personalizing user requirements; however, it really is a general process that can be used in any kind of development effort.

- User Stories

    - Are a succinct way of defining requirements in Agile.

    - Simplify the definition of the requirements in a language that can be easily understood by both developers and users.

- Break the requirements into small chunks of functionality that can be built incrementally.
- Follow the general format: *As a <role> I want <to be able to do something> so that <benefit>.*

- Epics
    - An Epic is a large user story.
- Product Backlog
    - The Product Backlog is a very important tool in an Agile project for maintaining a prioritized queue of project requirements.
    - It should be dynamic and should be continuously groomed as the project progresses.
    - Agile projects generally use an "iceberg" strategy for grooming (refining) the Product Backlog. The items that are near the top of the "iceberg" and are closest to going into development should get the most attention.
    - There should typically be about two to three sprints worth of stories at the top of the backlog that are well-groomed and ready to go into development in order to avoid a situation where the project team is waiting for work to do.
    - However, the Product Backlog Grooming (Refinement) does not need to be limited to two to three sprints in advance, and it is a judgment call of how far it is reasonable to plan into the future.

DISCUSSION TOPICS

Agile Planning Practices

1. What is rolling-wave planning? How is it different from a classical plan-driven project to an Agile project? Why would it make sense to use rolling-wave planning?

2. How would you develop a planning strategy for an Agile project? What factors would you consider in developing the strategy?

3. What is progressive elaboration, and why does it make sense in an Agile project?

4. What is "capacity-based planning"? How is it different in an Agile project versus a classical plan-driven project?

5. What is a spike, and how is it used? Provide an example.

Agile Requirements Practices

1. What is the role of a Business Analyst in an Agile project? When would a BA typically be needed and why?

2. Explain the concept of value-based functional decomposition and how it would be used. Provide an example of a real-world project and how that concept would be used in the project.

3. What is the "Five Whys" technique, and how would it be used? Provide an example of how it might be used from a real-world project.

4. What is the MoSCoW technique, and how is it used? Provide an example of how it might be used.

User Personas and Stories

1. What is a User Persona? Give an example of a User Persona in a project. How does it help to define the requirements?

2. What are the advantages of using User Stories to define requirements in an Agile project? Write an example user story for a project that you have been engaged in.

3. What are the characteristics of well-written User Stories? Provide an example of a User Story that is not well written and that violates these characteristics.

Product Backlog

1. What purpose does the Product Backlog serve? What are the characteristics of a well-organized Product Backlog?

2. Who is primarily responsible for the Product Backlog in an Agile/Scrum project?

NOTES

1. Winston Gonzalez, email comments to author on book review.
2. "5 Whys: Getting to the Root of a Problem Quickly," MindTools.com, http://www.mindtools.com/pages/article/newTMC_5W.htm.
3. MoSCoW Method, http://dsdm.org/content/10-moscow-prioritisation.
4. A Sample Persona, http://creativebeatle.files.wordpress.com/2010/12/sample-persona-from-interaction-design.pdf.
5. Bob Hartman, "New to Agile? INVEST in Good User Stories," *Agile for All*, May 14, 2009, http://www.Agileforall.com/2009/05/14/new-to-Agile-invest-in-good-user-stories/.
6. Winston Gonzalez, email comments to author on book review.
7. Ibid.

PART 2

Agile Project Management Overview

AGILE IS CAUSING US TO broaden our vision of what a *project manager* is and that will have a dramatic impact on the potential roles that a project manager can play in an Agile project. In fact, the role of a project manager at a team level in a typical Agile/Scrum project is undefined. That will cause us to rethink many of the things we have taken for granted about project management for a long time to develop a broader vision of what an *Agile Project Manager* is. Some of the project management functions might be integrated into other roles and performed by someone who is not a dedicated project manager and who doesn't have the title of "Project Manager."

Chapter 5 – Agile Development, Quality, and Testing Practices:

You might ask: "Why does a Project Manager need to know something about development practices?" In the past, the role of a project manager might have been somewhat limited to a coordination function to integrate the efforts of the different functional organizations that played a role in the project.

In many cases, the actual functional direction for the different functions involved (development, test, etc.) came from the functional managers who were responsible for those functions themselves, and the role of the project manager in providing functional direction may have been limited.

An Agile environment is different—cross-functional teams are much more integrated, and an Agile Project Manager needs to be a strong, cross-functional leader.

Chapter 6 – Time-Boxing, Kanban, and Theory of Constraints:

The typical way of modeling classical plan-driven projects is based on a fairly statically defined model composed of work-breakdown structures, Pert charts, Gantt charts, etc. An Agile approach is very different and is primarily oriented around managing the "flow" of requirements and tasks through an integrated development process. The tools that are most related to an understanding of managing "flow" are discussed in this chapter and include time-boxing, Kanban, continuous flow, and theory of constraints.

Chapter 7 – Agile Estimation:

An Agile estimation approach is very different from a classical plan-driven project estimation approach:

- A classical plan-driven project estimation approach is based on attempting to accurately pin down the requirements, costs, and schedule for the project upfront before the project starts and it is like a contractual commitment with the customer.

- An Agile estimation approach is based on much higher levels of uncertainty and that level of uncertainty needs to be openly and transparently shared with the customer based on a collaborative spirit of trust and partnership with the customer rather than an "arm's-length" contractual relationship.

Chapter 8 – Agile Project Management Role:

Agile is causing us to broaden our vision of what project management is, and that will have a dramatic impact on the potential roles that a Project Manager can play in an Agile project.

The image of a Project Manager is typically very heavily focused around a plan-driven development approach. That role will likely change dramatically in an Agile environment, and the formal role of a project manager, as we've known it, may even be eliminated, in some situations.

What is important; however, is not to preserve the role of a project manager, as we've known it, but learning to apply the discipline of project management in a much broader context and learning to blend classical plan-driven principles and practices with more Agile and adaptive principles and practices in the right proportions to fit a given situation.

Chapter 9 – Agile Communications and Tools:

In my opinion, an Agile Project Manager who doesn't know how to use one of the widely used Agile Project Management tools is equivalent to a traditional project manager who doesn't know how to use Microsoft Project.

- Agile tools are fundamentally different from traditional project management tools like Microsoft Project.

- The fundamental difference is that traditional tools like Microsoft Project are heavily oriented around defining and managing the structure of project activities while Agile tools are more oriented around managing the flow of project activities through a much more fluid structure.

Chapter 10 – Learning to See the Big Picture:

An Agile Project Manager needs to understand the "big picture" behind Agile to fully realize how it works and why it makes sense. The key to that is to develop a systems thinking approach to understand Agile principles behind the Agile practices at a deeper level. Knowledge of Complex Adaptive Systems is also useful to understand the "big picture" behind Agile.

Chapter 11 – The Roots of Agile:

An Agile Project Manager needs to understand Agile at a deeper level in order to apply it to different situations effectively. In order to develop a systems thinking approach, it is valuable to understand the roots of Agile and know how Agile thinking evolved. There are two significant areas that had a huge impact on development of an Agile approach:

- Total Quality Management had a strong influence on the quality management approach in Agile

- Lean Manufacturing had a significant influence on the way an Agile process works from a process efficiency perspective.

⑤ Agile Development, Quality, and Testing Practices

YOU MIGHT ASK: "Why does a project manager need to know something about development practices?"

- In the past, the role of a Project Manager might have been somewhat limited to a coordination function to integrate the efforts of different functional organizations that played a role in the project.

- In many cases, the actual direction for the different functions involved (development, test, etc.) came from the managers who were responsible for those functions themselves, and the role of the Project Manager in providing direction may have been limited.

   An Agile environment is different:

- Instead of a relatively loosely knit team of people from a variety of functional departments who might only work on a particular project on a part-time basis, an Agile team is typically dedicated to a project and should be much more tightly integrated.

- The methodology for an Agile project is much more of one well-integrated project methodology with all the functions (development, test, etc.) working more collaboratively and concurrently. Any decisions about how development and testing is done need to be integrated with the overall project management approach.

   Those factors require a much higher level of cross-functional leadership. Agile Project Managers can help provide that leadership if they have the cross-functional knowledge that is required, but it is more than a simple coordination function.

AGILE SOFTWARE DEVELOPMENT PRACTICES

This section provides an overview of some of the most important Agile software development practices that an Agile Project Manager should be familiar with.

Code Refactoring

In order to improve the reliability and maintainability of the software, code refactoring involves:

- removing redundancy
- eliminating unused functionality
- rejuvenating obsolete designs
- improving the design of existing software.

Agile approaches tend to emphasize quickly creating code to meet functional requirements *first* and then refactoring the code as necessary *later* to clean it up. Refactoring throughout the entire project life cycle saves time and increases the quality of the software.

In a classical plan-driven project, there is typically an emphasis on laying out the entire design and architecture of the project early in the project to avoid unnecessary rework later. In an Agile project, that approach may not be very realistic, for a couple of reasons:

1. It's very difficult in an Agile approach to define all the requirements in a sufficient level of detail at the beginning of the project to lay out the entire design upfront.
2. An Agile approach encourages changes as the project progresses in order to better adapt to user needs and requirements and attempting to define and control the design approach early in the project may restrict that ability to make changes.

The question then becomes—how do you avoid having code that is a complete mess and is unreliable and difficult to maintain as a result of so many unplanned changes?

- Fortunately, modern development tools have made it easier to implement code refactoring as the project progresses to more easily rework and clean up the code to make it more maintainable and reliable in an Agile environment.

- The use of design patterns and coding standards can also make this approach more practical to implement.

The truth is that code refactoring is just as important in a classical plan-driven project, but it often hasn't been done adequately because it was assumed that the design of the code was defined and stabilized upfront, and that just isn't a very realistic assumption in many cases, even in a non–Agile environment. The strengths and benefits of a code refactoring approach are that:

Strengths and benefits

- It encourages developers to put the primary focus on the functionality provided by the code first and clean it up later to make the code better structured and maintainable.

- It reduces the time required to produce functional code that can be available for prototyping and user validation.

Risks and limitations

- The amount of rework of the code required might be significant.

- The structure of the code might not be optimized around the most desirable architectural approach.

- Time pressures to complete the iteration may short-circuit the code-refactoring effort and allow poorly organized code to be released.

Continuous Integration

Continuous integration is the practice of frequently integrating new or changed software with the code repository and performing overall system integration testing throughout the project rather than deferring that effort until the end of the project.

- Continuous integration provides a way of early detection of problems that may occur when individual software developers are working on code changes that might conflict with each other.

- In many typical software development environments, integration might not be performed until the application is ready for final release.

I was involved in a large software development project in the early 1990s where a major electronics company invested over $150 million in the development of a very complex hardware/software switching system and the project had to be abandoned in the end because when it was time to integrate the project, the various components of the system could not be integrated, and the overall system would not work reliably. That is a perfect example of the critical importance of continuous integration.

Strengths and benefits

- Developers detect and fix integration problems continuously.

- Early warning of broken/incompatible code and of conflicting changes.

- Constant availability of a "current" build for testing, demo, or release purposes.

- Immediate feedback to developers on the quality, functionality, or system-wide impact of code they are writing.

- Frequent code check-in pushes developers to create modular, less-complex code.

- Metrics generated from automated testing and Continuous Integration focus developers on developing functional, quality code, and help develop momentum in a team.

Risks and limitations

- It might be difficult to implement on larger code development projects where the integration effort may be too complex to do as frequently.

- It requires a level of sophistication and close teamwork on the part of the team to make it work especially on large, complex projects.

- Resources and tools to automate the Continuous Integration, building, and testing process are essential and can be expensive.

Pair Programming

Pair programming is another Agile development process that is sometimes used to improve the quality and reliability of software. Pair programming is analogous to the way a pilot and a copilot fly a commercial airliner. In a commercial airliner, one of the pilots flies the airplane while the other pilot oversees the overall flying of the aircraft and provides support to the person who is actually doing the flying. Pair programming in a software development environment works in a similar fashion.

- One developer typically writes the code, and the other developer provides overall guidance and direction as well as support and possibly mentoring to the person writing the code.

- This technique might be used by two peer-level developers who rotate between the two roles, or it might be used by a senior-level developer to mentor a more junior-level developer.

- Pair programming is not as widely used as some other Agile development practices because the economics of dedicating two programmers to work together on the same task can't always be justified. In some cases, code reviews are substituted for pair programming to achieve a similar effect.

Strengths and benefits

- Design quality:
  - One developer observing the other person's work should result in better quality software with better designs and fewer bugs.
  - Any defects should also be caught much earlier in the development process as the code is being developed.

- The cost of a second developer that is required may or may not be at least partially offset by productivity gains.

- Learning and training. Sharing knowledge about the system as the development progresses increases learning.

- Overcoming difficult problems. Pairs are able to more easily resolve difficult problems.

Risks and limitations

- Work preference:

  - Some developers prefer to work alone.
  - A less-experienced, less-confident developer may feel intimidated when pairing with a more experienced developer and might participate less as a result.
  - Experienced developers may find it tedious to tutor a less-experienced developer.

- Costs:

  - The productivity gains may not offset the additional costs of adding a second developer.

Test-Driven Development

Test-driven development is a somewhat widely used Agile development practice. It is typically used in conjunction with unit testing by developers.

- Rather than writing some code first and then, as a second step, writing tests to validate the code, the developer actually starts by writing a test that the code needs to pass to demonstrate that it does the functionality that was intended and then writes code to make that test succeed.

- Of course, the test fails initially until the functionality has been implemented, and the objective is to do just enough coding to make the test pass.

- Development is done incrementally in very small steps—one test and a small bit of corresponding functional code at a time.

Strengths and benefits

- It encourages the development of small, incremental modules of code that can be easily tested and integrated with a Continuous Integration process

- It is well suited to testing of the design as it progresses and provides immediate feedback to the developers on how well the design meets the requirements

- It also encourages developers to write only the minimal amount of code necessary to pass a given test

Risks and limitations

- Test-driven development primarily addresses only unit testing of code modules—much more testing is typically needed at different levels of the application.

- Test-driven development emphasizes rapidly implementing software to provide a minimum level of required functionality and relies on later refactoring the code to clean it up to meet acceptable design standards. If that refactoring isn't done, the code might not be reliable and supportable.

Extreme Programming (XP)

At one time, Extreme Programming (commonly referred to as XP) was a more important Agile methodology, but XP is primarily just a collection of Agile development practices consisting of:

- User Stories
- Release planning
- Iteration planning
- Test-driven development
- Collective ownership of code
- Pair programming
- Continuous integration
- Ongoing process improvement.

It doesn't provide a management framework for how an Agile project should be managed, as Scrum does; it is more of a development process. Scrum does provide that management framework; and as a result, has become much more dominant and a more widely used Agile process. Sometimes XP might be used to define the development process used with Scrum, but more often than not, people take more of an *à la carte* approach to selecting the development processes that they use with Scrum.

AGILE QUALITY MANAGEMENT PRACTICES

For the same reason that cross-functional Agile Project Managers need to understand development practices, they also need to understand quality management practices.

Key Differences in Agile Quality Management Practices

Table 5.1 shows some of the key differences between the Agile approach to quality management and the approach typically used for quality management in more classical plan-driven projects.

TABLE 5.1   Key differences between Agile and traditional quality management

	Typical traditional plan-driven (Waterfall) quality management process	Typical Agile quality management process
Integration of testing with development	Testing is typically done sequentially with development and is done by a separate organization	Testing is done more concurrently with development and is well-integrated into the development team
Testing approach	There is a more reactive approach to finding and correcting defects	There is a more proactive approach to preventing defects
Responsibility for quality	Responsibility for the quality of the software is perceived as on the QA organization	The overall team has responsibility for the quality of the software
Regression testing	Because testing is deferred until the end of the project and code has stabilized, there may not be a need for a complicated regression testing approach	Because testing is done concurrently with development and the code is constantly changing, it is more essential to do regression testing as the project progresses

Definition of "Done"

A very important concept in Agile is the "Definition of Done."

In Agile, the goal of each sprint is to produce a "potentially shippable" product. Of course, "potentially shippable" can mean different things to different people, depending on the nature of the project.

In a large, enterprise-level project, it might be impossible or impractical to really release something to production at the end of each sprint, for a lot of reasons. For example, the results might not be sufficiently complete to deliver a complete subset of functionality that is required, or additional integration work might be needed to integrate the software with other related applications before it is fully released to production.

The important thing is that the team should have a clearly defined definition of what "done" means so that it is unambiguous and well understood by everyone on the team. One of the biggest difficulties on a software project is that there could be different interpretations of what "done" means.

- Is it when the coding is complete?

- Is it when the coding and testing are complete?

- Does it depend on user acceptance?

Stating the requirements for something to be considered "done" more clearly and crisply removes a lot of subjectivity of making an assessment of whether something is "done" or not and makes it clear what the team is committing to when it makes a commitment to complete some number of stories in a sprint.

The Definition of Done is defined by the team and it is typically defined as a checklist of items that need to be completed for a work item to be considered "done." The definition of "done" typically includes:

- Code review

- Tested by developers and QA

- Accepted by the product owner and other stakeholders if necessary.

The important thing is that the team takes complete responsibility for the quality of the software it develops and unambiguously defines its own standards of completeness—quality isn't someone else's responsibility.

The Role of Quality Assurance (QA) Testing in an Agile Project

That leads right into the question of: "What is the role of QA in an Agile project?" As with many things in an Agile environment, there is not necessarily a single right/wrong way of doing things, and there are different schools of thought on this:

- There's a somewhat idealistic view that everyone on the team should be capable of doing anything required (development or testing), and all developers should be totally responsible for the quality of the work that they produce rather than relying on someone else on the team to test it and provide feedback.

- In actual practice, that view is very difficult to implement, and there is value in having people with focused skills within a project team. There is also value in having an independent tester look at software other than the person who developed it.

Some people would say that having people focused on specific tasks and skills within a team inhibits teamwork. I don't agree with that point of view. The analogy I like to use is of an athletic team—on a high-performance athletic team you have people who are highly trained and skilled in playing particular positions on the team, but that doesn't inhibit them working together as a cohesive team.

- A good team should be cross-functional and collaborative and having people on the team representing different perspectives will ultimately strengthen the team.

- QA is a discipline that needs focus and training to do it well. To give it to a developer as just another thing for them to do might not be the best use of resources, and there's always value in having an independent observer look at someone else's software. A trained and specialized QA tester will typically see things that the primary developer might have taken for granted or overlooked.

The important points are:

- The team, as a whole, owns responsibility for quality—*it is not someone else's responsibility*.
- QA should be an integral part of the team, not a separate organization external to the team.
- QA testing is an important skill, and it is often worthwhile to have people on the team who are specialized and skilled in QA testing who are dedicated to planning and executing QA tests.

AGILE TESTING PRACTICES

Concurrent Testing

The biggest difference in Agile testing practices is that testing is integrated with development in each sprint in a very collaborative process where testers and developers take joint responsibility for the quality of the product that they produce. In actual practice, there are a number of ways that can be done:

For example, instead of waiting until the end of the sprint to turn over all stories to QA for testing, a better practice would be to have the testers begin testing the software as soon as it is sufficiently complete for testing.

That approach allows more concurrency of development and testing and avoids any last-minute surprises at the end of the sprint.

Acceptance Test-Driven Development

Acceptance test-driven development is an excellent approach that is used in Agile projects. It is very similar to test-driven development, but it is at a higher level of functionality. Test-driven development is done at the level of unit testing to validate the implementation of code, while acceptance test-driven development is done at a higher level of functional testing and tests the features and behaviors of the system that are observable by the user.

Acceptance test-driven development involves writing the acceptance tests for the functionality that must be provided in each story prior to starting development. This is usually done as part of the writing or grooming (refinement) of the stories, and it helps to build a more concise, common understanding of exactly what needs to be done to satisfy the user need for each story.

A big advantage of acceptance test-driven development is that it can eliminate the requirement for writing detailed functional specifications, it simplifies the requirements definition process, and it keeps everyone on the team clearly focused on the functionality that must be provided to satisfy the business need.

Repeatable Tests and Automated Regression Testing

Because testing is done concurrently with development in an Agile project and new functionality is being continuously added, it is essential for tests to be repeatable in an Agile project to ensure that new functionality or other development activity has not broken something that was already tested and completed. As any item of new functionality is added, a regression test is needed to repeat the testing of any items of functionality that were previously tested to ensure that nothing has been inadvertently broken by adding the new functionality.

As Agile projects get larger, it becomes impractical and perhaps even impossible to repeatedly manually run a very large suite of regression tests that could be in the hundreds or thousands of tests so there is a big advantage to automating regression tests so that they can be run repeatedly and efficiently on a frequent basis.

In an ideal case, automated regression tests can be run nightly to check the entire system as new functionality is being added. It would be very difficult, if not impossible, to do that frequently without automation for a complex system with a large number of individual regression tests to be run. And, of course, automating the regression testing frees up QA test resources to focus on more value-added tasks, and also ensures that it is done consistently each time it is run.

Value-Driven and Risk-Based Testing

Another consideration in planning the testing strategy in any project is to recognize that it is impossible to test 100% of everything you could possibly test, and some method should be developed to determine and prioritize what testing should be done and what testing can be minimized or eliminated. Fortunately, an Agile project makes that easier to do because there is a lot more direct communications with the users to determine what is important and what is not important, and the user can be directly engaged in testing the functionality of the software as it is being developed.

SUMMARY OF KEY POINTS

It's important for an Agile Project Manager to have a broad, cross-functional view of all aspects of a project including development and testing practices to allow him/her to ensure that the people, process, and tools in the project are very well integrated to maximize the overall efficiency and flow of the project and that they are also well aligned with achieving the business goals the project is intended to accomplish.

Agile Software Development Practices

An Agile project calls for a different approach for managing the development process that should be well integrated and well aligned with the overall Agile project management approach.

1. Code Refactoring

- It is often not possible to completely lay out and stabilize the design of a project early in the project.

- A more dynamic approach for evolving the design as the project progresses is needed.

- That also necessitates cleaning up the design as it is in progress to keep the software well-organized and well-structured.

2. Continuous Integration

Because the design is potentially changing throughout the project, a rigorous approach is needed to manage how changes are merged into the design, emphasizing Continuous Integration to ensure that the overall reliability of the design is still maintained.

3. Pair Programming

Pair programming is sometimes used in Agile projects to coach and mentor less-experienced developers and to provide higher levels of code quality and reliability. However, it is difficult to justify the economics of pair programming on all Agile projects, and sometimes a more limited form of code reviews can accomplish some of the same goals.

4. Test-Driven Development

Test-driven development is often used in Agile projects to simplify the development effort and to improve the reliability and quality of the software.

5. Extreme Programming (XP)

Extreme Programming (XP) is primarily a set of development processes, and it does not specify a project-level framework as Scrum does. Sometimes Extreme Programming (XP) might be used to define the development process used with Scrum, but more often, people take more of an à la carte approach to selecting the development processes that they use with Scrum.

Agile Quality Management Practices

1. Key Differences in Agile Quality Management Practices

The philosophy and approach for managing quality in an Agile project are very different from a classical plan-driven project.

- The most important differences are that testing is an integral part of the development effort and is done concurrently with the development effort.

- The team, as a whole, needs to own responsibility for quality—it is not someone else's responsibility.

- The orientation is also more proactive toward preventing defects rather than a typical reactive approach of finding and fixing defects after the fact.

2. Definition of "Done"

The definition of "done" is very important in an Agile project. It provides a very well-defined standard of completeness that the team commits to. If it is properly done, it removes a lot of subjective judgment of whether a development effort is really complete or not.

3. The Role of QA Testing

QA testing plays a different role in an Agile project. QA should be an integral part of the team and not a separate organization external to the team; however, that doesn't necessarily mean that developers should do their own QA testing. QA testing is an important skill, and it is many times worthwhile to have people on the team who are specialized and skilled in QA testing who are dedicated to planning and executing QA tests.

Agile Testing Practices

1. Concurrent Testing

Instead of waiting until the end of the sprint to turn over all stories to QA for testing, a better practice would be to have the testers begin testing the software as soon as it is sufficiently complete for testing. That approach allows more concurrency of development and testing and avoids any last-minute surprises at the end of the sprint.

2. Acceptance Test-Driven Development

Acceptance test-driven development is very similar to test-driven development, but it is at a higher level of functionality. A big advantage of acceptance test-driven development is that it can eliminate the requirement for writing detailed functional specifications, it simplifies the requirements definition process, and it keeps everyone on the team clearly focused on the functionality that must be provided to satisfy the business need.

3. Repeatable Tests and Automated Regression Testing

Because testing is done concurrently with development in an Agile project and new functionality is being continuously added, it is essential for tests to be repeatable in an Agile project to ensure that new functionality or other development activity has not broken something that was already tested and completed.

- As Agile projects get larger, there is a big advantage to automating regression tests so that they can be run repeatedly and efficiently on a frequent basis.
- In an ideal case, automated regression tests can be run nightly to check the entire system as new functionality is being added.

4. Value-Driven and Risk-based Testing

It is impossible to test everything. Fortunately, an Agile project makes it easier to determine what is important and what is not so important, and the user can be directly engaged in testing the functionality of the software as it is being developed.

DISCUSSION TOPICS

Agile Software Development Practices

1. Why is it important for a Project Manager to have some understanding of development practices in an agile project?

2. What is continuous integration, and why is it important in an Agile project?

3. What is test-driven development, and how is it different from acceptance test-driven development?

Agile Quality Management Practices

1. What is the biggest advantage of an Agile quality management approach over a conventional non-Agile testing approach? How is it different? What do you think it would take to make it work effectively?

2. What does the definition of "done" mean, and what is its importance in an Agile project? Provide an example of a definition of "done."

3. What is the role of a QA tester in an Agile project? How is it different from a conventional, non-Agile project?

Agile Testing Practices

1. Why is it important that tests are repeatable in an Agile project?

2. Explain what automated regression testing is and why it is important in an Agile project.

6 Time-Boxing, Kanban, and Theory of Constraints

THE TYPICAL WAY OF MODELING classical plan-driven projects is based on a fairly statically defined model composed of work-breakdown structures, PERT charts, Gantt charts, etc. Figure 6.1 shows an example Gantt chart and Figure 6.2 shows an example Pert chart.

Those modeling tools work fine for a heavily plan-driven approach where the requirements as well as the overall design approach for meeting those requirements can be defined prior to the start of the project. However, they can be very difficult or impractical to apply in a more dynamic and adaptive project approach where the requirements are much more uncertain and difficult to define upfront. In that kind of project, it is very difficult and perhaps impractical to define and manage the structure of the project, and it becomes far more important to manage flow.

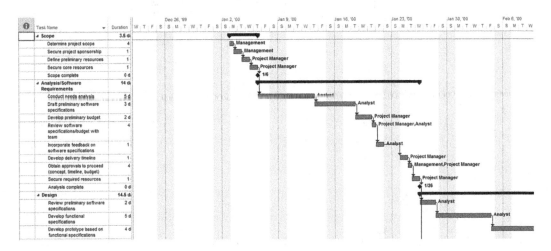

FIGURE 6.1   Example Gantt chart

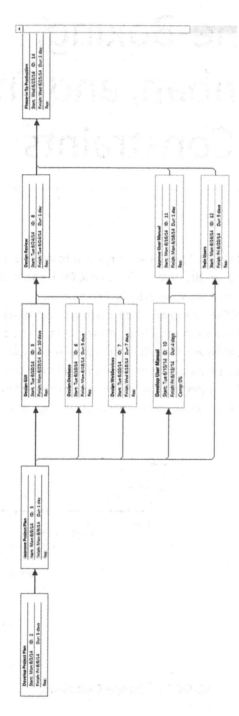

FIGURE 6.2　Example PERT chart

THE IMPORTANCE OF FLOW

Agile Project Management approaches are heavily based on a more dynamic model and on optimizing flow in order to maximize the efficiency and the throughput of the project team rather than statically managing the structure of the project tasks to achieve control and predictability.

■ The idea of flow originated with Lean Manufacturing and has been used in manufacturing to streamline production processes.

■ Flow is one of the most important Lean principles to understand to maximize the efficiency of any process.

There are a number of factors that contribute to maximizing flow.

Small Batch Sizes

The first factor to consider to optimize flow is small batch sizes. Think of trying to push a number of balls through a pipe having a fixed diameter that limits the capacity you can push through the pipe. Large balls can block the pipe and cause bottlenecks while a greater number of smaller balls are more likely to flow smoothly through the pipe. For that reason, Agile is heavily oriented on the idea of breaking up work into small chunks of effort to optimize flow.

Just-In-Time Production

Maximizing the efficiency of a process is also dependent on having the right materials available in just the right amounts just in time to start production.

■ In a manufacturing operation, it is wasteful to have huge inventories of raw materials waiting to start production, and it is also wasteful to not have a sufficient level of raw materials to start production.

■ Either of those situations results in waste or inefficiency and doesn't contribute to maximizing flow.

■ In a manufacturing process, flow is dependent on having the right raw materials that are needed available just-in-time to start production.

■ In a product development process, requirements are equivalent to raw materials and flow is dependent on having requirements defined just-in-time to start development.

Concurrent Processing

Concurrent processing is another major factor that contributes to maximizing flow. If activities can only be done sequentially and there is a bottleneck in one phase of the process, it will inhibit the overall flow in the process as a whole. For example, if development and testing are done sequentially and there is a limited number of QA testers, the overall flow may be limited by the availability of testers. If, on the other hand, the process is designed around doing testing and development concurrently and those resources are more coordinated in a much more collaborative fashion, there is likely to be a much more efficient overall flow to the process.

This chapter will discuss several topics related to flow:

- *Time-boxing* plays a role in developing a cadence for optimizing the product development flow based on short-fixed length sprints. Having a regular cadence and a relatively stable velocity is essential for Agile estimation.

- *Kanban* also provides an approach for optimizing WIP (work-in-process) flow.

- The *Theory of Constraints* is a well-known method developed by Eliyahu Goldratt for resolving bottlenecks that may inhibit flow.

Flow is one of the five Lean principles described in the 1990 *Lean Thinking* book by Womack and Jones.[1] We'll discuss the Lean principles in more detail in Chapter 11, Table 11.6.

TIME-BOXING

Classical plan-driven project managers are used to working with an approach called schedule-boxing or scope-boxing, which means expanding the size of the schedule interval to accommodate the size of whatever task is being performed. That method might work if you can accurately size the scope of the requirements upfront and adjust the schedule to fit the scope of the items, but it has some significant disadvantages and is difficult to apply to a continuous flow-based model where the requirements are less-defined.

The alternative Agile approach is breaking up tasks into small chunks that can be developed incrementally and using a fixed-length interval called a time-box for a sprint or iteration for development of those features.

- The scheduling approach with time-boxing becomes focused on "how many of these small, incremental features can be completed in a short fixed-length sprint?" rather than "how long does the development interval need to be to accommodate these features?"

- Naturally, to make this approach work, the individual features need to be small enough to fit inside of a fixed-length sprint.

Time-Boxing Advantages

There are a number of advantages to a time-boxing approach, which are summarized as follows:[2]

- **Focus:** The great advantage of time-boxing is you learn how to focus your attention on the job at hand for the specified period of time.

- **Increased productivity:** When you set a timer and work diligently and in a focused manner on only the task you have identified, you work smarter and harder, and you get more done.

- **Realization of time spent:** When you use time-boxing to get a job done, you realize how much time you might normally waste when working.

- **Time available:** Time-boxing makes you consciously aware of something you previously weren't consciously aware of—how much time you can give to a particular project or task.

Additional Time-Boxing Productivity Advantages

Beyond these advantages, there are a number of additional advantages of time-boxing that may not be immediately apparent. Time-boxing addresses two common productivity issues:

- **Parkinson's law** says, "Work expands so as to fill the time available for its completion." Fixing the time allowed eliminates wasted slack time that might be built into a scope-boxing approach.[3]

- **The Student Syndrome** refers to the phenomenon that many people will not start to fully apply themselves to a task until the last possible moment before a deadline. This leads to wasting any buffers built into individual task duration estimates.[4]

Time-boxing can be applied to classical plan-driven projects as well as Agile projects—it's a good development practice to keep a sustained pace of effort, but it is highly dependent on the culture and style of the organization:

The general concept of time-boxing has many applications beyond its role in optimizing the flow of work in an Agile project. In a broad sense, you can apply time-boxing to many different things to improve efficiency.

For example, setting a time-box limit on the length of a meeting is often very useful to improve productivity and efficiency. Instead of letting a meeting go on and on without any limit (which can easily happen), it's useful to set a time limit for the meeting to keep the meeting focused on producing an acceptable end result in a fixed amount of time.

THE KANBAN PROCESS

A *Kanban* process is a different kind of Agile process. A Kanban process is typically used to manage a queue of work to be done particularly when it is difficult to plan the work to be done such as a reactive process to manage customer service requests. It can be used to a limited extent for project work but it is more often used to manage the flow of work in a project-level process like Scrum. To understand Kanban, it is first necessary to understand the difference between a push process and a pull process.

Push and Pull Processes

A *push system* is totally planned-in-advance. An example would be a traditional manufacturing process:

- It starts with a manufacturing plan based on a *forecast* of demand that someone *thinks* will take place.

- Raw materials are ordered and held in inventory waiting to go into production.

- Production resources are planned in advance and allocated to meet the forecasted demand in the manufacturing plan.

In essence, raw materials are planned and queued and *pushed* through the manufacturing process. A classical Waterfall development process works on the same "push" principle. Requirements are forecast based on what someone *thinks* the customer needs, work is planned in advance, and resources are scheduled against the plan. Figure 6.3 shows what a push process looks like.

A *pull system*, on the other hand, is heavily demand-driven and less planned. An example would be a manufacturing process where furniture is custom-built to satisfy customer orders. In an ideal "pull" system:

- There may be a loosely defined plan to put capacity in place in anticipation of demand but no actual work is planned or scheduled until a customer order is received.

- Raw materials might be ordered only as needed to satisfy customer orders.

- Production resources would be allocated to work only as needed to satisfy customer orders.

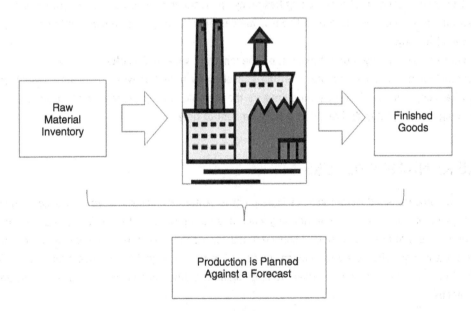

FIGURE 6.3   The push process

This is the inverse of a "push" process—in a "pull" process, customer orders are queued, and raw materials are *pulled* through the process to meet actual demand. Figure 6.4 shows what a "pull" process looks like.

What Is a Kanban Process?

All Agile processes are adaptive to customer demand to some extent rather than being totally planned; however, an ideal Kanban process is based totally on a "pull" process. It is totally un-planned and is totally reactive to customer demand. The word *Kanban* is derived from the Japanese words *Kan*, meaning visual, and *Ban*, meaning card or board. The idea originated from inventory demand cards that are sometimes used in a manufacturing system. In an automotive assembly line, for example:[5]

■ The person putting doors on cars only assembles enough doors to meet the demand for cars.

■ When his inventory of doors runs low, he sends a demand signal (Kanban card) back to the previous operation that builds the doors.

■ The person building the doors builds only enough doors to meet demand and, in turn, sends a demand signal for more materials to build the doors his supply or materials runs low.

In the real world, very few processes are totally unplanned (pull) and totally reactive. For example, in the manufacturing plant example, some amount of capacity needs to be planned to meet anticipated demand, but that capacity may be idle until it is needed to meet actual demand.

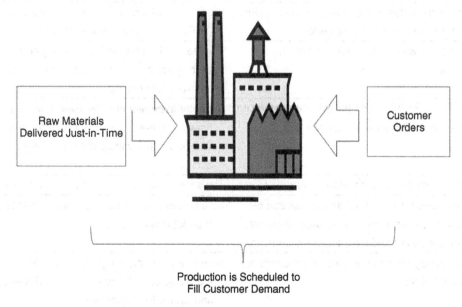

Production is Scheduled to
Fill Customer Demand

FIGURE 6.4   The pull process

Outside of a manufacturing environment, Kanban processes are used primarily for demand-driven processes that are very reactive and difficult and/or impractical to plan. Examples that would be good candidates for a Kanban process include:

- a call service center queue for handling customer service calls
- some repetitive development processes that are heavily demand-driven, such as a support function that creates reports for a business group
- an ongoing support process for providing continuous ongoing development of a product that includes bug fixes and minor enhancements.

Kanban is sometimes used for software application development processes that need to be highly adaptive and may be difficult to plan in advance. Typically, the flow of items through a Kanban process is shown in what is called a "Kanban board." Kanban boards will be discussed in a later section of this chapter.

Differences Between Scrum and Kanban

Table 6.1 shows a summary of the major differences between a Scrum process and a Kanban process.

TABLE 6.1   Differences between Scrum and Kanban

Typical Kanban process	Typical Scrum process
Kanban is generally based on a continuous flow model and work is not organized into sprints	Scrum uses time-boxed sprints to organize work incrementally
Items can be added to the work-in-process (WIP) any time the capacity is available	Items cannot be added to work-in-process (WIP) once a sprint is in progress
Estimation of work is optional in a Kanban process	The team estimates work at the beginning of each sprint and plans sprint capacity based on expected velocity
Work-in-process (WIP) can be organized into stages with WIP limits to manage flow	A similar approach may be used if desired; however, there is less of a need to manage WIP limits to organize the flow of work in Scrum
A Kanban board can be shared by multiple teams	The sprint backlog is owned by a single team
No prescribed roles—a Kanban process can be adapted to almost any kind of team structure. There are no defined roles in a Kanban team.	Scrum uses prescribed roles (Scrum Master, Product Owner, team)
Kanban is very streamlined, and no meetings are required at all	Scrum requires meetings to include Sprint Planning, Daily Standups, Sprint Review, and Sprint Retrospective

A Scrum process is actually a hybrid of a push/pull process:

- Work is planned to some extent in the Product Backlog and broken up into sprints and releases (push).

- Within a sprint, the work is planned, limited, and allocated to the team (push).

- However, it is demand-driven based on priorities set by the Product Owner, and the Product Owner has considerable ability to influence what is built (pull).

In one sense, it is a pull process because a certain amount of capacity is made available in the team to develop requirements and there is not a large backlog of requirements, as there would be in a Waterfall process.

- Requirements in a Scrum process are typically groomed to get ready for development more on a just-in-time basis, which is similar to a pull process.

- On the other hand, a Scrum process can be somewhat of a push process because there is nothing to prevent planning high-level requirements in advance of development in order to estimate the project cost and schedule at a high level.

A pure Kanban process is totally demand-driven (pull) and not planned:

- Only a limited effort is made to plan requirements in advance—for example, in planning a customer service response capability, it's impossible to predict the exact number and nature of the customer service calls that might come in, but a limited amount of planning is needed in order to do capacity planning and pipeline planning to put the appropriate capacity in place to handle the anticipated volume of calls.

- A Kanban process does not break up the work into sprints at all.

- It is a continuous flow model—work is taken into the process immediately based on priority as soon as resources are available to work on it.

The best example of a Kanban process is something like a customer service call queue. A certain number of customer service representatives are hired to answer calls in the expectation of customer demand, but the actual work in the process is totally reactive to customer demand and the utilization of the process is dependent on the level of demand that comes in and the time it takes to handle each call.

Work-In-Process (WIP) Limits in Kanban

A critical concept in Kanban systems is the idea of work-in-process (WIP). Kanban processes many times have stages that work goes through—an example is a customer service system designed to solve customer problems. That kind of system typically has different levels of support, depending on the nature and severity of the customer's problem, as shown in Figure 6.5.

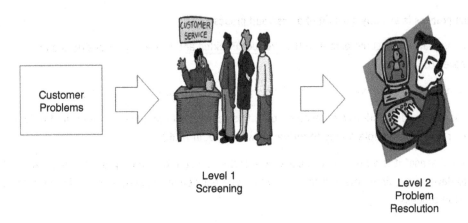

Customer
Problems

Level 1
Screening

Level 2
Problem
Resolution

FIGURE 6.5 **Levels of support in a customer service response system**

A Kanban system is designed to optimize the overall flow through the system and each stage in the process may have capacity limitations that limit the overall work in process.

- For example, in a typical customer service response system, there often is an initial level that is designed to screen calls and weed out problems that can be easily resolved without requiring more sophisticated resources.

- Only problems that cannot be resolved in that stage are routed on to a second or third level of support for further resolution depending on the complexity and difficulty of the problem.

In this situation, the size of the screening staff would typically be much larger than the more sophisticated problem resolution staff to optimize the flow and efficiency of the overall process. Each stage has a WIP limitation based on the capacity of the resource capacity in that stage and the flow through the process will be determined by which stage in the process is more limiting on the overall flow.

Kanban Boards

Kanban boards are tools that are used to visually show the flow of items through a Kanban process. For example, Scrum and other Agile processes use Kanban boards to manage the flow of items within a sprint. The Kanban board can be as simple as a whiteboard with stickies on it or 3 × 5 cards that are manually moved around the board to show progress through the flow.

Kanban boards can also be implemented using an automated online tool to track progress, as shown in Figure 6.6.

FIGURE 6.6   An example Kanban board in an online tool.

Source: VersionOne, Inc.

THEORY OF CONSTRAINTS

All Agile processes (especially Kanban) emphasize optimizing the flow through the entire process—bottlenecks can develop in each stage of the process, and a key concept that is extremely useful for optimizing the overall flow is the Theory of Constraints that was originally published by Eliyahu Goldratt in his book *The Goal* in 1984.[6] It involves five steps that are repeated over and over until you finally optimize the flow through the process:

1. **Identify:** The first step is to *identify* the stage or portion of the process that is the most critical constraint or bottleneck. For example, in the customer service system, it may be that there are not enough people to handle calls and do the initial screening and, as a result, people are waiting for a long time to get through.
2. **Exploit:** The second step is to do whatever you can to *exploit* or optimize that portion of the process to make it more efficient. For example, in the customer service response system, you might add an improved interactive voice response system that would provide some initial screening to relieve the load on the people answering the calls.
3. **Subordinate:** The third step is that, after you've done whatever you can to optimize the performance of that constraint, you should *subordinate* everything else in the process to work within that limitation. For example, if the number of people available to answer the phone is

the limitation, there is no point in having an excessive number of people in the second stage to resolve problems if those people are idle waiting for calls to come through.

4. **Elevate:** If the process flow through the overall system is still not sufficient, the next step is to do whatever is necessary to *elevate* the constraint. For example, in the customer service response system, you might add additional resources to answer the phone until that portion of the process is no longer the limiting constraint.

5. **Check for new constraint:** Finally, once you've relieved a bottleneck, another part of the process then will typically become the bottleneck. For example, in the customer service response system, if you have relieved the constraint of the front-end people answering calls, the people on the back end resolving problems may then become the bottleneck and you need to repeat the above steps with that as the new primary constraint.

Here's an example. In a company I worked in, we had over 100 legacy applications that needed to be upgraded to a new version of software, and we designed a Kanban flow with a number of different stages that each application would go through:

1. **Initial assessment by tech lead:** A senior-level technical lead would check out the application and do a preliminary survey of the application to identify any risks in the conversion process and flag those items to the project team. He would also establish coordination with the owners of that application and any other related applications to begin the conversion process.

2. **Conversion by offshore team:** An offshore project team was assigned to do the conversion process and to perform initial unit testing.

3. **QA testing:** The application was turned over to Quality Assurance (QA) for formal QA testing.

4. **Staging for release:** The application was staged for release to production and released.

The process flow is shown in Figure 6.7.

We were given a goal to convert more than 100 applications in about a year, and the plan was to do about 6 to 7 applications per month. We explored a number of options to try to accelerate this process; this is a good example of how the theory of constraints can be used:

1. **Identify:** The first step in the process is to *identify* the system constraint or bottleneck that is most critical for limiting the flow.

   - In this situation, the system constraint was the resources available to perform the QA testing.
   - There were a limited number of QA resources to do the testing that was required, and they would become quickly saturated with testing. The conversion effort wasn't that difficult for many of the applications and some of it could be automated but the testing effort was significant to ensure that the application functionality wasn't impacted by the conversion.

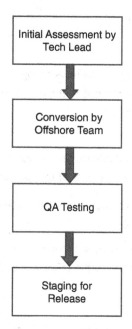

FIGURE 6.7   Example process for Theory of Constraints analysis

2. **Exploit:** The next step is to *exploit* the constraint. Given that the QA resources were the bottleneck, what can be done to optimize the use of those resources to improve the overall flow? We were able to do several things to optimize those resources, including planning the conversion process to spread the load among the QA resources we had as evenly as possible.

3. **Subordinate everything else:** Once we had done everything we could do to optimize the QA resources, it made no sense for the conversion process to go faster than QA could handle the applications. As a result, the application conversion effort needed to be slowed down to match the speed of the QA process. Otherwise, a very large queue of applications waiting to be tested would have piled up, waiting for QA.

4. **Elevate:** The next step to improve the flow would be to add more QA resources to *elevate* the constraint and reduce the bottleneck. In this particular situation, that was not an option.

5. **Check for new constraint:** Of course, if you add enough QA resources, at some point that will no longer be the most critical constraint and the bottleneck will move somewhere else in the process, such as the number of developers to do the conversion process. (We never really got to that point in this particular example.)

SUMMARY OF KEY POINTS

1. The Importance of Flow

The idea of flow originated with Lean Manufacturing and has been used in manufacturing to streamline production processes. Flow is one of the most important Lean principles to understand to maximize the efficiency of any process.

2. Time-Boxing

Time-boxing, as opposed to schedule-boxing or scope-boxing, is a more efficient way of managing flow in an Agile project for a number of reasons and can result in higher productivity.

- It's equivalent to passing work to be done through a pipeline and using uniform size containers to handle the elements in the flow.

- It provides a way of establishing a regular cadence on an Agile project, as well as measuring velocity, which can be used to measure the efficiency of the process and predict future performance.

3. The Kanban Process

Kanban is normally a continuous flow process. It is primarily a pull process that is totally demand-driven.

- Rather than work being planned and broken up into sprints, the work is taken in continuously from a prioritized queue of work to be done.

- The best candidates for Kanban are processes that are reactive and difficult to plan in advance. For example, managing the workload in a queue of customer service requests would be a good candidate for a Kanban process.

- Although a Kanban process and a Scrum process are different, many of the principles related to managing flow are the same in both processes.

Kanban boards are widely used in Scrum and many other processes as a visual way of tracking progress through a process.

4. Theory of Constraints

The Theory of Constraints is an approach for analyzing problems related to continuous flow. It provides a systematic way of identifying and removing bottlenecks to improve the overall flow of a process.

DISCUSSION TOPICS

The Importance of Flow

1. What are the factors that contribute to maximizing flow?

Time-Boxing

1. What is time-boxing? What is the alternative? Why is time-boxing generally considered a better approach for Agile projects?

2. When would it not work so well?

The Kanban Process

1. What are the most important differences between a Kanban process and Scrum? When would a Kanban process make more sense than a Scrum process?

2. What is a Kanban board? How is it used in Scrum?

3. Explain how WIP limits affect a Kanban process.

Theory of Constraints

1. Discuss an example of a problematic project situation and how you might have applied the theory of constraints to analyze the problem and find an appropriate solution.

NOTES

1. James P. Womack, and Daniel T. Jones, *Lean Thinking: Banish Waste and Create Wealth in Your Corporation* (New York: Free Press, 2003).
2. "Timeboxing." Agile Hardware, http://www.agilehardware.com/pages/Timeboxing.html.
3. "Parkinson's Law." http://en.wikipedia.org/wiki/Parkinson's_Law.
4. "Student Syndrome." http://en.wikipedia.org/wiki/Student_syndrome.
5. "Kanban Development Oversimplified." AgileProductDesign.com (April 20, 2009), http://www.agileproductdesign.com/blog/2009/kanban_over_simplified.html.
6. Eliyahu Goldratt, *The Goal: A Process of Ongoing Improvement* (Great Barrington, MA: North River Press, 2012).

⑦ Agile Estimation

AGILE ESTIMATION OVERVIEW

THIS CHAPTER IS DESIGNED TO PROVIDE A HIGH-LEVEL OVERVIEW OF AGILE ESTIMATION PRACTICES and how they are different from classical plan-driven estimation practices. References to other Agile estimation techniques that are outside the scope of this book will be provided at the end of this chapter.

What's Different about Agile Estimation?

Before discussing Agile estimation, it's important to have an understanding of what's different about the environment for an Agile project, because there are some very significant differences that affect how you would do any kind of estimation.

A classical plan-driven or Waterfall project is usually based on somewhat of a contractual relationship between the business users and the project team.

- Typically, the business users agree to some fairly well-defined requirements and sign off on them prior to the start of the project.

- The project team then commits to a cost and delivery schedule to meet those requirements and any changes in the scope of the requirements are controlled from that point forward.

- In order to manage the reliability of the estimates, some kind of disciplined process for managing changes to the requirements is implemented, and significant changes are considered to be an exception rather than the norm.

- If a significant change takes place, it normally triggers an assessment to determine the impact on the costs and schedule of the project and any initial estimates are recalculated and revised if necessary.

An Agile project has a very different model. First, the requirements are typically not necessarily well-defined in detail upfront; and, in addition, change is the norm rather than the exception.

Change is expected and encouraged as the project is in progress. Naturally, any estimates can only be as good as the requirements are defined and any change in the requirements might invalidate any initial estimates. For that reason, some people say that it is futile to even attempt to estimate the costs and schedule of an Agile project because the requirements are only going to change. I don't necessarily agree with that point of view—just because a project has some level of uncertainty, doesn't mean it is totally futile to estimate the costs and the schedule. But, of course the level of accuracy in the estimate needs to be understood in the context of the level of uncertainty associated with the project. Very few people get a blank check to do an Agile project without any expectations of what the cost and schedule might be.

An Agile project is based on a much more collaborative partnership between the business users and the project team and there is a much higher level of transparency and sharing of information related to the project, which should lead to a much higher level of mutual understanding of the risks and uncertainties in the project as well as a shared understanding of the level of accuracy in any cost and schedule estimates.

The most important thing is to properly set the expectations of the customer regarding whatever estimates are made:

■ A classical plan-driven project estimate might have been based on what was thought of as a relatively static and well-defined model of what the requirements are and how the project will be managed (which might or might not have been realistic).

■ An Agile estimate is typically based on a high-level and dynamic view of what the requirements are that is understood to be not extremely accurate and likely to change.

Naturally, those are not discrete, binary alternatives and there are many variations that involve a blend of those two approaches. Table 7.1 shows a summary of the differences in these two environments.

TABLE 7.1 Estimation in Agile vs. classical plan-driven processes

	Typical plan-driven or Waterfall estimation process	Typical Agile estimation process
Relationship of customer or business user to the project team	■ Arm's-length contract based on firmly-defined and signed-off requirements ■ Limited transparency and visibility of the user to project details and issues	■ Collaborative joint partnership model ■ High levels of transparency and sharing of information ■ Mutual understanding of the relationship of uncertainty and flexibility to estimates
Control over changes	Changes to requirements are controlled, and any significant changes are treated as an exceptional situation	Changes to requirements are encouraged and are the norm rather than an exception

Developing an Estimation Strategy

There are two common problems associated with estimation:

1. Analysis paralysis

- Spending too much time attempting to develop detailed estimates based on highly uncertain and unreliable information
- Delaying making commitments or delaying making any decision at all until the information required to make the decision has a very high level of certainty about it and all of the risks associated with a project are well-known and completely understood

2. Cavalier approach

- Not worrying about managing uncertainty and risk at all and just starting to do a project with very little or no upfront planning and estimation
- Assuming that whatever uncertainty and risk is inherent in the project will be discovered and somehow work itself out as the project proceeds

The right approach for a given project is somewhere in between those extremes and the approach selected needs to be consistent with the level of uncertainty in the project:

- The level of uncertainty in a project can vary widely, from a pure research and development initiative with very high levels of uncertainty to projects with very low levels of uncertainty (like building a bridge across a river).

- It is foolish to underestimate the level of uncertainty in a project and create unrealistic estimates based on very uncertain information.

- However, it can be equally foolish to not take advantage of information you do know and use it to make informed decisions.

The key is to honestly and objectively assess the level of uncertainty and risk in the project and reach agreement among all participants in the project (customer and project team) on an estimation approach that is consistent with that assessment.

Management of Uncertainty

Being able to deal with uncertainty effectively is a very critical skill for an Agile Project Manager. You have to be able to accurately assess the risks and uncertainty in a given situation and make informed decisions based on the best information that you have, knowing that it might not be a perfect decision. That is not really a new skill with Agile—it is a skill that all project managers should have; however, it is especially important in an Agile environment where the level of uncertainty may be significantly higher.

This process of managing uncertainty is similar to making a decision when you're betting on a card game. Many people are familiar with the game of blackjack where players compete against the dealer to get the highest value of cards without going over 21 points (see Figure 7.1):

- Face cards are worth 10 points each and aces can be worth either 1 or 11 points each.

- Players are each dealt two cards face up.

- The dealer is dealt two cards (one face up and one face down).

- The player has to decide to take more cards (called taking a hit), risk going over 21 points, and automatically losing; or to stay with his/her current cards and risk losing to the dealer because the dealer has a higher number of points.

Here's how I would go about analyzing this situation:

1. What information do I know for certain?

   - I have 16 points in my hand.
   - The dealer has more than 10 points in his/her hand.

2. What information is unknown or uncertain?

   - What is the dealer's second card?
   - What card will I be dealt next?

3. What is a reasonable assumption to make?

   - There is a high probability that the dealer's second card is another 10 or a face card, giving the dealer a total of 20.
   - If that is true, I will lose on 16 if I don't take a hit, so I should risk taking another card.

This is an example of making an informed decision based on incomplete information. It is something a gambler does all the time. It is a very important skill for an Agile Project Manager. An Agile Project Manager needs to be frequently making informed decisions in a very uncertain environment.

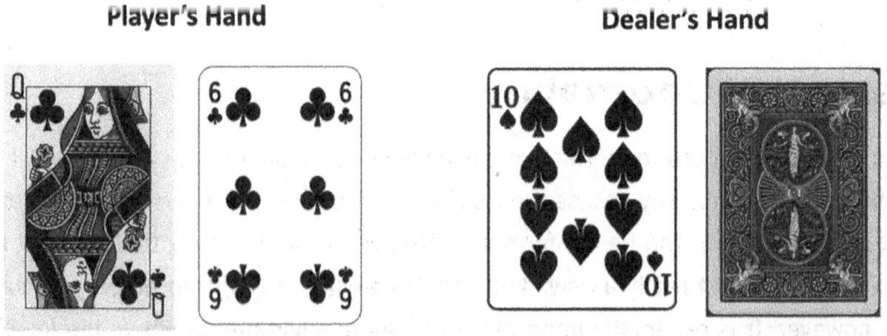

Player's Hand   **Dealer's Hand**

FIGURE 7.1   Example of a blackjack hand

AGILE ESTIMATION PRACTICES

Levels of Estimation

In an Agile project, there is no prescribed way to do estimation, and the exact approach could range from doing very little or no estimation at all to having several levels of planning and estimation based on the scope and complexity of the project, as shown in Figure 7.2.

The level of emphasis that you would put on each of these levels will also vary significantly, based on the nature of the project. Doing sprint-level planning and estimation is most essential, but any of the levels of planning and estimation just described are optional, depending on the nature of the project. For example, an Agile estimation approach could be as simple as having sprint-level estimates only without any project-level or release-level estimates at all.

The approach is totally dependent on the level of uncertainty in the project and the level of predictability and control that is desired. Of course, the level of predictability and control needs to be consistent with the level of uncertainty in the project. Attempting to achieve a very high level of predictability and control in a project that has very high levels of uncertainty may be futile.

Each of these levels of planning would naturally have a different planning horizon, different levels of uncertainty, and a different level of accuracy for any estimates associated with it. Naturally, the longer the planning horizon, the more the uncertainty will be and the less accurate the estimates will be.

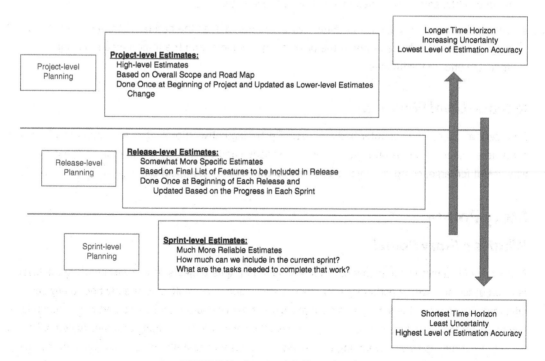

FIGURE 7.2   Levels of Agile estimation

Sprint-Level Planning

Sprint-level planning is the lowest level, has the shortest planning horizon, and it should have the highest level of certainty and the highest accuracy in the estimates. An experienced Agile team should be able to accurately predict the capacity that can be completed in a given sprint after the team has reached maturity and the velocity of the team has begun to stabilize.

- Before the team starts a sprint, the features to be included in that sprint should be known, and they should be understood well enough to make a reasonably accurate estimate of the level of effort required to complete those features.

- The details associated with those features may be further elaborated during the sprint, but features should not normally be added or subtracted from the items to be included in the sprint while the sprint is in progress.

Project-Level Planning

Project-level planning is the highest level of planning, has the longest time horizon, and has the highest level of uncertainty.

- Naturally, the estimates at this level can only be as accurate as the level of certainty in the requirements, so the estimates may not be very accurate.

- The level of accuracy depends on a number of factors such as the nature and complexity of the project, the level of uncertainty in the project requirements, and the importance of having schedule and cost estimates.

Release-Level Planning

Release-level planning is an intermediate level of planning between project-level planning and sprint-level planning and is optional, based on the scope and complexity of the project. It can be very useful for large, complex projects to break up the project into releases.

Story Points

What Is a Story Point?

A *story point* is a metric commonly used for estimation in Agile projects. It is a way of sizing the level of effort associated with a particular user story. A story point is normally not directly tied to any specific measure of time; it is simply a relative measure of the level of difficulty of a particular story relative to other stories, and a story point may not have the same meaning among different Agile teams.

For example, as I was developing the course materials for the Agile Project Management course based on this book, I used story points to estimate the level of effort required for each lesson. One story point was equivalent to one simple PowerPoint slide, and I used that as a reference point for sizing the level of effort associated with other slides and lessons, which is a relative sizing estimate.

Jeff Sutherland has written a great article to explain why story points are better than hours for Agile estimation:

> Story points give more accurate estimates, they drastically reduce planning time, they more accurately predict release dates, and they help teams improve performance. Hours give worse estimates, introduce large amounts of waste into the system, handicap the Product Owner's release planning, and confuse the team about what process improvements really worked . . .
>
> The stability of a user story is critical for planning. A three point story today is three points next year and is a measurable part of the product release for a Product Owner. The hours to do a story depend on who is doing it and what day that person is doing it. This changes every day.[1]

Story points are also a more relevant metric to measure completion of work: "Hours completed tell the Product Owner nothing about how many features he can ship and when he can ship them."[2]

In reality, it is very difficult to start a new Agile team using story points because they have no frame of reference to compare it to.

- In that case, I have sometimes made an arbitrary decision to let one story point be approximately one man-day of work but that is just an arbitrary decision to get started with.

- As the team gains more experience, they should be able to make relative comparisons among stories based on their past experience without a reference to any specific measure of time. An alternative to story points for teams who are not very experienced is *ideal days*. An ideal day is defined as follows:

> A unit for estimating the size of product backlog items based on how long an item would take to complete if it were the only work being performed, there were no interruptions, and all resources necessary to complete the work were immediately available.[3]

Once a team gains experience, story points become very easy and intuitive.

> A mature Agile team intuitively knows what a story point means in terms of the relative size of a user story compared to other stories that it has sized in the past, but how does a new team that perhaps even has people who are new to Agile get started with story points? Ideal days is a story point estimation scale that blends size with effort and degrades the backlog sizing process into a drawn-out time estimation exercise.[4]

A Fibonacci sequence is often used for story point estimates because it provides an appropriate level of discrimination between large and small estimates without attempting to very accurately discriminate large estimates.

■ A Fibonacci sequence is a series of numbers of the form of (1, 2, 3, 5, 8, 13, 21—where each number is the sum of the previous two numbers).

■ T-shirt sizes (Small, Medium, Large, XLarge, etc.) can also be used to create estimates.

■ Both story points and T-shirt sizes are designed to provide a very imprecise framework for making estimates that is appropriate to an Agile environment.

For an estimation approach based on story points to work with Agile teams, it is important for the teams to be stable. If people are rotated in and out of an Agile team, it will destabilize the team and make the estimation process very difficult.

How Are Story Points Used?

Story points are used in different ways based on the level of planning and estimation being done.

1. High-level product backlog grooming (refinement)

■ In grooming (refinement) the product backlog, stories are estimated in story points.

■ A voting technique is used to estimate the level of effort associated with each story in story points.

■ The relative size of the story points provides some information to the Product Owner to assess the business value against the level of effort.

■ Based on that assessment, the Product Owner will rank order the items in the product backlog to try to maximize the business value to be gained.

2. Tactical sprint planning

■ The project team determines the team's velocity in story points based on the number of story points completed in each sprint.

■ At the beginning of each sprint, the project team will make a determination of how many stories that they can take into the current sprint based on their previous velocity history.

■ In sprint planning, a more detailed estimate of the level of effort required for the tasks associated with each story may be used to validate the story point estimates.

3. Project-level and release-level planning

■ High-level story point estimates, combined with team velocity estimates, may also be used for developing project-level and release-level estimates.

■ For example, to develop a rough estimate of the time required to complete a project, it would be necessary to make a high-level estimate of the story points required for each story to be included in the project or release, total those story points, and then divide that by the velocity of the team in story points per sprint to determine how many sprints it will take to complete the project or release.

Other Relative Sizing Techniques

Agile estimation typically does not use highly precise, absolute, quantifiable estimates. More often, Agile estimates are much more imprecise and use relative sizing rather than absolute sizing. An example is T-shirt sizes where the size of a user story is estimated in terms of T-shirt sizes such as:

- Extra-small

- Small

- Medium

- Large

- Extra-large

That could be a quick and easy way to do a relative size estimate on a number of different user stories. Then, if desired, those estimates could be converted to an equivalent number of story points. I have seen this technique used to come up with a very quick estimate of the time required to do a complete project. The approach we used was this:

- We had a large number of user stories and put them on 3 x 5 cards.

- People on the team were asked to sort the cards into different piles based on approximate T-shirt sizes.

- The team was asked to review the piles and come to consensus on how they were ordered. For example, if someone did not agree with the sizing of a particular use story, they were asked to discuss it and try to resolve the difference of opinion.

- After the team agreed on the T-shirt size estimates, we assigned an equivalent number of story points to each T-shirt size and totaled the overall story points for all of the stories in the project.

Using that technique we were able to quickly come up with a very high-level estimate of the total story points required to complete the whole project.

- Then, based on the estimated velocity of the team, we were able to project how long it might take to complete the entire project.

- Naturally, this estimate is very imprecise and not very accurate, but it served the purpose it was intended for.

■ The company had committed to completing a very large project with only one Agile team. We needed to see if that was really feasible, and this effort showed that it wasn't feasible at all, and at least two teams would be needed to complete the project in any reasonable length of time.

What Is Planning Poker?

Planning poker is a commonly used approach to come up with story point estimates:

■ A story is presented to the team and is discussed to resolve any significant uncertainties or questions.

■ Each member of the team chooses a card representing his or her estimate in story points but does not reveal the estimate to others.

■ When everyone is ready to vote, all members of the team show their cards.

■ The team then discusses why they voted differently to try to converge on a single estimate that everyone agrees to.

Planning poker is based on the Delphi method. The Delphi method is a structured communication technique, originally developed as a systematic, interactive forecasting method that relies on a panel of experts coming to consensus on an estimate:

■ The experts answer questionnaires in two or more rounds.

■ After each round, a facilitator provides an anonymous summary of the experts' forecasts from the previous round as well as the reasons provided for the facilitator's judgments.

■ Thus, experts are encouraged to revise their earlier answers in light of the replies of other members of their panel.

■ It is believed that during this process the range of the answers will decrease, and the group will converge towards the "correct" answer. Finally, the process is stopped after a predefined stop criterion (e.g., number of rounds, achievement of consensus, and stability of results) and the mean or median scores of the final rounds determine the results.[5]

An online tool is available for implementing planning poker on the following website: www.planningpoker.com.

More Sophisticated Agile Estimation Techniques

As previously mentioned, this chapter is designed to provide a high-level overview of Agile estimation practices and how they are different. Here are some references on some more sophisticated Agile estimation techniques that are outside the scope of this book:

■ Average Cycle Time.[6]

■ Monte Carlo simulation.[7]

VELOCITY AND BURN-DOWN/BURN-UP CHARTS

The estimation process in an Agile project should be dynamic and continuously refined as the project progresses. An Agile estimate is based primarily on the velocity of the team, projecting that velocity into the future, monitoring the actual velocity against the projected velocity, and then adjusting the estimate as necessary.

Velocity

The concept of velocity is essential to Agile estimation. Velocity is defined as follows:

> Velocity is the measure of the throughput of an Agile team per iteration. Since a user story, or story, represents something of value to the customer, velocity is actually the rate at which a team delivers value. More specifically, it is the number of estimated units (typically in story points) a team delivers in an iteration of a given length and in accordance with a given definition of "done."[8]

Here is an example of how velocity can be used to develop an estimate for completing future work:

- Assume that a team has completed an average of 60 story points of work per sprint over the past three sprints, and each sprint is two weeks long.

- Suppose there is a total of 300 story points' worth of work to be completed in the current release, and the Product Owner wants to estimate when the remainder of the release will be completed.

- An estimate for completing the remaining in the release would be 300/60, or five sprints, and since each sprint is two weeks long, it will take approximately 10 weeks to complete the work remaining in the release.

It should be noted that the use of velocity as a metric has some inherent weaknesses:

- It is normally measured in terms of story points completed in a sprint, and story points are only a relative metric that has no absolute value and is only meaningful within a team.

- Because it is such a subjective metric, the data from this metric may not be very reliable.

Burn-Down Charts

Of course, any estimate is only an estimate, and it should be continuously adjusted as the project progresses as more becomes known about the actual velocity of the team and the level of effort remaining to be completed. A way of doing that is to measure the ongoing actual velocity of the team against the projected velocity and adjust the projected velocity based on actual results. An important tool for doing that is a burn-down chart.

> A burn-down chart is a chart often used in Scrum/Agile development to track work completed against time allowed. The x-axis is the time frame, and the y-axis is the amount of remaining work left that is labeled in story points, man hours, or the like. The chart begins with the greatest amount of remaining work, which decreases during the project and slowly burns to nothing.[9]

Figure 7.3 shows an example of a burn-down chart.

- In this chart, the y-axis shows the total work remaining to be completed in story points, which is 120 units at the beginning of the sprint and is projected to go down to zero by the end of the time interval.

- The time interval being measured could be a sprint, a release, or an entire project.

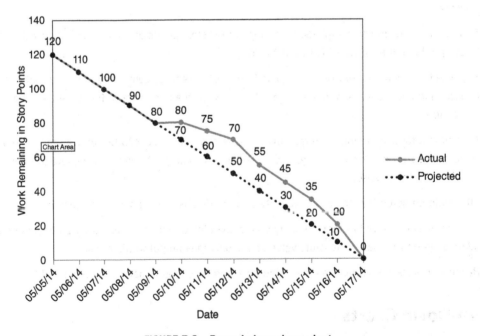

FIGURE 7.3 Example burn-down chart

■ The diagonal line between those two points shows the projected rate of completing work in the time interval.

The solid line shows the actual performance of the team as the sprint is in progress:

■ If the actual work remaining is above the projected line, work is proceeding slower than expected.

If the actual work remaining is below the projected line, work is proceeding faster than expected.

In Figure 7.3 the work started out very close to the projected schedule, then started to lag behind and finally caught up with the projected schedule by the end of the sprint. Burn-down charts are very useful for monitoring the actual progress against projected progress to continuously adjust the estimated completion of an effort.

Burn-down charts are typically measured in either hours of work remaining or story points of work remaining. Each of those metrics provides different information:

■ Hours of work remaining is difficult to track because it requires the people on the team to report time spent on tasks, as well as estimating the hours of work remaining, which can be difficult to do. Estimates of hours of work remaining may not be very accurate.

■ Story points are easier to track, and it's an all-or-nothing metric—the team gets full credit for all of the story points associated with a given story if the story has been complete and meets the definition of "done", or no credit if it has not been completed. That's probably a more reliable and more accurate metric.

Burn-Up Charts

A burn-up chart is essentially the mirror image of a burn-down chart.

> A burn-up chart is a graphical representation that tracks progress over time by accumulating functionality as it is completed. The accumulated functionality can be compared to a goal such as a budget or release plan to provide the team and others with feedback. Graphically the x-axis is time and the y-axis is accumulated functionality completed over that period of time. The burn-up chart, like its cousin the burn-down chart, provides a simple yet powerful tool to provide visibility to the sprint or program.[10]

Figure 7.4 shows an example of a burn-up chart. One of the advantages of a burn-up chart is that it provides a way to show a change in the scope of the projected work. In this example on May 13, 2014, there was an increase in the scope of work remaining from 80 to 100 story points, which raised the end goal from 120 to 140. (That shouldn't happen in the middle of a sprint, but it can

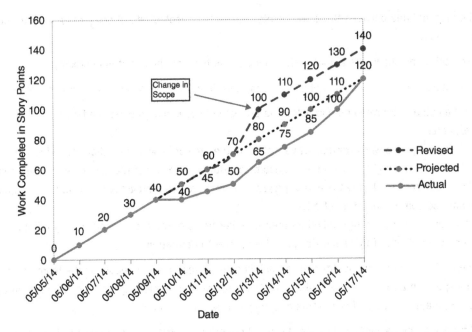

FIGURE 7.4   Example burn-up chart

easily happen in the middle of a release or project.) The chart shows that the team finished the sprint on the original goal of 120 but failed to meet the revised goal of 140.

SUMMARY OF KEY POINTS

1. Agile Estimation

Agile estimation is based on a very different environment than classical plan-driven project estimation approaches.

- A classical plan-driven model is based on somewhat of a contractual relationship between the customer and the project team and the project estimates are expected to be reasonably accurate to support that kind of relationship.
- Any Agile project estimate can only be as accurate as the requirements are known and there typically is lot of uncertainty in the requirements that would limit the accuracy of any estimate. For that reason, it is essential for an Agile project to be based more on a collaborative partnership between the business users and the project team as well as a shared understanding of the level of accuracy in any cost and schedule estimates.
- Management of uncertainty is an important skill for an Agile Project Manager—it is something that all project managers should have, but it is especially important in an Agile environment where the level of uncertainty may be significantly higher.

2. Agile Estimation Practices

There is no prescribed way to do estimation in an Agile project and the exact approach could range from doing little or no estimation at all to having several levels of planning and estimation based on the scope and complexity of the project. Whatever level of estimation is selected, the approach should be consistent with the level of uncertainty in the project and mutually understood by all participants in the project.

- Agile estimation practices are based on an understanding of velocity and flow and require breaking up requirements into small functional elements typically in the form of user stories that can be sized by the project team in terms of story points to estimate their difficulty.
- Typically, Agile estimation techniques are based on relative size estimates rather than absolute size estimates. Naturally, that is not a very precise way of estimation, but it is appropriate for use in many Agile projects.
  - A *story point* is a metric commonly used for estimation in Agile projects. It is a way of sizing the relative level of effort associated with a particular user story.
  - T-shirt sizes is another relative sizing technique.

3. Velocity and Burn-Down/Burn-Up Charts

An Agile estimate is based primarily on the velocity of the team, projecting that velocity into the future, monitoring the actual velocity against the projected velocity, and then adjusting the estimate as necessary.

- Burn-down charts are typically measured in either hours of work remaining or story points of work remaining. Each of those metrics provides different information.
- A burn-up chart is a graphical representation that tracks progress over time by accumulating functionality as it is completed.

DISCUSSION TOPICS

Agile Estimation

1. How would you go about determining the appropriate estimation strategy for a project?
2. Discuss an example of a real-world project and how you might have done it differently with a more Agile estimation approach.
3. What is the advantage of story points over estimating in hours?
4. Complete the Agile estimation exercise in Table 7.2 and answer the following questions:
   - Select two to three of the items in the list and write a user story to describe each one.
   - If the work is divided into two-week sprints and I have a velocity of completing 20 story points of work in each two-week sprint, how long will it take to complete this work? (Note that the items in the hierarchy are summed at different levels.)

TABLE 7.2 List of tasks for estimation

Title	Estimate
1. Introduction, Course Objectives, and Agile Overview	8.00
Introduction and Course Objectives	2.00
Introductions	1.00
Course Objectives	1.00
Agile Overview	6.00
What Is Agile?	3.00
Agile Perception versus Reality	3.00
2. Agile Fundamentals	17.00
Agile History, Values, and Principles	11.00
Agile Manifesto Values	3.00
Agile Manifesto Principles	8.00
Agile Benefits and Obstacles to Becoming Agile	6.00
Agile Benefits	3.00
Obstacles to Becoming Agile	3.00
3. Scrum Overview	34.00
Scrum Roles	3.00
Scrum Master Role	1.00
Product Owner Role	1.00
Team Role	1.00
Kanban Process Overview	10.00
What Is Kanban?	1.00
Differences Between Push and Pull Processes	2.00
Differences Between Kanban and Scrum	2.00
WIP Limits in Kanban	1.00
Theory of Constraints	2.00
Kanban Boards	2.00
Scrum Methodology	8.00
Time-Boxing	3.00
General Scrum/Agile Principles	10.00

NOTES

1. Jeff Sutherland, "Story Points: Why Are They Better Than Hours?," posted on May 16, 2013, http://scrum.jeffsutherland.com/2010/04/story-points-why-are-they-better-than.html.

2. Ibid.

3. Ideal day definition, http://www.innolution.com/resources/glossary/ideal-day.

4. Blog archives, "Getting Started with Story Points," posted by Alec Hardy, Feb. 3, http://agilereflections.com/tag/ideal-days/.

5. "Delphi Method," http://en.wikipedia.org/wiki/Delphi_method.

6. "Agile and Metrics – Cycle Time," Screenful.com, https://screenful.com/blog/software-development-metrics-cycle-time.

7. Web-based Monte Carlo simulation for Agile estimation, InfoQ, https://www.infoq.com/news/2019/04/monte-carlo-agile-estimation/.

8. Agile Sherpa, "Velocity," http://www.agilesherpa.org/agile_coach/metrics/velocity/.

9. Techopedia, "What Is a Burn-down Chart?" http://www.techopedia.com/definition/26294/burndown-chart.

10. Thomas F. Cagley, "Metrics Minute—Burn up Charts," May 9, 2011, http://tcagley.wordpress.com/2011/05/09/metrics-minute-burn-up-charts/.

8 Agile Project Management Role

AGILE IS CAUSING US TO BROADEN OUR VISION of what *project management* is, and that will have a dramatic impact on the potential roles that a Project Manager can play in an Agile project. The image of a Project Manager is typically very heavily focused on a plan-driven development approach. If you read many books on Project Management, you will find a lot of discussion on how to use:

- Project plans

- Work breakdown structures

- PERT charts and Gantt charts

- Microsoft Project to plan a project.

All those things are good things for a classical plan-driven Project Manager to know, but many of them can become irrelevant in an Agile environment, which calls for a very different approach:

- In an Agile environment, you may not have enough information upfront to develop detailed project plans and work breakdown structures, and it may be very difficult, if not impossible, to predict how all the project activities will be organized in advance in terms of PERT charts and Gantt charts.

- Agile calls for a much more fluid and dynamic approach that is optimized around maximizing business value in a much more unpredictable and uncertain environment.

That doesn't mean that the typical plan-driven approach to project management is obsolete and no longer useful, but it certainly should not be the *only* way to run a project. The challenge for Project Managers is going beyond a classical plan-driven project management approach and learning a much broader range of project management practices in a wider range of environments with potentially much higher levels of uncertainty. It is much more of a multidimensional and adaptive approach to project management.

- For example, the person responsible for the success or failure of a project in an Agile environment may be called a Product Owner rather than a Project Manager, but that person still needs to understand and utilize many principles and practices that have been associated with project management in the past.

- It's mostly a matter of applying the project management discipline and skills in a very different context that emphasizes a more adaptive approach to maximizing business value in a very uncertain environment, rather than a more plan-driven approach that attempts to maximize predictability and control.

To understand the potential role of an Agile Project Manager, we need to first get past some of the stereotypes and misconceptions that exist about both classical plan-driven project management and Agile. The following are some of the stereotypes that exist about project management:

1. Project Managers are very command-and-control-oriented

Project managers are noted for getting results and sometimes that means being assertive and somewhat directive to set goals and manage the performance of project teams.

- Many times, that behavior is expected of a Project Manager by the businesses that they operate in.
- In many companies, if a project team is underperforming, the Project Manager is the one held responsible and is expected to take corrective action to get the project on track.

2. Project Managers are rigid and inflexible and only know how to manage by the Waterfall methodology

For many years, Project Managers have been held accountable for managing the costs and schedules of projects, and we all know that to meet cost and schedule goals, you must control the scope of the project.

- That, in turn, requires a disciplined approach to defining and documenting detailed requirements and controlling changes, where changes become the exception rather than the norm.
- The emphasis on managing costs and schedules requires accurately defining the requirements up front, which leads to extensive use of plan-driven or Waterfall-style methodologies that are based on trying to define the project requirements in detail upfront before the project starts and controlling changes once the project is in progress.

3. Project Managers are just not adaptive and cannot adapt to an Agile environment

Like the other stereotypes, there may be some amount of truth in this stereotype, but it would be inaccurate to generalize and say that this is true of all Project Managers. Agile will require some considerable rethinking of the project management approach and some Project Managers are so heavily engrained in the classical plan-driven way of operating because it has been so widely accepted as the norm for such a long time that they may have a difficult time adjusting to an Agile project approach.

It should be apparent from the above that Project Managers are a product of the environment that they work in and what is expected of them. Developing a new image of a Project Manager may require some changes to the environment that Project Managers work in and what is expected of them.

- These are not insignificant challenges, but it certainly isn't impossible for a Project Manager with a classical plan-driven project management background to adapt to an Agile environment.

- A good Project Manager knows that you must adapt the project management approach to fit the problem rather than force-fitting every problem or project to a single approach.

AGILE PROJECT MANAGEMENT SHIFTS IN THINKING

The career path for a Project Manager who has a classical plan-driven project management background to become an Agile Project Manager is not easy because it can require some significant shifts in thinking from some of the accepted beliefs that many Project Managers have had for a long time. Making a shift in thinking is probably one of the most significant changes that a classical plan-driven Project Manager has to make in developing a more adaptive approach to operate in an Agile environment.

Emphasis on Maximizing Value Versus Control

The first major difference in an Agile Project Management role is the emphasis on maximizing value versus the classical plan-driven emphasis on predictability and control. Figure 8.1 shows the

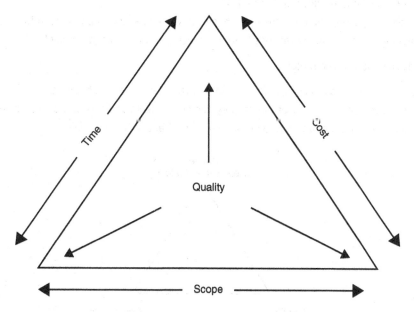

FIGURE 8.1   Classical plan-driven project management iron triangle

classical plan-driven project management iron triangle, which is based on an emphasis on control that has represented a primary focus of the project management profession for many years:

- The idea of it is to fix the scope of the project and put the primary emphasis on controlling the project cost and schedule by controlling changes to scope. A project was deemed successful if it met the original requirements within the budgeted cost and schedule.

- With those constraints, the three legs of the triangle are set, and you can't change any one leg of the triangle without affecting the other legs. For example, for a given scope, you cannot decrease the time of the project without also changing the cost.

    This model has been the predominant model in project management for a long time.

- The problem with this approach is that an excessive emphasis on managing the costs and schedule of a project can lead to a relatively rigid and inflexible style of project management that is not adaptive to uncertain or changing user needs.

- In many cases, it is just unrealistic to be able to completely define all the requirements for a project in detail. For that reason, many projects have met their cost and schedule goals but failed to deliver the business value required by the user.

    The project management triangle in an Agile project is much more complicated because the constraints may not be as rigidly fixed, change is the norm rather than the exception, and there are many more trade-offs to be made in a much more dynamic environment. So, it really is a challenge to apply some level of project management discipline in that kind of environment and it requires a very different approach, as shown in Figure 8.2.

    The impact of putting more emphasis on business value is a significant shift in the focus of decision-making and the primary responsibility for success or failure of the project:

- In a classical plan-driven project:
    - The primary decision-making is focused on managing costs and schedules.
    - The Project Manager is one of the primary decision-makers and has primary responsibility for the success or failure of the project in meeting cost and schedule goals.

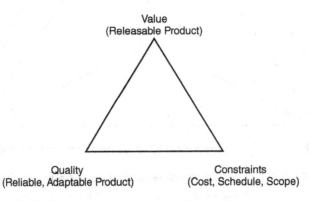

FIGURE 8.2   Value-based project management approach

- Since the requirements are assumed to be relatively fixed, the business involvement can be limited once they're signed off on the requirements, and only significant changes in requirements need to be escalated to the business for approval.

- In an Agile project, the primary decision-making is focused on maximizing business value, and the Product Owner, as a representative of the business, is the key decision-maker. Since requirements are never completely signed off, the business and the Product Owner need to be much more continuously involved in the project and are held responsible for the success or failure of the project in providing the required business value.

In an ideal world, as I've mentioned, there is typically no role for a Project Manager in a small, single-team Agile project, and the Product Owner takes on most of the responsibilities that would normally be held by a Project Manager. However, in the real world, in many companies, Product Owners are not well prepared to take on that responsibility, and a Project Manager may be needed to play a supporting role to assist the Product Owner in fulfilling that responsibility.

> Project Managers are noted for getting results and driving projects to successfully achieve cost and schedule goals, so it does require somewhat of a shift in thinking to focus on delivering value as the primary goal.

Emphasis on Empowerment and Self-Organization

In addition to the shift in decision-making responsibility, there is also a significant shift in the style of how the project is managed to more empowerment of the individuals on the project team to plan and organize their own work without too much detailed direction, as shown in Figure 8.3.

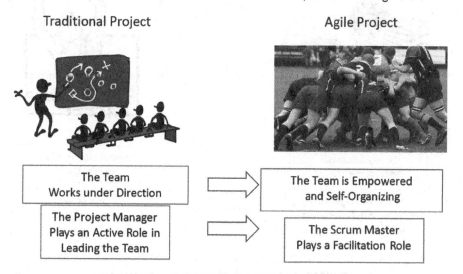

FIGURE 8.3   Impact of empowerment and self-organization.

Source: Alison Bowden/Adobe Stock.

- In a classical plan-driven project, the Project Manager typically might play a very active role in organizing and leading the team to the extent of organizing, planning, and tracking of tasks required to successfully complete the project.

- In an Agile project, the team is expected to be self-organizing. The team reaches agreement with the Product Owner on the goals of each sprint and takes responsibility for planning and organizing tasks that are needed to complete those goals.

In an ideal world, the Scrum Master takes over the project management function in leading the team, but it is a different kind of leadership; the Scrum Master is regarded as a servant leader; he/she plays a facilitation role to help the team plan and organize the activities rather than being overly directive.

In the real world, many teams have difficulty becoming completely self-organizing and more active leadership is needed to steer them in the right direction. Also, the Scrum Master may or may not have all the project management skills needed to take an active leadership role, help the team organize its activities, and remove obstacles to making the team productive. For that reason, sometimes companies will combine the role of a Scrum Master with a Project Manager and call it an Agile Project Manager.

Limited Emphasis on Documentation

One of the important roles played by a Project Manager in a classical plan-driven project is in developing and managing the documentation required by the project because requirements documents and specifications are important deliverables in themselves (Figure 8.4). In fact,

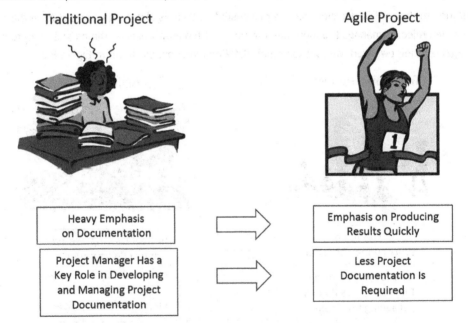

FIGURE 8.4   Limited emphasis on documentation

in a classical plan-driven project, many of the project deliverables are heavily associated with documentation and would not be considered complete without the required documentation.

In an Agile project, documentation plays a much more limited role, and the emphasis is shifted to producing tangible and demonstrable results with much more limited documentation.

- As noted in Chapter 3, there is a popular misconception that Agile projects do not require documentation at all, but that is not the case. Instead, any documentation should provide value in some way to the project.

- An important role for a Project Manager in an Agile project is to implement tools that help the team operate more efficiently and many times that will include online electronic tools that take the place of documentation.

Managing Flow Instead of Structure

A very big change in the role of an Agile Project Manager is a shift in emphasis from managing structure to managing flow:

- In a classical plan-driven project, Project Managers are heavily oriented toward planning and organizing project activities in the form of work breakdown structures, Gantt charts, PERT charts, and others that are all oriented around the structure of the project.

- In an Agile project, many of those things become somewhat irrelevant because you may not have sufficient information in the front-end of the project to define the project structure. In an Agile project, structure is considerably simplified and is understood to be much more fluid. The role of a Project Manager shifts to managing flow to optimize the efficiency of the project. That role is somewhat of an Agile Coach role.

  This is also consistent with the focus on helping the team operate most efficiently.

POTENTIAL AGILE PROJECT MANAGEMENT ROLES

The question you might ask at this point is, where does that leave the Project Manager in a typical Agile/Scrum project? If a lot of the planning and decision-making responsibility are shifted to the Product Owner, there is less emphasis on active team leadership, and less emphasis on documentation; what's left for the Project Manager to do?

The role of a Project Manager in an Agile project is not well defined. Officially, there is no role for a Project Manager at the team level in an Agile project. However, there are a number of potential roles that a Project Manager with the right skills and training for the Agile Project/Program Manager role can perform.

Making Agile Work at a Team Level

In a typical, pure Agile project, the tasks associated with making Agile work at a team level typically involve training the team members in Agile principles and practices and then coaching them in successfully applying those principles and practices in a real-world project. (Hybrid Agile projects that require blending an Agile approach with some level of classical plan-driven project management will be discussed later.)

■ Agile relies very heavily on the skill and judgment of the individuals performing the process, and it can take a lot of training, coaching, and mentoring to get the people on an Agile team to a level of full proficiency.

■ Those tasks are many times handled by an Agile Coach. Rachel Davies, author of *Agile Coaching*, explains what an Agile Coach does:[1]

In a nutshell, an Agile coach helps teams grow strong in applying Agile practice to their work. It takes time to adopt these changes so you can't do this effectively as a seagull consultant or trainer who swoops in to deliver words of wisdom and then makes a sharp exit. You need to spend time with a team to help them to become more aware of their workflow and how to collaborate effectively.

How is being a coach different from a team lead or Project Manager job role? Well, it's not incompatible. The difference is that these roles have a wider set of project- and company-specific responsibilities, such as reporting progress, performance appraisals, etc. I notice that the pressure to deliver can distract from a focus on process improvement. Whereas, if you work solely as an Agile coach, you can make this your sole focus because you don't have responsibility for project deliverables and administrivia.

Being a coach is also different because it's a transitory role not tied into project duration. Your goal is for the team to become self-coaching and adept in applying Agile, then you move on. That doesn't limit Agile coaches to introducing Agile into organizations and establishing new Agile teams. The majority of the teams that I coach are already applying Agile techniques and seek coaching because they want to boost their performance and proficiency in Agile software development.

Can a Project Manager add value at that level? It is certainly worthwhile for an Agile Project Manager to have some Agile coaching knowledge and skills, but being an Agile Coach is a specialized role and it takes a level of focus to do it well. An Agile Project Manager should certainly understand the role of an Agile Coach, at least to the extent of learning to recognize performance issues in an Agile team that might require the skills of an Agile Coach; however, it might be very difficult and awkward in many cases for a Project Manager to try to perform the role of an Agile Coach.

Although there might be no one with the formal title of "Project Manager" at the team level in an Agile project, a number of project management functions and skills are still needed—they are just distributed among the other roles in an Agile project rather than being done by one single person with the title of "Project Manager." For example, the Scrum Master has responsibility for team leadership and facilitation, and the Product Owner has overall responsibility for the success of the project and for planning and prioritizing the requirements of the project.

A well-trained Agile Project Manager can help put in place the right people, process, and tools in a Scrum project to make the project work most efficiently and will help the team blend the right level of classical plan-driven project management with Agile principles and practices into their process and roles. However, that is a very different role than a classical plan-driven project management role and can require some significantly different skills and training.

Hybrid Agile Project Role

As I've mentioned, I don't believe it is a binary choice between Waterfall and Agile; there are a number of alternatives between those extremes, as shown in Figure 8.5.

■ For example, many IT services companies do software application development under contract for clients. Naturally, their clients are not going to write a blank check for a software development project without some expectations of the cost, and it would be irresponsible for an IT services company to commit to a cost estimate without some level of detail of what the requirements are.

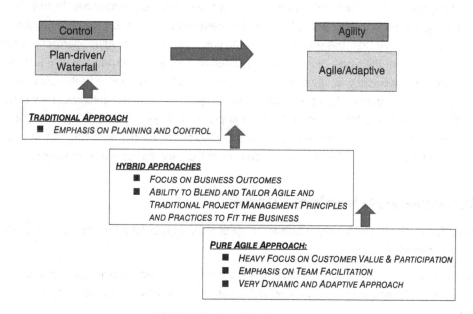

FIGURE 8.5   Hybrid Agile approach

■ Of course, the level of those estimates can range from a firm, fixed-price contract at one extreme based on more detailed requirements to a fairly rough budgetary estimate and very high-level requirements at another extreme.

■ The important thing is that both the IT service provider and the customer have a mutual understanding of the level of accuracy in the estimates.

This is a very challenging role for an Agile Project Manager and may involve:

■ Blending Agile and classical plan-driven project management principles and practices in the right proportions to develop a hybrid management approach when required.

■ Taking a more iterative and adaptive approach to classical plan-driven project management projects.

There are a number of situations where a hybrid Agile approach might be needed. Here are a few examples:

1. **Agile contracts:** Many contracting scenarios require a commitment to some kind of cost and delivery schedule yet still require some flexibility in defining the requirements. At first, those two things might seem to be mutually incompatible, but they are not—this book provides more detail on a methodology I have used to manage a large government contract, and I will also go over a case study of a defense contractor (General Dynamics, UK Limited), who also successfully developed such an approach.

2. **Regulatory requirements:** In fields such as in the medical or pharmaceutical industries, you might need to show traceability over requirements and sufficient control over testing and release processes to ensure that the overall process complies with any regulatory control requirements. At first, this might also seem to preclude any kind of Agile approach; however, that is also not the case—there are ways to intelligently blend Agile and classical plan-driven project management principles and practices in the right proportions to fit the situation.

3. **Large, complex projects:** As projects are scaled to larger, more complex enterprise-level projects, there is also a need for a significant amount of project management. Such projects might involve multiple teams and possibly significant levels of coordination with other activities such as training, support, and operational cutover. We will discuss this scenario separately.

Enterprise-Level Implementation

Above the team level, there is a much broader range of opportunities for a Project Manager to add value at an enterprise level:

■ Managing large, complex enterprise-level projects that require multiple teams and may also require a significant level of coordination and integration with activities outside of the project team.

- Aligning an Agile development process with a company's overall business objectives and culture and developing an overall strategy that blends Agile and classical plan-driven project management principles and practices in the right proportions to fit the company's projects and business environment.

- Planning and leading Agile transformations at the enterprise level.

Those tasks are a more logical extension of a Project Manager's skills and are more likely to utilize the value-added that a Project Manager can provide. Let's look at these roles more closely.

1. Enterprise-Level Project/Program Management Role

There is a potential role for an Agile Project/Program Manager at the enterprise level in:

- Providing project/program management of large, complex initiatives requiring additional planning and management above the team level and helping to align team-level activities with the company's business goals.

- Planning and leading enterprise-level Agile transformations designed to align with the company's business objectives, particularly those that require a blended approach of Agile and classical plan-driven project management principles and practices.

Both of these roles will require a number of skills, including program management skills as well as an understanding of how to blend Agile and classical plan-driven project management practices together at the enterprise level to fit a company's business. At the enterprise-level, the shift in thinking that a Project Manager needs is similar to the team-level role; however, at the enterprise-level, the role may be more of a program management role and, for that reason, it may require more of a strong active leadership role than just a facilitation role.

An example of such a large-scale enterprise-level project is Harvard Pilgrim Health Care. There is a case study on Harvard Pilgrim Health Care later in Chapter 28. The project involved almost completely replacing most of Harvard Pilgrim's legacy systems over a five-year period involving about 100 Agile teams.[2]

The project used a hybrid approach based on a combination of classical plan-driven project management and an Agile development process. It was further complicated by the fact that a large percentage of the development effort was outsourced to another company. Some of the more significant success factors were:

- developing a very collaborative relationship with the company that was contracted to do the development effort, since most of the development effort was outsourced;

- having an integrated hybrid methodology to manage the effort;

- strong leadership with early planning and well-trained people.

The results were very significant. During the five-year duration of this project, which involved a massive redesign of many of Harvard Pilgrim's core business systems, Harvard Pilgrim continued to be named #1 for member satisfaction and quality of care for over nine consecutive years. That's an amazing accomplishment, considering the number of moving parts involved in this effort that all had the potential for impacting Harvard Pilgrim's business.

2. Aligning an Agile Approach with a Company's Business

A senior-level Agile Project/Program Manager could play a role in helping a company align an Agile development approach with its business. Agile is not something that should be done mechanically, and it is important that the Agile approach is well aligned with a company's business and culture:

- An effective implementation of Agile depends on a set of people who have a common understanding of the principles behind Agile and who are also driven by a common set of values.

- A consistent set of principles and values is essential to build cross-functional teamwork and cohesion among everyone involved, both within the team and the business sponsors who are directly engaged with the team.

An Agile development process is not likely to be as effective if it is done mechanically without a common understanding of the values and principles behind it among both the people on the Agile team as well as the people who directly interact with the team or are impacted by the project, such as business sponsors and stakeholders.

- This often causes some level of conflict if the company's culture is not well aligned with an Agile development process.

- If there is a misalignment between the values and principles of Agile and the culture of the company, it may require trying to overcome those differences.

Changing the company's culture so that it is more conducive to Agile is desirable, if that's possible; however, it's often not that simple:

- A company's culture should be shaped around whatever makes sense to drive their primary business and that may or may not be in alignment with an Agile development process.

- Changing the culture of a company is not easy to do and is typically not something that can be done quickly.

For those reasons, it is often necessary to find a compromise, at least in the short term, between an ideal world where the company has adopted a complete Agile model from top-to-bottom, and a more pragmatic approach that recognizes the need to blend an Agile development approach with the company's existing culture and business processes. Corporate culture and enterprise-level Agile transformations will be further discussed later in this book.

3. Integration with Enterprise-Level Management

Many companies have a well-established existing Project Management Office (PMO) to define and manage the implementation of their project management structure and perhaps other levels of enterprise management, such as project/product portfolio management that you just can't unravel overnight. At an enterprise level, many of these companies may have a command-and-control style management approach that would benefit from being replaced with a more sophisticated and more adaptive management approach that is more consistent with an Agile development process, but it may take time to make that transformation.

The strategy that an Agile Project Manager might take in integrating a project management strategy with the company's business may be very different from one company to the next. The challenge for an Agile Project Manager is in helping the company determine how much, if any, of the company's existing enterprise-level management approach it makes sense to keep in place to provide the right balance of control and agility. This decision is very much related to the previous discussion about aligning an Agile approach with the company's business.

Aligning an Agile Project Management approach with a company's business and integrating the approach with existing enterprise-level management processes will be discussed in Chapter 18, "Scaling Agile to an Enterprise Level," and Chapter 20, "Adapting an Agile Approach to Fit a Business."

Using Agile Concepts in Non-Agile Projects

Even if you never manage a true Agile project, knowledge of Agile concepts is likely to significantly enhance the effectiveness and project management skills of someone who only has a classical plan-driven project management background. Almost any kind of project can benefit, to some extent, from a more flexible and adaptive project management approach. Here are a few examples that can go a long way to improving a classical plan-driven project without becoming 100 percent Agile:

- developing a more collaborative approach with the business users

- putting more emphasis on maximizing business value

- taking a more iterative approach to deliver value incrementally

- reducing unnecessary documentation and overhead.

Some other specific Agile techniques can also be useful to apply to a non-Agile project such as:[3]

- incorporating Daily Standups (DSU) meetings to keep the project team members focused on the project objectives

- adding Kanban boards to track the project progress and/or adopting Kanban practices as needed

■ adopting the practice of doing a retrospective on a given cadence so that the project team can discover opportunities to improve its performance quickly rather than waiting for a post-mortem at the end of the project.

AGILE, PMI®, AND PMBOK®

It's been very clear to me for the past 10 years that the project management profession was on the cusp of a very significant change as a result of the influence of Agile, but the Project Management Institute (PMI) has been slow to implement changes to reflect that:

■ **Prior to about 2013:** PMI did not even recognize Agile as a legitimate form of project management.

■ **2013:** In 2013, PMI® released the PMI-ACP® certification, which for the first time, was a formal recognition of the importance of Agile to the project management community. However, at that time, Agile and classical plan-driven project management were still treated essentially as separate and independent domains of knowledge with little or no integration between the two.

■ **2017:** In 2017, PMI released the *Project Management Body of Knowledge (PMBOK® Guide)*, version 6, which didn't really go very far to integrate a focus on Agile Project Management at all, but PMI also released *The Agile Practice Guide* as a separate document.

■ **2019:** PMI announced the acquisition of the Disciplined Agile framework which was a major move by PMI to becoming a player in the enterprise-level Agile space.

■ **2021:** PMI released the *Project Management Body of Knowledge (PMBOK® Guide)*, version 7, which laid the foundation for developing a more integrated approach to project management that included Agile as well as classical plan-driven project management.

Because this change is so significant and requires rethinking a lot of things that have been taken for granted for a long time, it has taken PMI a long time to make this kind of major change in direction. However, we're finally seeing some very significant results with PMBOK® version 7.

Prior PMBOK® Versions

Prior to PMBOK® version 7, PMBOK® was heavily based on a classical plan-driven approach to project management and an explicit knowledge philosophy.

■ It attempted to develop a fairly complete, explicit, and prescriptive base of knowledge of how to do project management.

■ The idea was that you could define a process and checklist telling you what to do and how to do it for almost every conceivable aspect of project management.

- That leads to a somewhat mechanical approach to project management that just doesn't work well in an environment with a lot of uncertainty.

- In that kind of uncertain environment, standardized and prescriptive checklist just don't work well, and a significant amount of judgement and skill is needed to implement a much more flexible and adaptive approach to project management.

For example, prior to PMBOK® version 7, PMBOK® defined a number of major project knowledge areas with a checklist of tasks to consider in each knowledge area:

- Project Integration Management

- Project Scope Management

- Project Time Management

- Project Cost Management

- Project Quality Management

- Project Human Resource Management

- Project Communications Management

- Project Risk Management

- Project Procurement Management

- Project Stakeholder Management

What's Different about PMBOK® Version 7?

PMBOK® version 7 has significantly moved away from a prescriptive plan-driven approach to much more of a principles-based approach. Instead of defining specific process areas with checklists for each, PMBOK® version 7 consists of two major parts:

1. *The Standard for Project Management.*
2. *A Guide to the Project Management Body of Knowledge.*

Part 1 The Standard for Project Management

The first part (*The Standard for Project Management*) is more focused on defining the principles that a Project Manager should follow and leaves it up to the Project Manager to interpret those principles in the context of the situation he/she is in. That is much more consistent with a flexible and adaptive approach to project management. PMI® seems to have recognized that a highly prescriptive approach no longer works in all situations. Instead, PMBOK® version 7 defines some much broader principles as follows:

- Be a diligent, respectful, and caring steward

- Create a collaborative project team environment

- Effectively engage with stakeholders

- Focus on value

- Recognize, evaluate, and respond to system interactions

- Demonstrate leadership behaviors

- Be tailor-based on context

- Build quality into processes and deliverables

- Navigate complexity

- Optimize risk responses

- Embrace adaptability and resiliency

- Enable change to achieve the envisioned future state.

Instead of attempting to tell someone explicitly what to do and how to do it, PMBOK® version 7 focuses on some higher-level principles which must be interpreted in the context of the situation.

It also puts a lot more emphasis on creating value rather than the classical, narrow project management emphasis on planning and control – The most important, whole first part of PMBOK® version 7 is entitled "A System for Value Delivery." It also puts a lot more emphasis on the "softer" aspects of project management such as people skills and leadership skills.

These should help to close the gap significantly between the Agile and project management communities; however, it can be a gut-wrenching change for many Project Managers who have been trained in a classical plan-driven project management approach and never known any other approach.

Part 2 A Guide to the Project Management Body of Knowledge

The second part (*A Guide to the Project Management Body of Knowledge*) is more focused on tangible and explicit results (performance) that a Project Manager is expected to accomplish. It is somewhat similar to the process-orientation of prior PMBOK® versions, but instead of positioning it in a process orientation of explicitly telling you what to do and how to do it, it focuses on the results to be accomplished which, I believe, is a very healthy change.

In other words, instead of saying "here's the process you need to do and here's a checklist of how to do it," it focuses on the results that need to be produced and offers some guidelines for how to produce those results.

There is also a whole section on "tailoring" that recognizes the need to fit the project management approach to the nature of the project.

The Difference Between Explicit and Tacit Knowledge

A major challenge in this area has been the difference between explicit and tacit knowledge. In the past, PMI relied heavily on explicit knowledge to try to define all the project management principles and practices in PMBOK®. Agile is based much more on tacit knowledge which is less explicitly defined.

With PMBOK® version 7, instead of taking an explicit knowledge approach of defining guidelines and checklists to consider in almost every conceivable project management situation, PMI has taken more of a tacit knowledge approach based on principles.

To better understand this comparison, it is essential to understand the difference between tacit and explicit knowledge. Here's a good definition of these two different kinds of knowledge:[4]

- *Explicit knowledge* is codified knowledge found in documents, databases, etc.

- *Tacit knowledge* is intuitive knowledge and know-how, which is:

  - rooted in context, experience, practice, and values
  - hard to communicate—it resides in the mind of the practitioner
  - the best source of long-term competitive advantage and innovation
  - passed on through socialization, mentoring, etc.—it is not handled well by systems that try to document and codify that knowledge.

This site goes on to define another category of knowledge, called embedded knowledge, which is defined as follows:

> *Embedded knowledge* refers to the knowledge that is locked in processes, products, culture, routines, artifacts, or structures. Knowledge can be embedded formally (e.g., through a management initiative to formalize a certain beneficial routine), or informally as the organization applies the other two knowledge types.[5]

An Agile approach relies very heavily on tacit knowledge. It's impossible to tell someone explicitly and exactly what to do in a very uncertain environment and it often requires a certain amount of trial-and-error and adaptation to find the right approach.

For example, Scrum is based on an iterative approach to continuously refine both the solution and the process for arriving at the solution as the project is in progress. That requires much more skill and judgement to adapt an approach to fit the problem and the environment.

An explicit knowledge approach of attempting to tell someone exactly what to do and how to do it in every conceivable project management situation no longer works in the complex, modern environment.

SUMMARY OF KEY POINTS

1. Agile Project Management Shifts in Thinking

There are several significant shifts in thinking that someone with a classical plan-driven project management background may have to make to develop the more adaptive approach that is needed in an Agile environment:

- Emphasis on maximizing value versus control

- Emphasis on empowerment and self-organization

- Limited emphasis on documentation

- Managing flow instead of structure

2. Potential Agile Project Management Roles

The role of a Project Manager in an Agile project is not well defined; and officially, there is no role for a Project Manager at the team level in an Agile project. However, there are a number of potential roles that a Project Manager with the right skills and training for the Agile Project/Program manager role can perform. These roles include:

- Making Agile work at a team level

- Managing hybrid Agile projects

- Enterprise-level project/program management role

However, even if a Project Manager is not actually involved in a pure Agile project, many of the general concepts and principles behind Agile are applicable, to some extent, to almost any project. For that reason, learning these principles and developing a more adaptive approach are probably of benefit for any Project Manager.

3. Agile, PMI®, and PMBOK®

PMBOK® has historically been heavily written for a plan-driven, document-centric approach to project management. The philosophy behind PMBOK® has also been very different from an Agile approach. Up until PMBOK® version 7, it has been more oriented to an explicit knowledge approach—based on the idea that you can codify a checklist of the things you need to do in almost every imaginable project management situation.

- An Agile approach is more oriented to a tacit knowledge approach. Instead of attempting to define an explicit checklist of things to do in every possible situation, Agile tends to define some very broad-based principles, values, and practices that you need to interpret in the context of the situation you're in.

- PMBOK® version 7 has made a significant shift away from the former explicit process-based approach to much more of a principles-based approach. Instead of attempting to tell a Project

Manager what to do and how to do it in almost every conceivable project management situation, it focuses on understanding principles that must be interpreted in the context of the situation. That approach is much more consistent with a flexible and adaptive approach to project management.

DISCUSSION TOPICS

Project Management Shifts in Thinking

1. What do you think is the most important shift in thinking that a Project Manager needs to make to operate in an Agile environment?

2. What would be the most difficult change? Why?

Potential Agile Project Management Roles

1. What do you think is the most likely role for a Project Manager to play in an Agile project?

2. What is the current methodology in your organization? What role can you play in moving the needle towards agile project management?

Agile and PMBOK®

1. Discuss how you might take a different approach to a project based on a broader understanding of the principles in PMBOK® version 7.

NOTES

1. Blog, "Agile Coach: Understanding the Role," posted by Jessica Thornsby, February 25, 2011, http://jaxenter.com/agileagile-coach-understanding-the-role-35038.html.
2. Charles Cobb, *Managed Agile Development: Making Agile Work for Your Business* (Outskirts Press, 2013).
3. Winston Gonzalez, email comments to author on book review.
4. "The Different Types of Knowledge," KMT – An Educational KM Site (2013), http://www.knowledge-management-tools.net/different-types-of-knowledge.html.
5. Ibid.

⑨ Agile Communications and Tools

AGILE COMMUNICATIONS PRACTICES

TWO OF THE MOST IMPORTANT VALUES behind Agile are *openness and transparency* (see Chapter 3). Many classical plan-driven projects in the past limited the sharing of information to carefully controlled channels of communication and bad news was sometimes hidden from view in order to present a favorable image of progress. An important role of a Project Manager in that environment has been to manage the flow of information.

Since the flow of information in an Agile project is more open and transparent, information can be allowed to flow much more easily and automatically using the concept of an *information radiator*. For that reason, the role of a classical plan-driven Project Manager to manage the flow of that information is less essential. In addition to the general importance of openness and transparency, there are a number of other reasons why communication is extremely important in an Agile project:

- Information is rapidly and dynamically changing in real time throughout the project and needs to be shared efficiently.

- Good communications are essential to support close collaboration among everyone on the project team as well as people who may be peripheral to the project team, especially if the team cannot be co-located, that is, working in the same area, which is often the case.

- Sharing information with the customer and business sponsor is also essential for the same reasons to support a close and collaborative partnership relationship.

As Agile projects become larger and more complex, using tools to help distribute information quickly and efficiently becomes essential.

Information Radiators

An *information radiator* is a large and highly visible display of critical team information that is typically located in a place where the team and others can see it constantly, and it is continuously updated either during the Daily Standup meetings or even more frequently in real time as work is completed. It could take on several different forms:

- In some cases, it is actually a large whiteboard or a complete wall of a room that is used to track progress.

- It could also be in electronic form by using one of a number of different Agile project management tools so that a much broader audience can view the information online.

The concept of an information radiator in Agile is consistent with the value of openness and transparency:

- Everyone on the team is aware of the work of all other members of the team and how it contributes to the goals of the team, which is consistent with promoting a strong and unified teamwork approach.

- The Product Owner and other stakeholders outside of the team are also aware of progress and issues that might impact progress, which is consistent with promoting a spirit of partnership and customer collaboration.

- A whiteboard approach, using a physical whiteboard and colored "stickie" notes, has been widely used in Agile projects for a long time, but it has some significant limitations.
  - It doesn't work well with distributed teams.
  - It doesn't provide the ability to easily roll up information across multiple teams.
  - It is difficult to keep the information organized and it also doesn't provide the capability to sort and report on the information in different ways.

For those reasons, many Agile projects are moving to more widespread use of online tools. Figure 9.1 shows an example of an information radiator using an online tool.

A big advantage of information radiators is that each person on the project team can individually update the status of tasks that he/she is responsible for in near real time and all of that information will be aggregated in the information radiator for sharing with others both inside and outside the project team.

That reduces the role that a Project Manager has provided of aggregating and reporting project status information. An information radiator is typically used in conjunction with the Daily Standup meeting to discuss and review the work being done by the team.

If an online tool is used, the information can also be rolled up at a number of different levels across multiple teams and even across multiple projects.

The use of information radiators is a very powerful aspect of Agile, and if it is implemented correctly:

- It promotes openness and transparency which help build stronger and more effective teamwork. That is consistent with Agile Manifesto principle #4, "Business people and developers must work together daily throughout the project."

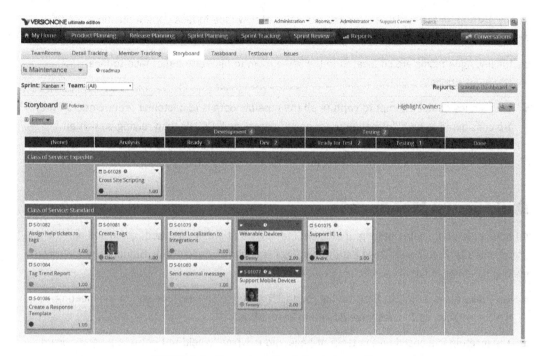

FIGURE 9.1 Example online information radiator.

Source: VersionOne, Inc.

- It promotes customer collaboration and helps to build a spirit of trust and partnership between the development organization and the business users.

- It can remove a huge burden on management that are attempting to control all aspects of an operation. In one operation, a senior manager commented on how much Agile has relieved the management team from resolving day-to-day issues that the project team now takes responsibility for, and it has enabled the company to focus on much higher-level strategic goals.

Face-to-Face Communications

Agile heavily emphasizes face-to-face communications wherever possible over other forms of communication.

> The most efficient and effective method of conveying information to and within a development team is face-to-face conversation.[1]

However, that does *not* mean that documentation and other forms of communication are eliminated, and this principle needs to be adapted to fit the situation, particularly in the case of widely distributed teams. Here are some examples of how face-to-face communications are used in Agile projects:

- Ideally, Agile teams are co-located in the same room to facilitate direct face-to-face communications.

- User stories do not attempt to capture all the possible details of customer requirements—they are considered to be a "placeholder for communications." Rather than relying on written documents and specifications for requirements, user stories are very brief and succinct and rely on face-to-face communication to clarify and further define what is needed.

Daily Scrum Meetings

All of us have participated in long, inefficient meetings that are not focused, go on and on, and don't use people's time very well. Daily Scrum meetings (sometimes called Daily Standups) are a way of making meetings much more efficient in an Agile project.

- They take place daily, but they are limited to 15 minutes, and each person in the meeting typically stands up to encourage people to keep it brief.

- The meeting is focused on the tasks at hand and is typically held in front of a progress board.

- The meetings are structured so that each person typically answers three primary questions:

  - What did you accomplish yesterday?
  - What are you going to accomplish today?
  - What obstacles are in your way?

Naturally, the idea of a daily Scrum might need to be modified to fit the environment, particularly with distributed teams and the format of these three questions is only a suggestion and might be different in actual practice. New teams may also need some coaching on how to make these meetings meaningful and productive.

Distributed Teams

Many Agile communications practices (e.g. Daily Standups) are based on the assumption that the team is co-located. However, in many situations, it isn't possible to have that level of co-location and the communications practices need to be adjusted to support distributed teams. Distributed teams may not be in the same geographical location and could even be in different time zones, which is often the case with offshore development teams. In that case, you have to adjust the communications strategy to fit distributed teams:

- Video conferencing might be used in lieu of face-to-face communications.

- Online status boards might be used in lieu of physical wall boards.

■ Wikis might be used for sharing of information and online communications tools such as Slack, Zoom, Webex, etc. will likely play an important role.

■ Daily Scrum meetings might need to be extended or supplemented with other meetings to provide more communications.

Managing communications with offshore teams in an Agile project can be very challenging. Some of the challenges are:

■ geographic separation

■ cultural and language differences

■ time zone differences.

I was in a situation where the entire development team was in India and only the customer-facing people were in the United States. Informal meetings next to the water cooler will not work in this situation. But even in an Agile environment where the team is co-located, a brief Daily Standup might not be sufficient. If it is the only (or primary) form of communication with the team throughout the day, it can be very difficult to limit the Daily Standup to just a 15-minute session. You must learn to adapt the communication practices to the situation.

AGILE PROJECT MANAGEMENT TOOLS

Agile Project Management tools are very different than classical plan-driven project management tools in several respects as shown in Table 9.1:

TABLE 9.1   Comparison of Project Management Tools

	Classical plan-driven project management tools	Agile project management tools
Primary focus	Focused on managing a relatively static project *structure* with Gantt charts, PERT charts, etc. and managing progress against predefined project budgets and schedules	Focused on managing the overall *flow* of work in the project and maximizing the efficiency of producing value as quickly as possible
Emphasis	Control of information and limiting information to those with a need to know	Sharing information and making all information available to a broad audience
Tool management	The information in the tool is typically managed and controlled by a Project Manager	Everyone on the team has the ability to update information in the tool related to their area of work
Example	Microsoft Project	Jira

In the original edition of this book, I tried to go into more detail about Agile Project Management tools and to show some specific examples of tools that were in use at that time; however, that information quickly became out-of-date because this area changes so frequently. For that reason, I am going to limit the discussion in this edition to a more general description of Agile Project Management tools.

You might think that tools play a limited role in Agile due to the Agile Manifesto value of "Individuals and interactions over processes and tools," however, tools still play a very important role in Agile, especially as you start to scale projects to large, complex enterprise-level solutions and that's the area where you are most likely to find an Agile Project Management role.

There are many tools that are used in Agile, ranging from very simple tools that are oriented around simple, team-based activities to more complete tools that include team-level capabilities but also go well beyond that level and provide a capability for scaling projects to enterprise-level projects.

In my opinion, an Agile Project Manager who doesn't know how to use one of the widely used Agile Project Management tools is equivalent to a classical plan-driven Project Manager who doesn't know how to use Microsoft Project.

Benefits of Agile Project Management Tools

At an enterprise level, Agile Project Management tools provide some significant value:

1. **Ability to fully engage all members of the team in the process**: An Agile Project Management approach is very different from a classical plan-driven project management approach.

   - In a classical plan-driven project, the Project Manager is primarily responsible for planning the project as well as managing and reporting progress.
   - Microsoft Project is well designed to support that role because it is a stand-alone desktop tool and most of the information flows through the Project Manager.
   - In an Agile project, the responsibility for planning and managing the effort is distributed among everyone on the team, and Agile Project Management tools are designed with that in mind.

2. **Ability to update progress in real time and rapidly view status and issues (information radiators):** An Agile project is also fast moving, and it's very important to be able to update information quickly in real time and for anyone to be able to view that information easily. That's the concept of an information radiator.

3. **Ability to scale projects to an enterprise level and provide a standardized way of reporting across projects:** As I've previously discussed, there are different kinds of information radiators.

   - Small, single-team Agile projects might use a simple Kanban board on a whiteboard with index cards or "stickies" on it to manage and track progress. That method works OK for small teams but falls apart quickly for projects with teams that are not co-located or projects that require multiple teams.

■ Most of the Agile Project Management tools provide at least the capability for an online Kanban board to enable sharing of information across distributed teams but go beyond that to provide an enterprise-level view of all major projects in the organization.

Characteristics of Enterprise-Level Agile Project Management Tools

As an example, VersionOne has developed a nice summary of the required criteria for selecting a robust, enterprise-level Agile project management tool, which is reproduced here with permission from VersionOne.[2] Note: since the original edition of this book was published, VersionOne has been acquired by Digital.ai.

Full Agile Process Life Cycle Coverage

Agile management systems must span the full range of processes from integrated customer feedback through portfolio/product planning, code integration, testing, and delivery.

■ Product, release and iteration planning and tracking

■ Agile portfolio management, including epic boards, epic ranking, rapid epic breakdown, and reporting

■ Strategic goals, functional rollups (themes), goal assignment, impediment tracking, and retrospectives

■ Consolidated requirement backlog and defect repository

■ Test management (manage, track, and execute tests across multiple stories, defects, and distributed projects)

■ Integrated product road-mapping linked to releases/iterations.

Team and Customer Collaboration

Collaboration capabilities should promote teamwork and expedite communication between team members, teams, and organizations.

■ Social media-style communication portal for project teams

■ Contextual collaboration includes links to work items, team members and dedicated Team Rooms

■ RSS feeds and email notifications for receiving message and alerts

■ Cross-project planning, tracking and reporting for distributed team members

---

- Board Views are customizable, filterable, and support flexible color coding to convey information

- Integrated customer Idea Management platform (facilitating customer engagement, collaboration, and prioritization).

Visibility, Reporting, and Analytics

Agile management solutions must include concise views of data, work estimates, actuals, and trends.

- Executive-level dashboards with best practice metrics

- Advanced planning, including: release forecasting, "what-if" analysis, and work-load balancing

- Team burn-down, burn-up, cumulative flow, and test trend reporting

- Epic boards/Storyboards with drag-and-drop tracking of story cards, work-in-process limits, aging and cycle-time reporting

- Roll-up reporting for teams working across projects and projects with multiple teams

- Best-practice Agile dashboards: Project, Sprint, Program, Team Member, and role-based dashboards

- Custom reports and graphs created via web-based, wizard-driven interface

- Agile visualizations—Epic bubble charts, Hierarchy/Tree charts, Relationship Visibility, Release Dependency Mapping.

Simplicity and Ease of Use

Use a tool with a simple user interface and built-in process navigation.

- Built-in Agile process navigation with customizable dashboards for team members to track their projects and work

- Drag-and-Drop ranking and whiteboard-style release and iteration planning

- Drag-and-Drop epic, story, task, and test boards with customizable workflow processes and configurable card data

- Multi-select options for actions such as: Move; Close; Reopen; Delete; and Rank

- Multi-level estimation at the story/defect level and the effort tracking level

- TeamRoom™—dedicated team-based environment supporting the daily activities of development teams

- PlanningRoom™—dedicated environment for program-level managers to collaborate in focused planning sessions.

Configurable Workspaces, Process, and Terminology

Agile Tools should guide you through the Agile process and help you implement "Agile your way." Custom workspaces and terminology support unique process configurations without sacrificing visibility, reporting, and analytics across the portfolio.

- Drag-and-drop epic, story, task, and test boards with customizable workflow and configurable card data
- Customizable methodology templates (XP, Scrum, DSDM, Kanban, etc.)
- Extensive customization options, including boards, fields, lists, value, grids, etc.
- Customizable folder structure for nested project/release hierarchy
- Customizable and filterable board views that support color-coded visual indicators.

Agile Portfolio and Program Management

Agile tools should grow with you as your needs grow, including portfolio-level planning for your strategic initiatives, program-level coordination and project-level story delivery with full traceability from high-level epics to development-level tasks.

- Portfolio-level business initiatives, release rollouts, progress and organizational velocity mapped against a timeline
- Program-level Epicboards, Epic Bubble charts, epic ranking, planning, and roll-up reporting
- Integrated product roadmapping linked to releases/iterations
- Advanced planning, including: release forecasting, "what-if" analysis, and workload balancing
- Cross-project team planning, tracking, and reporting
- Release Dependency Diagram to better prioritize story completion
- Extensive support for Scaled Agile Framework® (SAFe™) PPM methodology (Planning, Process, Metrics, Reports, etc.).

Deployment, Security, and Integrations

Agile tools should provide: flexible deployment options for ANY team size and ANY Agile methodology; application-, role- and project-level security; and open integrations for simplified deployment and customization.

That's a long list of capabilities and, naturally not all of those capabilities may be required in a particular implementation, but it's a good list, and it's always easier to take away something that's not needed rather than adding something that's not there.

SUMMARY OF KEY POINTS

Agile Communications Practices

1. **Communications:** Communications in an Agile project are extremely important to support openness and transparency and they are essential to support teamwork and rapid and efficient coordination among the people on the project team, as well as close collaboration and partnership with the customer and Business Sponsor.

2. **Information radiators:** Information radiators are widely used in Agile projects to disseminate information rapidly and efficiently in a very dynamic and fast-paced environment. The simplest form of information radiator is a whiteboard with colored "stickies" to track progress of work. Online tools have a number of advantages over whiteboards and can be more effective in many environments for a number of reasons, such as the ability to easily roll up information across multiple teams and the ability to support distributed teams that cannot be co-located.

3. **Face-to-face communications:** Agile emphasizes face-to-face communications; however, in the real world, communications practices frequently need to be adapted to fit teams that cannot be co-located. In that situation, there are a number of alternatives that can be used in lieu of or to supplement direct face-to-face communications. Those alternatives include video conferencing, online status boards, and Wikis.

Agile Project Management Tools

1. **Different orientation:** Agile project management tools have a very different orientation than classical plan-driven project management tools.

   - Classical plan-driven project management tools are heavily oriented around planning and managing the structural aspects of a project (Gantt charts, Pert charts, dependencies, etc.).

   - Agile project management tools are much more organized around planning and managing flow and the classical emphasis on structure is considerably simplified or not needed at all.

2. **Collaborative team approach:** Agile project management tools are also designed around a collaborative team approach where each individual on the team has direct access to the tool for planning and tracking their own work rather than all work being coordinated and managed by a Project Manager.

3. **Scalability of tools:** Many Agile project management tools are very scalable and offer capabilities all the way from simple single-team Agile projects to complex, large-scale enterprise-level capabilities.

DISCUSSION TOPICS

Agile Communications Practices

1. What are the major differences between communications practices in an Agile project and a conventional, non-Agile project? Why are communications so important in an Agile project?

2. What is an information radiator? What are the advantages of online tools for managing communications in an Agile project?

3. What are some of the challenges associated with distributed and/or offshore teams, and how would you go about resolving them?

Agile Project Management Tools

1. What Agile tool capability do you think is most important in a typical Agile project? Why?

2. What do you think is the most important factor in choosing an appropriate tool for an Agile project environment? Why?

NOTES

1. "Principles behind the Agile Manifesto," http://www.agilemanifesto.org/principles.html.
2. "VersionOne Tool Evaluator Guide," http://www.versionone.com/pdf/agiletoolevaluator.pdf.

(10) Learning to See the Big Picture

AN AGILE PROJECT MANAGER NEEDS TO UNDERSTAND Agile at a deeper level in order to apply it to different situations effectively. The key to that is to develop a "Systems Thinking" approach to understand Agile principles and practices at a deeper level. In order to develop that kind of systems thinking approach, it is also valuable to understand the roots of Agile and how Agile thinking evolved. The roots of Agile will be discussed in Chapter 11.

SYSTEMS THINKING

What Is Systems Thinking?

BusinessDictionary.com defines systems thinking as follows:

> Practice of thinking that takes a holistic view of complex events or phenomenon, seemingly caused by myriad of isolated, independent, and usually unpredictable factors or forces. Systems Thinking views all events and phenomenon as "wholes" interacting according to systems principles in a few basic patterns called systems archetypes. These patterns underlie vastly different events and phenomenon such as diminishing returns from efforts, spread of contagious diseases, and fulfillment in personal relationships.
>
> Systems Thinking stands in contrast to the analytic or mechanistic thinking that all phenomenon can be understood by reducing them to their ultimate elements. It recognizes that systems ("organized wholes") ranging from soap bubbles to galaxies, and ant colonies to nations, can be better understood only when their wholeness (identity and structural integrity) is maintained, thus permitting the study of the properties of the wholes instead of the properties of their components.[1]

In the context of Agile, *systems thinking* means understanding the principles behind the methodology rather than just focusing on the mechanics of how the methodology works and understanding how those principles interact with the overall project and business environment that they are part of. Why is systems thinking important? It allows you to see things in an entirely different perspective:

- You see the whole rather than the pieces and understand their relationship. In an Agile implementation, you see the business as a large ecosystem and see the development process as only one component of that ecosystem and you begin to better understand how the two are interrelated.

- Within an Agile development process, you begin to better understand how all the components of that process work together to make the overall process more effective and instead of following the process rigidly and mechanically, you see it as a much more dynamic process where each component of the process may need to be adjusted to fit the situation.

Binary thinking is the antithesis of systems thinking. Instead of seeing the real complexity that is inherent in many situations, people who engage in binary thinking are sometimes looking for a simple, cause-effect explanation for something that isn't really very simple at all:

- They tend to see the Agile values and principles in black-and-white terms, as absolute statements, rather than relative statements that need to be interpreted in the context of the situation as they were intended to be.

- They see the relationship of Agile and more classical plan-driven approaches as either-or, mutually exclusive choices (either you're Agile or you're not) and they may see these approaches as competitive with each other rather than seeing them as potentially complementary.

That sort of narrow thinking has led to many stereotypes, myths, and misconceptions about what Agile is, and also about what classical plan-driven project management is. We need to rethink what Agile is as well as rethink what classical plan-driven project management is to see them in a new light as potentially complementary rather than competitive approaches. Systems thinking is the key to that.

How Is Systems Thinking Used in Organizations?

System thinking is closely related to the idea of a learning organization. BusinessDictionary.com defines a *learning organization* as follows:

> An organization that acquires knowledge and innovates fast enough to survive and thrive in a rapidly changing environment. Learning organizations (1) create a culture that encourages and supports continuous employee learning, critical thinking, and risk taking with new ideas, (2) allow mistakes, and value employee contributions, (3) learn from experience and experiment, and (4) disseminate the new knowledge throughout the organization for incorporation into day-to-day activities.[2]

Systems thinking provides a mechanism to understand the dynamics behind how an organization works at a deeper level. The culture of a *learning organization* creates an environment where that information is used for ongoing, continuous improvement. Adopting a *systems thinking* approach and becoming a learning organization are two of the most important aspects of achieving enterprise-level agility.

COMPLEX ADAPTIVE SYSTEMS

What Are Complex Adaptive Systems?

An empirical process control model is an example of a "Complex Adaptive System."

An Agile Team or a Scrum Team is also an example of a complex adaptive system (CAS) where one to many Agile teams work together towards the same goal, learning and adapting as needed using feedback from its environment and interconnected parts. CAS also can self-correct during good and bad times.[3]

An understanding of Complex Adaptive Systems is important because it fundamentally defines how an Agile approach works. The following is a definition of Complex Adaptive Systems:

> As early as 1997, Kevin Dooley, defined a Complex Adaptive System (CAS) as a group of semi-autonomous agents who interact in interdependent ways to produce system-wide patterns, such that those patterns then influence the behavior of the agents. In human systems at all scales, you see patterns that emerge from the interactions of agents in that system. Thoughts, experiences, perceptions interact to create patterns of thought.[4]

A good example of a complex adaptive system is an ant hill. When ants are building an ant hill, I'm guessing that they don't start out with a grand plan of what the finished ant hill is going to look like—it just evolves and these thousands or millions of ants all work collaboratively together as a self-organizing team without a lot of direction. And, when the ant hill is finished, it may not be a perfect-looking structure, but it satisfies the purpose it was intended for, and it will evolve as needed to fit the overall environment that it is part of (Figure 10.1).

If it is done correctly, a Complex Adaptive System approach is a very powerful approach to project management.

- It's like the ant hill example, I suspect that there aren't designated project managers among the ants building the ant hill to plan and delegate the work to individual ants.

- The individual ants somehow just work together naturally and collaboratively to build the ant hill without a lot of explicit direction.

FIGURE 10.1   (a) Ants at work; (b) typical large ant hill.
Source: (a) Violet Kaipal/Adobe Stock; (b) BestPhotoStudio/Adobe Stock.

- That's exactly the way an Agile project works. If it is done right, an Agile team doesn't need a lot of explicit direction to be told exactly what to do and how to do it.

- With some very general direction, they should be able to figure out what needs to be done to accomplish the goals that have been established.

Peter Fryer has developed an excellent explanation of how Complex Adaptive Systems have evolved:

> For many years scientists saw the universe as a linear place. One where simple rules of cause and effect apply. They viewed the universe as a big machine and thought that if they took the machine apart and understood the parts, then they would understand the whole. They also thought that the universe's components could be viewed as machines, believing that if we worked on the parts of these machines and made each part work better, then the whole would work better. Scientists believed the universe and everything in it could be predicted and controlled.[5]

We need to recognize the dynamic nature of Complex Adaptive Systems in our projects and recognize that those systems need to evolve as necessary to fit the environments that they are part of. Many classical plan-driven projects fail to recognize and take advantage of the dynamic nature of the project teams that are part of the project; and, as a result, fail to utilize the full power of the teams.

Characteristics of Complex Adaptive Systems

Complex Adaptive Systems Are Emergent

Rather than being planned or controlled, the agents in the system interact in apparently random ways. They are also part of the environment that they operate in and need to change to ensure best fit to the overall business environment.

Complex Adaptive Systems Are Sub-Optimal

Another characteristic of Complex Adaptive Systems is that they are sub-optimal:

- A Complex Adaptive System does not have to be perfect in order for it to thrive within its environment.

- It only has to be slightly better than its competitors and any energy used on being better than that is wasted energy.

- A Complex Adaptive System, once it has reached the state of being good enough, will trade off increased efficiency every time in favor of greater effectiveness.[6]

This is also true of an Agile process—an Agile process is based on the understanding that absolute perfection is not necessarily the most appropriate goal. Ideally, an Agile process seeks to meet the requirements for a product in terms of minimal marketable features and only going beyond that level if it provides business value.

Complex Adaptive Systems Have Requisite Variety

Ideally, Complex Adaptive Systems also have what is called requisite variety:

- The greater the variety within the system, the stronger it is. In fact, ambiguity and paradox abound in Complex Adaptive Systems, which use contradictions to create new possibilities to co-evolve with their environment.

- Democracy is a good example in that its strength is derived from its tolerance and even insistence in a variety of political perspectives.[7]

Here's how this pertains to Agile—a strong Agile team is able to successfully integrate people of different perspectives into a very cohesive cross-functional team. A team of "yes men" who all think alike is typically not that strong.

Complex Adaptive Systems Have Connectivity

> The ways in which the agents in a system connect and relate to one another are critical to the survival of the system, because it is from these connections that the patterns are formed, and the feedback disseminated.[8]

In relation to an Agile process, this is extremely important, the agents in an Agile process are the people on the team as well as the outside stakeholders who interact with the team. The success of the process is critically dependent on the relationships and interaction among all these people. Ideally, those interrelationships are very dynamic and based on simple rules without being overly controlled or structured.

Complex Adaptive Systems Have Simple Rules

The next property is that Complex Adaptive Systems have simple rules:

- Complex Adaptive Systems are not complicated.
- The emerging patterns may have a rich variety; but like a kaleidoscope, the rules governing the function of the system are quite simple.
- A classic example is that all the water systems in the world, all the streams, rivers, lakes, oceans, waterfalls, etc. with their infinite beauty, power, and variety are governed by the simple principle that water finds its own level.[9]

Agile methodologies like Scrum are not overly prescriptive. In fact, Scrum is generally considered to be a "framework" and not a methodology. It provides some very simple rules to follow rather than defining an explicit and prescriptive methodology.

Complex Adaptive Systems Are Iterative

Complex Adaptive Systems are also iterative:

- Small changes in the initial conditions of the system can have significant effects after they have passed through the emergence-feedback loop a few times (often referred to as the butterfly effect).
- A rolling snowball, for example, gains on each roll much more snow than it did on the previous roll and very soon a fist-sized snowball becomes a giant one.[10]

An Agile process is both incremental and iterative:

- An "incremental" process develops the overall solution in small "chunks" rather than attempting to develop the entire solution all at once.

- An "iterative" process might start with a prototype of what is needed and then continue to refine that prototype based on user feedback until it satisfies the user requirement.

Complex Adaptive Systems Are Self-Organizing

Complex adaptive systems are self-organizing:

> There is no hierarchy of command and control in a Complex Adaptive System. There is no planning or managing, but there is a constant re-organizing to find the best fit with the environment.[11]

An Agile team is self-organizing. That means that no one gives the team explicit direction on how they should complete their work. They are given a general requirement and they are expected to figure out the best way for the team to accomplish that requirement without a lot of explicit direction.

Complex Adaptive Systems Operate on the Edge of Chaos

Systems exist on a spectrum ranging from equilibrium to chaos:

> - A system in equilibrium does not have the internal dynamics to enable it to respond to its environment and will slowly (or quickly) die.
>
> - A system in chaos ceases to function as a system.
>
> - The most productive state to be in is at the edge of chaos where there is maximum variety and creativity, leading to new possibilities.[12]

A popular model that is used to visualize this in an Agile environment is the "Stacey Complexity Model," shown in Figure 10.2.

The level of uncertainty in a project is a major determinant of the complexity level. In a typical project, there are two dimensions of uncertainty:

- **Requirements uncertainty**—how uncertain are the requirements in the project?

- **Technology uncertainty**—how uncertain is the technology that is required to implement the solution?

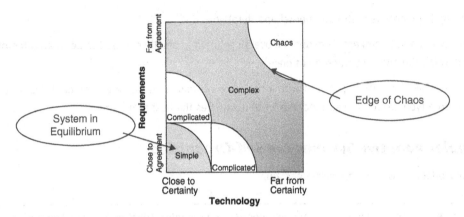

FIGURE 10.2   Complexity Theory

Classical plan-driven projects where predictability is important are not well designed for high levels of uncertainty. Sometimes, in order to achieve predictability, both the requirements and technology needed to implement the solution are simplified to reduce the level of uncertainty in the project. That might impact the quality of the overall solution and tend to suppress the creativity and innovation that might be needed.

Complex Adaptive Systems Are Nested Systems

Most systems are nested within other systems and many systems are systems of smaller systems. It is important to optimize the performance of a project team within the broader context of the business system that it is part of rather than just within a narrower project development context.

SUMMARY OF KEY POINTS

Systems Thinking

1. Systems thinking is very important in order to see Agile principles and practices in a holistic sense and in the context of how they fit with the overall business objectives of an enterprise. In the context of Agile,

   ■ Systems thinking means understanding the principles behind the methodology rather than just focusing on the mechanics of how the methodology works.

- It means understanding how those principles interact with the overall project and business environment that they are part of.

Complex Adaptive Systems

An understanding of "Complex Adaptive Systems" is important because it fundamentally defines how an Agile approach works. Complex Adaptive Systems are dynamic organisms that evolve and adapt to meet a particular challenge. Some of the key properties and characteristics of a Complex Adaptive System are:

- Emergent
- Sub-optimal
- Of requisite variety
- Have connectivity
- Follow simple rules
- Iterative
- Self-organizing
- Operating on the edge of chaos
- Nested systems.

DISCUSSION TOPICS

Systems Thinking

1. Discuss an example of a problematic situation where you might have used systems thinking to analyze the situation and determine an appropriate solution.

Complex Adaptive Systems

1. Identify two or three systems that occur in nature that would be considered Complex Adaptive Systems.

2. How does the idea of a Complex Adaptive System help to explain the difference between a classical plan-driven project management approach and an Agile approach?

3. How would you use the idea of Complex Adaptive Systems in a project management approach?

NOTES

1. Human Systems Dynamics Institute, "Complex Adaptive System," https://www.hsdinstitute.org/assets/documents/5.1.1.4.complex-adaptive-system-04may16.pdf.

2. Ibid.

3. Winston Gonzalez, email comments to author on book review.

4. Peter Fryer, "What Are Complex Adaptive Systems?" http://www.trojanmice.com/articles/complexadaptivesystems.htm.

5. Ibid.

6. Ibid.

7. Ibid.

8. Ibid.

9. Ibid.

10. Ibid.

11. Ibid.

12. Ibid.

11 The Roots of Agile

AN AGILE PROJECT MANAGER NEEDS TO UNDERSTAND the roots of Agile and how Agile thinking evolved, in order to gain a deeper understanding of why it makes sense. The roots of Agile go fairly deep, but there are two major sources that had the most impact on its development:

■ Total Quality Management (TQM) was probably the strongest factor in influencing the Agile approach to quality.

■ Lean Manufacturing was probably the biggest factor in influencing Agile process thinking.

  Each of those influences will be discussed in this chapter.

INFLUENCE OF TOTAL QUALITY MANAGEMENT (TQM)

Total Quality Management had a significant impact on how quality is managed in an Agile environment:

■ Many of the Agile principles related to quality have their roots in the philosophy of Total Quality Management (TQM). The TQM philosophy originated from the ideas of W. Edwards Deming and others. Although TQM was originally developed for a manufacturing organization, it's fairly easy to see how those same principles are equally relevant to a software development environment.

■ Dr. Deming was an American statistician who was credited with the rise of Japan as a manufacturing nation. His principles transformed the Japanese automotive industry into developing high-quality products that gained a significant market share against American automotive manufacturers in the 1970s and 1980s (see Figure 11.1).

  Dr. Deming's original 14 points can be summarized into five major areas that have a significant impact on an Agile project management approach. The following is a summary of some of Dr. Deming's original 14 points that the TQM philosophy is based on that form the roots of today's Agile approach

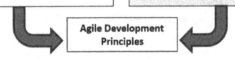

Total Quality Management (TQM) Principles	Lean Manufacturing (Lean) Principles
• Cease Dependence on Inspection and Build Quality into the Product • Emphasis on the Human Aspect of Quality • Need for Cross-functional Collaboration and Transformation • Importance of Leadership • Ongoing Continuous Improvement	• Focus on Customer Value • Map the Value Stream • "Pull" versus "Push" • Importance of Flow • Respect for People • Perfection (Continuous Improvement)

Agile Development Principles

FIGURE 11.1   The roots of Agile practices

for software development. According to Deming, "The 14 points all have one aim, to make it possible for people to work with joy."[1]

1. Cease dependence on inspection.
2. Emphasis on the human aspect of quality.
3. The need for cross-functional collaboration.
4. Importance of leadership.
5. Ongoing continuous improvement.

Each of these points and how it impacts an Agile Project Management approach will be discussed in the following sections (Tables 11.1–11.6).

Cease Dependence on Inspection

TABLE 11.1   Deming's principles: cease dependence on inspection

Deming's Principle	■ Point #3 "Cease dependence on inspection by building quality into the product in the first place."
Manufacturing problems prior to TQM	**Manufacturing TQM approach**
Prior to TQM, manufacturers relied heavily on quality control inspectors to detect problems at the end of the assembly line. The problems with that approach are:	A better approach is to go upstream in the process, find the sources of error that contribute to the defects, and eliminate those sources of error at the source
■ If problems are not detected until the end of the assembly line, it can be expensive to go back and rework or scrap the product at that point. It also can have a significant impact on cycle time to wait for a product to be reworked before it can be shipped.	■ That approach results in better quality because the quality is designed into the product from the beginning, and defects are prevented before they happen

(Continued)

Manufacturing problems prior to TQM	Manufacturing TQM approach
Limitations	**Advantages**
■ Any inspection approach like this relies heavily on sampling; and, with a limited number of inspectors, it's very time-consuming to do a very large sample size. For that reason, it's difficult to fully test every possible product or situation that could result in a defect. As a result, some defects are bound to slip through and result in very poor quality, as seen by the customer.	■ Reduced costs because it relies less on inspection to find problems
	■ Greater pride of ownership by the workers who are primarily responsible for producing the product
■ Quality is seen as the responsibility of the inspectors, who are perceived as the *enforcers*. As a result, the people who are producing the products may not feel fully responsibility for the quality of the products they produce.	■ Higher levels of quality because an emphasis on quality is "baked into" the design of the product
■ It is expensive to employ enough quality control inspectors to perform this function completely, and even with lots of inspectors, it still might not be very effective.	

Implications for Agile Development

■ An Agile development approach recognizes and incorporates this principle.

■ Instead of performing QA testing sequentially with development and relying heavily on a typical quality assurance approach to detect problems after development has been completed and sending the product back for rework to fix defects (bugs), it is far more effective to make quality an integral part of the development process, do it concurrently with development, and prevent the bugs (defects) from happening at the source.

■ This approach also results in reduced costs because it relies less on inspection (QA) to find problems after the product development is complete.

■ Agile makes producing quality products the responsibility of the team developing the product, which typically includes QA testers. It's not someone else's responsibility (like a separate QA department) to ensure that the product is free of defects. This increases pride of ownership of everyone on the team and results in much higher quality products.

Emphasis on the Human Aspect of Quality

TABLE 11.2   Deming's principles: emphasis on the human aspect of quality

Deming's Principles	
	■ Point #6: "Institute training on the job."
	■ Point #8: "Drive out fear."
	■ Point #12: "Eliminate barriers to pride of workmanship."
	■ Point #13: "Institute education and self-improvement."

Manufacturing problems prior to TQM	Manufacturing TQM approach
In the early days of manufacturing, *sweatshops*, where people worked under oppressive conditions, were quite common. It's obvious that people who are overworked and who aren't respected and recognized for the importance of the work they do are probably not going to be highly motivated and are likely to produce a much lower quality product. In particular,	Manufacturing companies learned a long time ago to pay attention to the human aspects of producing quality products:
■ People who are fatigued from working very long hours in a less-than-ideal working environment are prone to make errors	■ Improving working conditions and automating menial, repetitive tasks as much as possible, rather than relying on people to perform those tasks, uses the skills of people more effectively and removes a major source of errors and defects
■ People with narrowly defined and repetitive jobs, who only have a limited responsibility for a small portion of the overall product (such as putting the bolts on a wheel on an automobile), may have difficulty feeling ownership for and taking pride in the overall product they are producing. That may affect the quality of the product they produce.	■ Engaging people at all levels of production through *quality circles* and other mechanisms so that they feel responsibility for the quality of the overall products they produce, rather than just their own particular role in producing the product, will ultimately lead to more pride of ownership and higher-quality products
■ People who are primarily motivated by fear of the consequences of *not* performing a task well are generally not going to work as effectively as people who are *positively* motivated to produce a high-quality product because they take pride in their work	■ Providing training and education to all employees, building their skills to perform the process at a higher level, and empowering them to recognize and suggest opportunities for improvement in the process are essential for ongoing process improvement
■ No one wants to be a *cog in a wheel*. People want to see a higher purpose and value in the work they do, and they want to be respected and recognized for their work.	

(Continued)

Implications For Agile Development

An Agile software development approach recognizes this by:

- Making respect for people a very important value.

- Fully engaging everyone on an Agile development team as an equal contributor to the success of the product.

- Putting a high level of emphasis on the training and skill of individuals to exercise good judgment in how the process is executed rather than relying on highly prescriptive, predefined processes to tell people what to do and how to do it.

- Substituting positive motivation and leadership for classical command-and-control management. The Scrum Master on a team is a facilitator, not a directive manager, and should empower everyone on the team to be fully engaged in the process.

- Eliminating sweatshops and Death March projects where people are forced to work excessive amounts of overtime to meet arbitrary schedule commitments and instead working at a *sustainable pace.*

The Need for Cross-functional Collaboration and Transformation

TABLE 11.3 Deming's principles: the need for cross-functional collaboration and transformation

Deming's Principles	■ Point #2: "Adopt the new philosophy." ■ Point #9: "Break down barriers between departments." ■ Point # 14: "The transformation is everyone's job."

Manufacturing problems prior to TQM	Manufacturing TQM approach
Prior to TQM, responsibilities were typically split across different organizations (engineering, production, quality, etc.). As a result: ■ Each organization was typically focused on their individual objectives, and the responsibility for the overall effectiveness and quality of the process was fragmented	Moving to a TQM approach required a major shift in thinking using *systems thinking* to see the whole business from a much broader process perspective (rather than a hierarchical organizational perspective) and to develop a well-integrated cross-functional approach across all of the organization. Although this can be difficult to achieve, the results are significant:

(continued)

(*Continued*)

Manufacturing problems prior to TQM	Manufacturing TQM approach
■ It can be very difficult to break down these barriers to develop a much more integrated and unified approach	■ It eliminates conflicting goals among organizations and develops an integrated cross-functional approach that leads to much higher levels of productivity and efficiency
	■ It enables everyone in the company to see an overall vision of how the products and projects they're producing provide value to leverage the company's business success and how their individual role contributes to that goal

Implications For Agile Development

■ An Agile development approach addresses this by emphasizing self-sufficient and autonomous cross-functional teams as the focal point for responsibility and decision making to break down organizational barriers at the project level.

■ Even that doesn't go far enough, in many cases. A major problem with the implementation of an Agile process is that it is often perceived as just a software development process owned by the development organization. People incorrectly ignore the need for organizational transformation and senior management commitment to make it an integral part of the way the business operates to make it fully successful.

Importance of Leadership

TABLE 11.4   Deming's principles: importance of leadership

Deming's Principle	■ Point #7: "Institute leadership—Supervision should help people and machines do a better job. Supervision of management is in need of overhaul as well as supervision of production workers."

Manufacturing problems prior to TQM	Manufacturing TQM approach
Classical command-and-control styles of management may not be very effective, even in a manufacturing environment. There are a couple of significant problems in that approach: ■ It doesn't fully empower and engage employees to take an active role in processes that are more self-managed	Truly inspirational leaders help people see the higher-level purpose and vision for their work. As a result, people ■ Are much more highly motivated ■ Feel more ownership of the products they produce

Manufacturing problems prior to TQM	Manufacturing TQM approach
■ It can be very demotivating to employees to work in that kind of environment where their skills are not recognized and valued and without more effective leadership	■ Work more effectively ■ Use their initiative with less need for direct supervision

Implications for Agile Development

■ In a software development environment where creativity and innovation are extremely important, classical command-and-control management styles may stifle that initiative and creativity.

Ongoing Continuous Improvement

TABLE 11.5 Deming's principles: continuous improvement

Deming's Principles	■ Point #1: "Create constancy of purpose toward improvement of products and services with the aim of becoming competitive, staying in business, and providing jobs." ■ Point #5: "Improve constantly and forever. Constantly improve quality and productivity in order to constantly decrease costs."

Manufacturing problems prior to TQM	Manufacturing TQM approach
Processes that rely heavily on a reactive approach based on correction of defects for quality control (rather than a more proactive approach based on prevention of defects and continuous improvement) will not significantly improve their defect rate because they have not removed the root cause of the defects.	■ With the advent of TQM and Six Sigma in the 1980s and early 1990s, companies learned that they could "push the envelope" much further into developing higher levels of quality that were never believed possible before that time ■ It required a totally different and much more systemic approach to eliminate the source of defects and prevent them from happening in the first place, rather than fixing them after they've happened

Implications for Agile Development

■ Continuous improvement is a major focus of all Agile methodologies. The need for continuous improvement is done through retrospectives at the end of each sprint or iteration. Because the iterations are relatively short, learning and process improvement can take place rapidly.

INFLUENCE OF LEAN MANUFACTURING

Where TQM provides a foundation of how to integrate quality into the design of products, Lean Manufacturing principles (see Figure 11.2) complement and go beyond that by developing a stronger focus on *maximizing customer value and providing guidance on how to improve and stream-line processes to eliminate wasteful inefficiencies.* As with TQM, Lean was originally developed for a manufacturing environment, but it's fairly easy to see how those same principles are equally relevant in a software development environment.

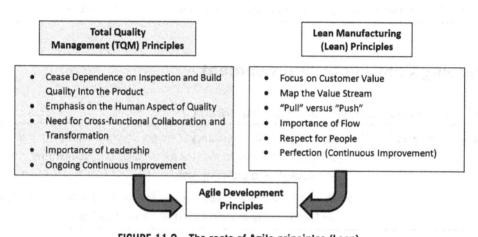

FIGURE 11.2   The roots of Agile principles (Lean)

Lean Manufacturing or Lean Production, which is often simply known as Lean, is defined as:

> A systematic approach to identifying and eliminating waste through continuous improvement by flowing the product at the demand of the customer.[2]

Lean considers the expenditure of resources for any goal other than the creation of value for the end customer to be wasteful and a target for elimination.

- Value is defined as any action or process that a customer would be willing to pay for.

- Agile is based on taking that same thinking from Lean Manufacturing and applying it to a software development process.

- It involves looking at a software development process and making critical decisions about whether each activity in the process adds value.

There are three kinds of work in any process:

1. **Value-added:** Process steps that produce value that either the customer is willing to pay for or that are essential to directly meeting customer requirements.
2. **Non-value-added:** Process steps that are not directly required to produce customer value but are required for other reasons, such as meeting regulatory requirements, company mandates, and legal requirements.
3. **Waste:** Process steps that consume resources but produce no value in the eyes of the customer.

Applying these concepts to a software development lifecycle model requires evaluating the various steps in the process and making a judgment of whether they really produce value in the eyes of the customer or not. Dr. David Rico provides a definition of *lean systems engineering* as follows:

Lean: Thin, slim, slender, narrow, adequate, or just-enough; without waste.

- A **customer-driven** systems engineering process that delivers the maximum amount of **business value**

- An economical, systems engineering way of **planning** and **managing** the development of complex systems

- A systems engineering process that is **free of excess waste**, capacity, and non-value-adding activities

- **Just-enough**, just-in-time, and right-sized systems engineering **processes**, **documentation**, and **tools**

- A systems engineering approach that is **adaptable to change** in customer needs and market conditions.[3]

INCOSE[4] has developed a list of systems engineering enablers to support the six key principles of Lean. Each of these topics is discussed in more detail in the following section.

TABLE 11.6 Lean principles

Lean principle	Enablers
Customer value	Focus on delivering customer value: ■ Use a defined process for capturing requirements focused on customer value ■ Establish the value of the end product or system to the customer (What are the business objectives?) ■ Frequently involve the customer
Map the value stream	Use a well-defined methodology for executing projects: ■ Poor planning is the most notorious reason for wasteful projects ■ Plan to develop only what needs developing ■ Plan leading indicators and metrics to manage the project
Pull	Tailor the process to the risks and complexity of the project to achieve maximum efficiency: ■ The *pull principle* promotes the culture of tailoring tasks and pulling them and their outputs based only on legitimate need and rejecting others as waste ■ Pull tasks and outputs based on need and reject others as waste
Flow	Eliminate bottlenecks that are likely to obstruct or delay progress: ■ In complex programs, opportunities for the progress to stop are overwhelming, and it takes careful preparation, planning, and coordination effort to overcome them ■ Clarify, derive, prioritize requirements early and often during execution ■ Frontload the architectural design and implementation ■ Make progress visible to all ■ Use the most effective communications and coordination practices and effective tools
Respect for people	Build an organization based on respect for people: ■ Nurture a learning environment ■ Treat people as most valued assets

(*Continued*)

Lean principle	Enablers
Perfection	Strive for excellence and continuous improvement in the software development processes:
	■ Use lessons learned from past projects for future projects
	■ Develop perfect communication, coordination, and collaboration policy across people and processes
	■ Use effective leadership to lead the development effort from start to finish
	■ Drive out waste through design standardization, process standardization, and skill-set standardization
	■ Use continuous improvement methods to draw best energy and creativity from project teams

Customer Value

Many businesses focus heavily on financial results rather than customer value as a primary goal. Without understanding the cause-and-effect relationships that drive those financial results, they are always in reactive mode. When the financial results for the current quarter go down, there's a lot of scrambling around to figure out what went wrong and what caused that to happen and then trying to fix the problem. That's a very reactive approach.

■ A more proactive approach is to develop an understanding of the factors that drive financial results for your business and focus on those factors.

■ If you focus on customer value and the factors that influence customer value and you do that successfully, the financial results should follow; it's a much more proactive, reliable, and consistent approach for managing a business successfully.

■ The idea is that managing the inputs to a process and managing the process itself are always a better approach than simply trying to manage the outputs of a process.

As an example, I once worked for a company that was known for putting enormous pressure on project teams to meet schedule deadlines. If a project was unsuccessful, there was a good chance that the whole project team might be fired. The problems with that approach should be very apparent—taking a brute force approach to try to force the outputs of a process to produce the desired results without an understanding of how the process works or the factors that influence the outputs is not likely to be very effective.

Map the Value Stream

An important principle of both Lean Manufacturing and Lean Software Development is to map the value stream.

■ This process basically involves starting from the point that the product or service is delivered to the customer (either internal customer or external customer) and working backward from that point to map all the process steps that led up to fulfilling that customer value.

■ The next step is to identify and differentiate steps that produce value to the customer from steps in the process that produce no value to the customer and may constitute waste.

An important factor is to identify *waste*. Mary Poppendiek has translated the seven wastes found in a manufacturing process to the equivalent seven wastes found in software development (Table 11.7).[5]

Note that these same areas of waste are also found in hardware development. Naturally, in order to manage the sources of waste in a process, you have to have a somewhat defined and documented process that is followed fairly consistently. This is one area where Lean and Agile may tend to pull people in somewhat different directions.

Pull

One of the key differences associated with Lean is the difference between a *push approach* and a *pull approach*—this difference is also found in most Agile approaches.

■ In a classical manufacturing process approach, production output is forecast, inventory is stocked at various points, and raw materials are then pushed through the process to fulfill that forecast.

■ This type of process has been the predominant approach used in manufacturing for many years to maximize the efficiency of the production equipment used in the process.

TABLE 11.7   Waste in Lean

The seven wastes of manufacturing	The seven wastes of software development
Inventory	Partially done work
Extra processing	Extra processes
Overproduction	Extra features
Transportation	Task switching
Waiting	Waiting
Motion	Motion
Defects	Defects

This approach has three potential deficiencies:

1. The forecasting process requires an intelligent guess at what the customer demand will be well in advance of when it is actually expected to be delivered from production. This is very difficult to do accurately and is fraught with lots of potential problems.
2. The process is very difficult to adjust—if there is a change in customer demand, it can take a considerable amount of time and effort to re-plan the entire process to adjust to that change. There also may be a considerable lag associated with restocking material to support the revised plan.
3. If the forecast is wrong, there are significant potential risks, including:

   - *Winding up with a significant amount of unusable inventory that might have to be scrapped.* In a software project, unusable inventory translates to extra features that no one needs or is likely to use that complicate the product and cause unnecessary maintenance if they are not removed.
   - *Not having sufficient inventory to fill customer demand if the forecast is wrong.* In a software project, this translates to not having the right features to satisfy customer needs.
   - *Having to stockpile or store inventory beyond the originally planned duration.* In a software project, this translates into undesirable product management overhead. "I've often seen this administrative burden lumped onto the project manager who must now wade through bloated scope matrices, backlogs of change requests and unwieldy specifications."[6]

Most classical plan-driven product development processes such as the Waterfall process are based on a similar *push process*.

- In a software development process, requirements are equivalent to raw materials in a manufacturing process.

- All the requirements are gathered upfront and are pushed through the rest of the development process in a sequential fashion just like a manufacturing assembly line.

- In a classical plan-driven product development process, the *push approach* may have the advantage of optimizing the utilization of the resources in a development process if the requirements are relatively certain and known in advance, but that is often not the case.

- If there is uncertainty in the requirements, it can result in serious potential problems and inefficiencies. Those potential problems and inefficiencies are similar to those associated with a push manufacturing process:

The product requirements for a new product development effort are essentially a *forecast* of the requirements for a product or application that a customer will need in the future. These requirements may be very uncertain and not very well defined.

- Attempting to forecast (or guess at) product requirements well in advance of when the product will actually be deployed and used is even more risky than forecasting production output in a manufacturing process.

- In addition to the normal risks of forecasting customer demand, the customer may not really know what he/she wants without seeing the product and seeing first-hand how it works. For that reason, the requirements might easily change over that time.

- The process also can be difficult to adjust when changes in customer requirements occur. Typically, many assumptions are made about what the customer requirements are, and elaborate plans, resource assignments, and documentation will be created to support those assumptions.

- There is also a significant risk that the assumptions in the requirements are wrong and don't reflect the real needs of the customer. This may not be discovered until the final product is ready for final acceptance testing.

Inaccurate or changing customer requirements may result in a significant amount of lost time and effort to replan around a different set of requirements and assumptions. It might also require substantial rework. If a typical change control system is used, the process for doing that may be very cumbersome and difficult. Both Lean and Agile approaches avoid these problems:

- By deferring the resolution of uncertain requirements until a decision is required (when that particular requirement is *pulled* into development for further processing).

- By avoiding guessing the requirements upfront. Waiting until more information is known will typically result in better decisions and avoid many of the problems associated with inaccurate and changing requirements.

- Lean and Agile systems openly acknowledge and are built on the assumption that requirements are uncertain and are likely to change as the project progresses. Many classical plan-driven processes do not recognize or acknowledge an appropriate level of uncertainty that is actually inherent in the requirements and attempt to superimpose a rigid control model on top of a very uncertain environment.

A *pull* system works by producing only the required amount to meet demand at each stage. In a manufacturing system, this would be characterized by a just-in-time production scheduling system. Many of the ideas for Lean Manufacturing came from the Toyota Production System and Kanban.

In an Agile software development process, there is a direct analogy between Kanban cards and User Story cards. User Stories are high-level descriptions of capabilities that the system needs to provide. The following is an example of a User Story:

> As a banking customer, I need to be able to withdraw funds from my account through an ATM machine so that I can get direct access to the funds easily.

User Stories are typically defined early in the project to identify the capabilities that the system must provide, with a sufficient level of detail to do only a rough estimate of the level of effort associated with each. The details of how the User Story will be implemented will generally be deferred until it is time to do the design:

- Instead of attempting to define all of the requirements in detail, typically only the high-level requirements are defined upfront to the level of user stories without a lot of detail.

- Instead of treating all requirements equally, the requirements are prioritized based on their value to the customer. After they are prioritized and broken up into releases and/or iterations, the most important requirements, which are at the top of the list and ready for development, get developed first.

- Once the developer has picked up a user story to begin working on, he/she will then work directly with the user to *pull* more detail as needed to fill that requirement.

A story card is equivalent to a Kanban card in a manufacturing system and describes one particular feature that a user needs. As in the manufacturing system, physical cards may or may not be used—there are computerized tools that will automate this task and eliminate the use of physical story cards if desired. However, in many cases, physical cards may actually be used and put on a board; each developer picks up a card to start working on it, similar to the way a Kanban card works in a factory.

A Kanban development process is sometimes used as an alternative to Scrum in situations that need to be more reactive, such as managing a queue of customer service requests.

- In a Scrum process, there is some level of upfront planning to at least identify the items in the product backlog at a high level before the project starts, and additions to the backlog items are not allowed once an iteration has started.

- From that perspective, a Scrum process is only *partially* based on *pull*. At a high level, requirements are *pulled* through the process based on priority; however, within an individual iteration, once the iteration has started, requirements are *pushed* through the process.

A Kanban process is designed to be much more reactive and responsive to customer demand. Table 11.8 shows a summary of the differences between a Kanban development process and Scrum.[7]

The key difference is with regard to how the two processes manage flow.

- A Kanban process is *totally* pull and allows demand for new items to take place at any time, *including during the middle of a sprint* if capacity is available (note that there is actually no concept of a "sprint" in Kanban).

- A Scrum process is *mostly* pull. The product backlog is considered to be dynamic and can be adjusted as needed to meet customer needs within the constraints that (a) the capacity to handle the demand is fixed and (b) additions are not allowed once an iteration has started.

TABLE 11.8   Scrum versus Kanban

	Scrum	Kanban
Primary flow	Time-boxed iterations (sprints)	Continuous flow
Primary metric	Velocity	Lead time
Work-in-process (WIP)	WIP limited indirectly (per sprint)	WIP limited directly (per workflow state)
Addition of new items	No new items added to the current sprint	Add new item as capacity is available
Roles	Three prescribed roles (Product Owner, Scrum Master, team)	No prescribed roles
Status board	Scrum board reset between sprints	Kanban board is persistent

Flow

Flow is one of the most important Lean principles to understand to maximize the efficiency of any process. There are a number of factors that contribute to maximizing flow, which include:

- small batch sizes

- just-in-time production

- concurrent processing.

   Each of these topics is discussed in the following sections.

Small Batch Sizes

In a manufacturing process, it is well known that small batch sizes are much better for optimizing the flow of a process than large batch sizes. If large batch sizes are used, bottlenecks can easily develop at various points in the process, and material winds up waiting at those bottleneck points, creating waste. There are at least a couple of types of *waste* associated with large batch sizes:

- *Excess material inventory* is used in the process, creating unnecessary inventory cost, space for storage, and additional handling costs. Mary Poppendiek uses an example of the construction of the Empire State Building in New York. The Empire State Building was the tallest building in the world and was built in a total of 20 months, including demolition of existing buildings and planning and design of the new building.[8]

  - One of the most serious constraints was that there was only a limited amount of vacant real estate in the area to store the materials needed for the building. This meant that the flow of the project had to be very carefully planned.

- The building was built in iterations of a few floors at a time, and the arrival of material had to be scheduled meticulously to have just the right materials available at the right time to maximize the flow.

- If the inventory is *perishable*, it can go stale and become unusable. By perishable, I'm not necessarily referring to fruits and vegetables. Dell Computer is a good example.

  - Dell builds computer systems for customers out of a variety of different components (main boards, storage devices, graphic cards, etc.), and those components become obsolete quickly and are constantly being replaced by newer versions.
  - Using small batch sizes and building systems on demand as customers need them reduces the risk of winding up with too much obsolete inventory of components in the pipeline.[9]

The other major advantages of small batch sizes are:[10]

- It reduces the end-to-end cycle time (Little's Law of Queuing).[11]

- It makes waste a very hard problem to ignore as any waste in a small batch size system will cause much larger problems than when you've got inventory to hand to smooth it over. You then have to confront and fix the waste. The visual metaphor often employed is that a stream running low uncovers the rocks on its bed.

Attempting to define all the requirements for a product or application upfront in a classical plan-driven development process is analogous to attempting to process large batch sizes in a manufacturing process. It's impossible to work on all the requirements at once, so bottlenecks develop at various points in the process and requirements wind up waiting to be processed. Having an excess of requirements sitting around waiting to be processed is similar to having excess inventory in a manufacturing process:

- *There are handling costs* associated with managing those requirements waiting to be processed— they have to be well documented and tracked or they may be forgotten and left out of the design. There is also a certain amount of overhead associated with managing changes to these requirements.

- *The requirements are also perishable*—if they wait for a long time to be processed, they could easily become obsolete. If that happens, either someone winds up designing and building a product or application based on obsolete requirements, or unnecessary labor is consumed in redefining and rewriting the requirements.

An iterative development process is analogous to small production batch sizes in a manufacturing operation. By breaking up the requirements into the smallest possible units of business functionality (user stories), the overall development process is likely to flow more smoothly and avoid the bottlenecks associated with classical plan-driven development processes. Of course, in actual practice, there are limits to how far it is practical to break down the requirements to optimize flow. For example:

1. **Requirements management considerations:** From a requirements management perspective, it may be necessary to group requirements into feature sets that are interrelated with each other. Those feature sets might be bigger than the effort that can be realized in a single iteration. That requires some compromises between:

   ■ an idealized approach where individual sets of requirements are completely processed immediately in each iteration; and

   ■ a more realistic hybrid approach where some of the requirements might be *pipelined* for processing and spread across more than one iteration.

   The idea of buffering some of the requirements that cannot be fulfilled immediately is called story pipelining.

> Story pipelining is often seen by purist Agile practitioners as strictly un-Agile as it violates the oft-held view that working software is the only thing that represents value and anything less is a cop-out. In practice, however, pipelining is often the best way to balance agility with the realities of real-world delivery constraints. You gain a measure of project progress that's still closer to real doneness in the eyes of the customer, you retain a lifecycle that encourages regular and frequent feedback, but you also recognize that complex software systems have a gestation period.
>
> Pipelining particularly recognizes requirements that are just really difficult to implement: thorny user interactions, challenging external systems dependencies, etc.[12]

2. **Testing/release and configuration management considerations:** Testing and release management considerations might require compromises from the ideal flow model:

   ■ Testing might want a functionality set to test against that's a relatively complete subset of functionality and will be stable for the duration of the test cycle.

   ■ Release and configuration management might have similar needs.

Both of these issues can be overcome but may require some rethinking of how the process works and very strong coordination of test planning with the rest of the development effort.

Just-in-Time Production

Having large quantities of raw materials waiting to be processed is inefficient because they become bottlenecks. A much more efficient process is based on just-in-time processing, when the raw materials arrive just at the right time that they are needed in production. A large business requirements document in a software development environment is equivalent to a large pile of raw materials in a manufacturing environment. Instead of developing large requirements documents that wait to be processed, it is generally more efficient to develop requirements just-in-time as needed in the software development process. For example:

- Early in the process, high-level requirements should be sufficient to do whatever level of planning is needed at that point.

- The elaboration of requirement details can be deferred until later in the process when those particular requirements are ready to enter development.

Concurrent Processing

Concurrent processing is another well-known way to improve flow in any process. In a manufacturing process, bottlenecks are much more likely to develop if there is only one path through the system and everything is sequential than if there are parallel paths available and some work can be done concurrently on different paths.

In a product development process, there are typically greater opportunities for concurrent engineering to improve the flow through the process. Here are a few examples:

- Requirements development can be overlapped with design instead of being sequential, and quality testing can also overlap with design instead of following it sequentially. This requires a much more collaborative, cross-functional approach to development, which can be difficult to achieve, but the potential payoff can be significant.

- Design teams can work on multiple iterations concurrently. This requires breaking up the design effort into iterations and requires some coordination among design teams:

> Concurrent engineering is especially useful when dealing with '"unprecedented" requirements . . . This can provide cover in situations where you have multiple integration approaches to choose from and don't know which is best (e.g. which will deliver the right performance, availability, or conform to other non-functional requirements). Similarly, challenging User Interface (UI) problems are sometimes best tackled using concurrent engineering (e.g. prototyping multiple solutions and testing them with real, representative users) in a "survival of the fittest" approach.[13]

Respect for People

In the early days of automotive manufacturing, processes were designed so that the people performing those processes did not require high levels of skill.

- An individual working on an assembly line could be assigned a small, repetitive task such as installing a tire on a car, which required only a minimum amount of skill and training.

- The primary requirement for higher levels of skill and training could be limited to a relatively few people who were responsible for designing and managing the overall process and training the workers to perform each task.

    There are several problems with that approach:

- No one really takes responsibility for the overall quality of the complete vehicle.

- This approach might rely heavily on quality control inspectors at the end of the line to try to find defects and send the vehicle back for rework if necessary.

- It can be a dehumanizing experience for anyone to perform that kind of limited, repetitive task.

- This approach doesn't take advantage of the complete range of skills and judgment of the people performing the tasks.

    Manufacturing processes have long recognized the need to respect and empower the people performing the processes as much as possible.

- In a manufacturing process, having people take pride in workmanship is extremely important for achieving high levels of quality and productivity.

- The need for fully utilizing the capabilities of people and motivating them is even more significant in a software development process, where the overall effectiveness of the process is so critically dependent on the performance of the people.

    Both Lean and Agile methodologies seek to eliminate those problems by empowering individuals and the team as a whole to take responsibility for the overall quality of their work.

    Many classical plan-driven development processes have been modeled on early manufacturing processes, where a process defines in detail the work to be done and how it should be done, and the process requires a lower level of skill to perform relatively well-defined tasks. Agile methodologies are generally much less well defined and rely heavily on the skill and training of the people performing the process to use appropriate levels of judgment and tailor them to a particular project, task, and business environment. That is a key reason why respect for people is so important in an Agile environment.

Perfection

The principle of perfection in Lean Manufacturing is similar to the TQM principles associated with ceasing reliance on inspection and ongoing continuous process improvement to remove defects. Defects in Lean are seen as a major source of waste:

- It takes a lot of resources to inspect for defects, which wouldn't be necessary if the defects were eliminated at their sources.

- Rework and scrap can result from defective products if those defects aren't discovered until the product is at the end of the assembly line (or at the end of the development process in a software development environment).

Lean Manufacturing also emphasizes continuous improvement to eliminate the waste caused by defects, just as TQM emphasizes it for improving product quality.

There is also a direct relationship with the principle of respect for people.

- In many cases, the people performing the process are the first ones to recognize opportunities for improvements in the process to prevent defects and/or to make the process more efficient.

- Unfortunately, many times they are not empowered to suggest or make those changes.

Lean and Agile methodologies recognize that and therefore are not rigidly defined or prescriptive—they provide some fundamental principles and practices that are common to most projects and are expected to be tailored to a given situation. And the people performing the process have a significant role in the design and management of the process. Naturally, it requires more skill to make good judgments about how to tailor a process to fit a business and project environment.

The word "perfection" is perhaps a misnomer. "I think Taichi Ohno (1998) suggested that we must aim for perfection through Kaizen (Continuous Improvement) knowing that perfection is nearly impossible to achieve."[14]

PRINCIPLES OF PRODUCT DEVELOPMENT FLOW

Don Reinertsen wrote a widely read book on the principles of product development flow, summarized here:[15]

1. Economics

Take an economic view. Many times, there is a point of diminishing returns associated with improvements in a process—understanding the economic impact is a critical factor in optimizing a process. For example, it is generally best to defer decisions on product features as long as possible; however, some decisions should be made early and should not be significantly deferred because of their economic impact.

Example: Increasing innovation should not be an end in itself and it reaches a point of diminishing returns at some point and begins to impact other proxy variables such as quality.

2. Queues: Actively Manage Queues

Agile development processes are based more on a continuous flow process as opposed to heavily plan-driven processes that are more of a large-scale batch process. A continuous flow product development process operates most efficiently when queues are managed.

Example: Developing requirements far in advance that sit in a queue waiting for development can be inefficient and wasteful because:

- The requirements may change prior to going into development and much of the effort involved in developing the requirements might have been wasted.

- Speculation in the requirements that are done too far into the future can result in erroneous assumptions that make their way into development without being questioned.

3. Variability: Understand and Exploit Variability

Reducing variability will many times improve efficiency but that isn't always the case. For example:

- Breaking up large requirements into smaller ones that are of a more uniform size reduces variability and can improve flow, however, at some point further attempts to reduce variability do not have economic value.

- Forecasting errors are a major source of variability . . . we can reduce this variability by forecasting at shorter time horizons.

- Design reuse reduces variability.

4. Batch Size: Reduce Batch Size

Large batch sizes tend to cause bottlenecks and inhibit flow. Reducing the batch size by breaking up requirements into small, independent user stories can significantly improve flow.

Examples of batch size inefficiencies include:

- Project scope: More is taken on in a single project than is truly necessary.

- Project funding: The entire project is conceived and funded as a single large batch proposal.

- Requirements definition: the tendency to define 100% of the requirements before the project starts.

5. WIP Constraints: Apply WIP Constraints

Use work in process (WIP) constraints to manage overall flow. For example,

- Control the number of projects taken on at any one time to avoid oversaturating development resources.

- Use specialized resources wisely to maximize their impact on overall flow.

6. Control Flow Under Uncertainty: Cadence and Synchronization

Reinertsen defines cadence[16] as follows:

> Cadence is the use of a regular predictable rhythm within a process. This rhythm transforms unpredictable events into predictable events. It plays an important role in preventing variability from accumulating in a sequential process . . .

Having a repeatable cadence improves the efficiency of the product development process and allows synchronizing a predictable development process with a much more unpredictable flow of requirements. Examples of the use of synchronization include:

- concurrent development on multiple paths at the same time

- concurrent testing of multiple subsystems.

7. Fast Feedback: Get Feedback as Fast as Possible

Fast feedback can lower the expected loss by truncating unproductive paths more quickly or raise the expected gain because we can exploit an emergent opportunity by rapidly redirecting resources.

Fast feedback, combined with selecting appropriate measures of performance, enables rapid learning and ongoing continuous improvement.

8. Decentralize Control

The final principle that Reinertsen has identified deals with decentralized control:

> Sophisticated military organizations can provide very advanced models of centrally coordinated, decentralized control. There is an impression that military organizations seek robotic compliance from subordinates to the orders of superiors. In reality, the modern military focuses on responding to a fluid battlefield, rather than executing a predetermined plan. It views war as a domain of inherent uncertainty, where the side that can best exploit uncertainty will win.[17]

SUMMARY OF KEY POINTS

Influence of Total Quality Management (TQM)

1. TQM and the thinking behind it revolutionized the quality and competitiveness of the automotive industry. TQM taught us to:

   - *Eliminate defects at the source*: Develop a more systemic approach to designing quality into the process that produces products to eliminate the defects at the source, rather than constantly finding and fixing the same or similar defects over and over again.

   - *Recognize the human aspects of quality*: Recognize the human aspects that are essential to build quality, such as engaging people at all levels so that they feel responsibility and ownership for the quality of the overall products they produce and empowering them to recognize and suggest opportunities for improvement in the process.

   - *Develop a cross-functional approach*: Break down barriers between departments, eliminate conflicting goals among organizations, and develop a much more integrated cross-functional approach that leads to much higher levels of productivity and efficiency, together with a more collaborative approach to quality and process improvement.

   - *Recognize the importance of leadership*: Eliminate classical command-and-control management in favor of inspirational leadership to empower people and to help them see the higher-level purpose and vision for their work.

   - *Strive for ongoing continuous improvement*: Commit to an ongoing continuous improvement effort to constantly find opportunities to improve processes.

Influence of Lean Manufacturing

1. The concept of Lean originated in manufacturing and has had a significant impact on many industries. The focus of Lean Manufacturing is on elimination of waste and improving operational efficiency rather than simply improving quality; however, Lean does recognize quality defects as an important form of waste that should be eliminated. Lean taught us:

   - *Customer value focus*: Focus on producing customer value and eliminate all unnecessary tasks that do not add value to the customer.

   - *Map the value stream*: Map the value stream to understand the process flow and identify any opportunities to eliminate non-value-added steps.

   - *Pull*: Use a pull approach rather than a push approach to plan and manage production capacity to meet demand.

   - *Flow*: Use the principles of flow, such as just-in-time processing, to optimize the efficiency of the overall process.

- *Respect for people*: Recognize and be sensitive to the human aspects of quality. People need to be respected and properly motivated to develop high-performance teams that produce high-quality products.

- *Perfection*: Use a systemic approach to identify the source of waste and defects in a process and use a continuous incremental improvement approach to perfect the process.

These same principles are adaptable to a product development environment.

Principles of Product Development Flow

1. A classical plan-driven management approach is heavily based on managing and controlling the structure of a project with such things as work breakdown structures, Pert charts, and Gantt charts. An Agile project is based on a much more fluid and adaptive process where structure is much less important and managing the flow of items through the process is much more important.

2. Understanding the principles of product development flow is very important to optimize the efficiency of an Agile project. The key principles of product development flow outlined by Don Reinertsen are:

- Economics: Take an economic view.

- Queues: Actively manage queues.

- Variability: Understand and exploit variability.

- Batch size: Reduce batch size.

- WIP constraints: Apply WIP constraints.

- Cadence and synchronization: Control flow under uncertainty.

- Fast feedback: Get feedback as fast as possible.

- Decentralized control: Decentralize control.

DISCUSSION TOPICS

Influence of Total Quality Management

1. Discuss an example of a problematic situation where the TQM principles might have been used to improve the overall quality of the work. What were some of the issues that impacted the quality of the overall product or service that the company produced? How could the impact of those issues have been reduced or eliminated?

2. How do you think that an understanding of the principles behind Total Quality Management might influence an Agile Project Management approach?

Influence of Lean Manufacturing (Lean)

1. Discuss an example of a situation where Lean thinking could have been used to improve the overall efficiency of the process. What was some of the waste involved in the process? How could it have been reduced or eliminated? What principles of Lean did you use to make that assessment?

2. How do you think that an understanding of Lean principles might influence an Agile project management approach?

Principles of Product Development Flow

1. Analyze the following situations and identify the appropriate principles of product development that might be important to consider in defining an appropriate solution to each:

   ■ A project is taking much longer than originally anticipated: it was originally planned to take only 6 months, it has now gone on for over 18 months, and there still doesn't seem to be an end in sight.

   ■ Projects are being delayed because users are frequently changing requirements in the middle of the project.

   ■ The quality assurance organization is becoming a bottleneck and scheduled releases are being delayed waiting for QA testing.

   ■ A company has just finished a massive implementation of Agile throughout the entire company but senior executives are not satisfied with the results. They feel that they have lost some visibility into projects and are not sure the projects are well aligned with their business goals.

   ■ An Agile development team is repeatedly missing its sprint goals because it has overcommitted the amount of work that can be done in a sprint.

   ■ The company CEO insists on personally making decisions in what he considers to be the most critical projects to the company's success.

NOTES

1. Alexander Laufer, *Mastering the Leadership Role in Project Management: Practices that Deliver Remarkable Results* (Upper Saddle River, NJ: Pearson Education, 2012).
2. Lean Manufacturing Guide, http://www.leanmanufacturingguide.com/.
3. David F. Rico, "Lean and Agile Systems Engineering," http://davidfrico.com.
4. International Council on Systems Engineering, Lean Systems Engineering Working Group, http://cse.lmu.edu/about/graduateeducation/systemsengineering/INCOSE.htm.
5. Tom Poppendiek and Mary Poppendiek, *Lean Software Development: An Agile Toolkit* (Reading, MA: Addison-Wesley, 2003), p. 4.

6. Erik Gottesman, email comments to author on book review.
7. Kanban Development Oversimplified, http://www.agileproductdesign.com/blog/2009/kanban_over_simplified.html.
8. Tom Poppendiek and Mary Poppendiek, *Lean Software Development*, op. cit., p. 102.
9. Ibid., p. 12.
10. Martin Burns, email comments to author on book review.
11. "Principle: Little's Law," http://www.factoryphysics.com/Principle/LittlesLaw.htm.
12. Erik Gottesman, email comments, op. cit.
13. Ibid.
14. Winston Gonzalez, email comments to author on book review.
15. Don Reinertsen, *The Principles of Product Development Flow* (Redondo Beach, CA: Celeritas Publishing, 2009).
16. Ibid.
17. Ibid.

PART 3

Agile Project Management Planning and Management

PART 2 PROVIDED AN OVERVIEW of Agile Project Management. In Part 3, we will go into much more detail on Agile Project Management planning and management practices.

Chapter 12 – Hybrid Agile Models:

A hybrid Agile approach is one that blends plan-driven principles and practices with Agile (adaptive) principles and practices in the right proportions to fit a given situation. Each has their own strengths. Mixing these together in the right way can fit the context better than just narrowly using only one of them and will provide the "best of both worlds":

- the predictability of a plan-driven approach, plus

- the flexibility and adaptivity of an Agile approach.

Chapter 13 – Value-Driven Delivery:

A classical plan-driven project was deemed to be successful if it delivered well-defined project requirements on time and under budget. The problem with that is the assumption that the "well-defined requirements" accurately reflect what the customer really needs.

There have been many projects that have met their cost and schedule goals but failed to deliver an acceptable level of customer value. It's a trade-off between maximizing the efficiency of meeting cost and schedule goals and the overall effectiveness of the solution.

Delivering value to the customer should always be the most important goal. Meeting cost and schedule goals certainly has some value, but it may not be the most important value in a particular project. The value of meeting cost and schedule goals needs to be balanced against delivering a solution that provides an acceptable level of business value to the customer.

Chapter 14 – Adaptive Planning:

Many people seem to think of planning as an "all-or-nothing" proposition—either you develop a highly-detailed plan, or you do no planning at all. This doesn't need to be the case. The right solution is to fit the planning approach to the nature of the project. Planning in an uncertain environment is a difficult thing to do and definitely requires more skill, but it definitely can be done.

Chapter 15 – Agile Planning Practices and Tools:

In this chapter, we will discuss some Agile planning practices and tools that can be used to implement an adaptive planning strategy. These include:

- Product/Project Vision

- Product Roadmaps

- Exploratory 360 Assessment

- Agile Functional Decomposition

- Agile Project Charter.

Chapter 16 – Agile Stakeholder Management and Agile Contracts:

Stakeholder management is probably one of the most important aspects of managing any project successfully. A project is not successful unless it meets stakeholder expectations:

- A classical plan-driven project sometimes assumes that meeting the defined requirements within the approved budget and schedule defines success; however, it's dangerous to assume that requirements documents accurately reflect the needs of all important stakeholders.

- Agile projects sometimes assume that the Product Owner totally represents all stakeholder interests but that is not necessarily the case. The role of the Product Owner is designed to make the project execution more efficient by putting a representative of the business users directly in contact with the project team, but it is very dangerous to assume that eliminates the need to talk directly to stakeholders outside of the project team.

Chapter 17 – Distributed Project Management in Agile:

There's a common misconception that "project management is not consistent with Agile. A contributing factor to that misconception is that there is normally no one called a "Project Manager" at the team level in an Agile project and the conventional view of project management is that it requires a project manager.

Another factor is people have a very narrow definition of what project management is that is heavily associated with planning and control over project costs and schedules.

There is actually a lot of "project management" going on in an Agile project, even though you may not find anyone with the title of "Project Manager."

- It's a different kind of project management with an emphasis on maximizing business value rather than the traditional project management emphasis of providing predictability on costs and schedules when delivering well-defined requirements.

- The functions that might normally be performed by a single person called a "Project Manager" have been distributed among the different members of the Agile team.

12 Hybrid Agile Models

I'VE SEEN MANY ARTICLES THAT POSITION "AGILE" AND "WATERFALL" as two binary and mutually exclusive alternatives with no middle ground between the two. Instead of thinking of what people commonly call "Agile" and "Waterfall" as individual discrete methodologies, it is more accurate to see it as a continuous spectrum of approaches from heavily plan-driven at one extreme to heavily adaptive at the other extreme.

A hybrid Agile approach is one that blends plan-driven principles and practices with Agile (adaptive) principles and practices in the right proportions to fit a given situation. Each has their own strengths. Mixing these together in the right way can fit your context better than just narrowly using only one of them.

As an example, some years ago, I was responsible for managing a very large development program for a US Federal Government agency. It had a fixed-price contract associated with it and a fairly aggressive delivery schedule. However, the customer wanted some level of flexibility in the details associated with defining requirements, so I created an approach for this situation that was very successful that I have called the "Managed Agile Development" process which will be discussed in Chapter 24 of this book. The Managed Agile Development process is an excellent example of a hybrid Agile process that is relatively easy to implement.

It's important to note that a given hybrid Agile approach might even require a different blend of approaches for different phases of the project. For example, a very Agile approach might be used to develop an initial prototype to better define and validate the design followed by a more plan-driven approach for the remainder of the development effort.

WHY WOULD YOU USE A HYBRID AGILE APPROACH?

There are two situations when a hybrid Agile approach would be most useful:

Fit for Purpose

The most common usage of a hybrid model is to fit the methodology to the nature of the project:

- For projects that have a lower risk profile, use plan-driven approaches to look for lower costs.

- For higher-risk projects, use iterative techniques to repeat activities until issues are revealed and resolved.

- For projects needing aggressive delivery, incremental techniques will deliver something sooner, to ensure customer engagement.

- Finally, in order to navigate complex environments, Agile techniques may have a higher initial overhead, but it might be worth it for the overall outcomes.[1]

Many people make the mistake of performing a methodology mechanically. They think they need to do it religiously and "by the book" (that's true of both Agile and other non-Agile methodologies).

- The right approach is to fit the methodology to the nature of the problem rather than force-fitting all problems and projects to a given methodology (Agile or non-Agile).

- It takes more skill to do that, but it definitely can be done.

- It requires understanding the principles behind the methodology and why they make sense in a given situation rather than doing a given methodology mechanically.

   If you think of methodologies as being rigid and prescriptive,

- It will be difficult to see how two seemingly disparate methodologies could be blended together in the right proportions to fit a given situation.

- On the other hand, if you understand the principles behind the methodologies at a deeper level, it is much easier to see how they could be complementary to each other rather than competitive.

As a Transition to a Full Agile Approach

Another major reason for using a hybrid Agile approach is to transition from a plan-driven environment to a more Agile (value-driven) environment.

- Many teams are not able to make the switch to Agile ways of working overnight.

- The larger the organization, the more moving parts, the longer it will take to shift.

- If you've lived in a plan-driven world for several years, then Agile methods will look and feel very different. As a result, your initial foray into the Agile world will be a messy amalgamation of both. That's okay. You're using specific techniques to move in a direction that you want to go to.[2]

WHAT ARE THE BENEFITS OF A HYBRID AGILE APPROACH?

General Benefits of a Hybrid Agile Approach

A hybrid Agile approach can provide the best of both worlds and combine some of the benefits of a plan-driven approach with an Agile approach (see Table 12.1).

Other Benefits of a Hybrid Agile Approach

The following is a summary of some other important benefits of a hybrid Agile approach:

1. **Time-to-market can be significantly improved by:**
   - ordering requirements and grouping them into releases and iterations
   - using an incremental delivery process to deliver functionality early
   - finding efforts can be more concurrent and less sequential where possible.

TABLE 12.1 Comparison of classical plan-driven and Agile approaches

Classical plan-driven approach	Agile (value-driven) approach
A well-defined process that is consistently implemented and is somewhat predictable	Flexibility and adaptableness that are optimized around an uncertain business environment
High-level milestones provide a framework for managing user expectations and integration with other activities outside the Agile development environment	Emphasis on creativity and innovation to maximize the business value of the solution
Scope and cost of projects can be easily and effectively managed	Ability to make changes easily provides higher user satisfaction resulting from direct user feedback and inputs

2. **There is also an opportunity to improve risk management:**

   ■ There is an ability to detect risks earlier by structuring the project to focus on high-risk items first and isolate them by dedicating a "spike" to evaluating a particular risk.

   ■ Once a risk has materialized, there is an ability to dynamically adapt to the risk without extensive re-planning.

3. **There is an opportunity to significantly improve productivity and efficiency:**

   ■ The team is empowered to tailor the process to fit the project.

   ■ The process can be optimized for the project.

   ■ Unnecessary paperwork and overhead are reduced.

   ■ More empowered teams can create a highly motivated environment.

   ■ A collaborative, cross-functional approach can improve communications and decision-making.

WHAT'S DIFFERENT ABOUT A HYBRID AGILE APPROACH?

There is no standardized, "cookbook" step-by-step solution for a hybrid Agile approach. It requires a fair amount of tailoring to fit the approach to the nature of the project—that may require a lot more skill, but it definitely can be done.

■ The major difference is to add a layer of planning and management over an Agile development process like Scrum.

■ Scrum is probably the most widely used Agile methodology in the world and it does provide an excellent framework for defining an Agile project, but it really primarily defines a process for a team-level and sprint-level process.

■ It doesn't explicitly specify any additional layers of management that might be needed on top of that for larger and more complex enterprise-level projects.

   Most Scrum projects will have project-level planning where at least the high-level goals and objectives of the project are defined; however, project-level planning can be as "thick" or "thin" as necessary depending on the nature of the project.

■ The project-level planning forms an "envelope" around the team and sprint-level planning process defined by Scrum; however, the project-level planning would typically be at least somewhat dynamic.

■ Rather than defining a rigid upfront plan that might not change for the entire duration of the project, it would be expected that lessons learned from the team and sprint-level planning process would be fed back into the project-level planning process to make adjustments to the project-level plan as necessary.

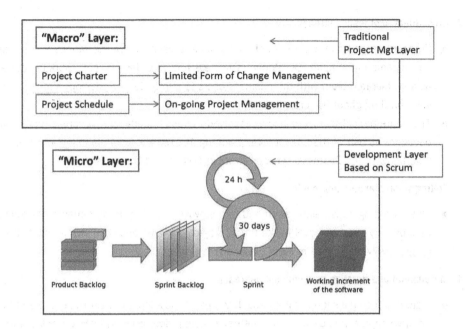

FIGURE 12.1   Hybrid Agile process

Figure 12.1 shows what a hybrid Agile process might look like. It consists of two major parts:

1. a "macro" level where the project planning and management are done;
2. a "micro" level development process based on Scrum.

Key Differences from a Plan-driven (Waterfall) Approach

The following is a summary of some of the major differences between a hybrid Agile approach and a classical plan-driven (Waterfall) approach.

1. **Partnership with the business sponsor and business users**

   - A high-level understanding of the project scope and high-level requirements exists in the form of user stories at the macro-level.

     - The level of detail in the requirements definition may vary depending on the nature of the project; however, it is normally not necessary to specify all the details of how a solution will be developed in the project-level planning.

     - The Business Sponsor/Users and the project team will jointly own responsibility for further defining the details of the solution as well as optimizing it to make it successful in the micro-level process.

2. **Cross-functional team approach**

- This requires more of a cross-functional team approach, where the various functional participants in the team (Developers, Business Analysts, Testers, etc.) work concurrently as an integrated team with the business users to jointly define, develop, and test the solution throughout the project.

- The entire project team will also be involved in making collaborative, cross-functional decisions as part of the project-level planning and release-level planning activities together with the Business Sponsor, Product Owner, and Business Users.

3. **Rolling-wave planning approach**

- This uses progressive elaboration and rolling-wave planning. Requirements are generally defined only to a high level in the initial project planning and then requirements are progressively elaborated in more detail as the project progresses.

4. **Incremental and iterative development approach**

- Instead of using a sequential approach to design, develop, and test the entire solution, an iterative approach is used to incrementally design, develop, and test portions of the overall solution.

Key Differences from an Agile Approach

The following is a summary of some of the major differences between a hybrid Agile approach and a pure Agile approach. Some people might argue that these items are just natural extensions to an Agile approach, but they're not explicitly defined.

1. **Project inception planning**

- At the front end of the project before the project starts, there might be an inception planning phase to develop at least a high-level plan for the project.

- The level of inception planning would be dependent on the level of uncertainty in the project. If there is a high level of uncertainty in the project, the level of inception planning might be limited.

2. **Project management**

- As the project is in progress, there would be a level of project management to monitor progress against goals and milestones of the project defined in the inception planning.

- The level of project management could be "thick" or "thin," depending on the level of uncertainty in the project and the relationship with the business.

3. **Change management**

- At least some minimal form of change management might be needed to control changes in scope.

- There should be enough slack in the project plan to absorb minor changes in the development process at the macro layer without changing the high-level project plan; however, more significant changes that might impact the project plan need to be reconciled with the project plan.
- The objective is not to limit changes but to understand what impact they might have on the overall plan.

CHOOSING THE RIGHT APPROACH

As I've previously mentioned, there is not just one way to do a hybrid Agile approach. There is a spectrum of approaches ranging from heavily plan-driven at one extreme to heavily adaptive at the other extreme with lots of alternatives in between. A big advantage of a hybrid Agile approach is that it allows the approach to be tailored to fit the situation rather than force-fitting a project to one of those extremes.

Most Important Factors to Consider

1. Level of uncertainty

There are two dimensions of uncertainty in a project:

- **Requirements uncertainty:** how certain are the requirements for the project and how likely are they to change?
- **Technology uncertainty:** how certain is the technology approach for implementing the project? How likely is it to change?

A project that has a high level of uncertainty would generally lean toward a more flexible and adaptive approach.

2. Stakeholder management

Managing the expectations of the business users and stakeholders is another important consideration:

- If the stakeholders are not willing to take an active and collaborative partnership role in the project, the project would likely lean toward a more plan-driven approach.
- If the stakeholders have definite expectations about the cost and schedule of the project that must be met, that will also lean more toward a more plan-driven approach.

3. Training and sophistication of the project team

The level of training and sophistication of the project team is also an important consideration:

- A hybrid Agile approach requires a level of training and sophistication of the project team.
- If the team hasn't been trained and proficient in Agile, it might favor more of a plan-driven approach.

Other Factors to Consider

1. Level of risk and risk management approach

An Agile approach may or may not be suited to a high-risk project.

- On the one hand, an Agile approach can potentially adapt to risks much more dynamically without extensive re-planning; but, on the other hand, that takes some skill and sophistication to do that.
- If a project has a high level of risk and the project team does not have a significant level of proficiency in taking a dynamic approach to managing risks, a more plan-driven approach might be more appropriate.

2. Geographic distribution of the project team

Another factor is the geographic distribution of the team(s). Agile works best if teams can be co-located and the more distributed the teams are, the more difficult it might be to implement a more agile approach.

3. scope and complexity of the project

The scope and complexity of the project are another major factor.

- Agile can be difficult to scale for projects requiring multiple teams and it can require a level of sophistication to do that.
- If the organization is not well prepared to implement a sophisticated model for scaling Agile for large, complex, multi-team projects; a more plan-driven approach might be needed.

SUMMARY OF KEY POINTS

Hybrid Agile Model

1. A hybrid Agile model has the potential to provide the best of both worlds by blending some of the elements of a pure Agile approach with some level of overall project management.

2. There is not just one way to implement a hybrid Agile model and there is potentially a broad spectrum of hybrid Agile approaches ranging from heavily plan-driven at one extreme to heavily adaptive at the other extreme.

3. It is important to fit the approach to the nature of the project and the most important factors to consider are:

- the level of uncertainty in the project
- stakeholder management
- the level of training and sophistication of the project team.

DISCUSSION TOPICS

Hybrid Agile Models

1. You have been assigned to manage a project in which there is a contractual relationship with the client where the client has definite expectations of the cost and schedule of the project. Is it possible to use a hybrid Agile approach in this project? How would you go about doing it?

2. How would you go about implementing some form of change management in a hybrid Agile project?

3. How does inception planning work in a hybrid Agile project if there is a high level of uncertainty in the project?

NOTES

1. Agile Alliance, "Why Would You Use a Hybrid Agile Approach?," https://www.agilealliance.org/what-is-hybrid-agile-anyway/.
2. Ibid.

13 Value-Driven Delivery

A CLASSICAL PLAN-DRIVEN PROJECT WAS DEEMED TO BE SUCCESSFUL if it delivered well-defined project requirements on-time and under-budget. The problem with that is the assumption that the "well-defined requirements" accurately reflect the "value" that the customer really needs.

In most projects, "value" has a much broader meaning than simply delivering well-defined requirements within a budgeted cost and planned schedule. There have been many projects that have met their cost and schedule goals but failed to deliver an acceptable level of customer value. It's a trade-off between maximizing the efficiency of meeting cost and schedule goals and the overall effectiveness of the solution.

In some cases, "business value" might be measured in financial terms such as Return on Investment (ROI) or Internal Rate of Return (IRR). In other words, did the project produce an acceptable financial return for whatever it cost to do the project? However, in many cases, the value that a project produces is not always quantifiable in financial terms. In many cases, it is a more subjective measurement that the Business Sponsor (the person paying for the cost of the project) believes that the project produced a valuable result considering the costs and effort that went into the project.

> According to Peter Drucker: "There is nothing quite so useless as doing with great efficiency, something that should not be done at all."[1]

VALUE-DRIVEN DELIVERY OVERVIEW

Delivering value to the customer should always be the most important goal. Meeting cost and schedule goals certainly has some value, but it may not be the most important value in a particular project. The value of meeting cost and schedule goals needs to be balanced against delivering a solution that provides an acceptable level of business value to the customer.

In today's world,

- There is generally a much higher level of uncertainty and volatility that makes it difficult to create well-defined requirements for a project upfront.

- An excessive emphasis on planning and control to meet cost and schedule goals can stifle the creativity and innovation that is needed to maximize the value of the solution.

> We have to accept that we live in a world of uncertainty that makes it very difficult, if not impossible for users to define in advance, what they need. Henry Ford said: "If I had asked people what they wanted, they would have said 'faster horses.'"[2]

What's Different about Value-Driven Delivery?

A major difference between a classical plan-driven project and an Agile project is that in a classical plan-driven project, no value is typically delivered until the very end of the project as shown in Figure 13.1.

In an Agile project, the requirements are prioritized, and value is delivered much more incrementally. Figure 13.2 shows how value and cost might compare in a typical Agile project.

- Cost is generally fairly linear and is a function primarily of the costs of operating the project team.

- If project requirements are properly prioritized with the highest value items first, value typically rises quickly at the start of the project, but as the project progresses, the incremental value of adding additional features and capabilities tends to diminish.

- At some point the project reaches a point of diminishing returns where the incremental value no longer exceeds the incremental cost of developing that value.

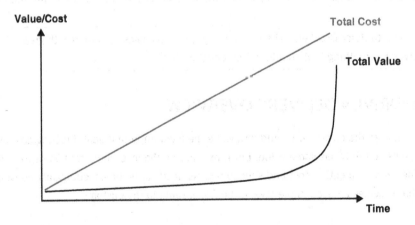

FIGURE 13.1   Typical classical plan-driven project cost-value curve

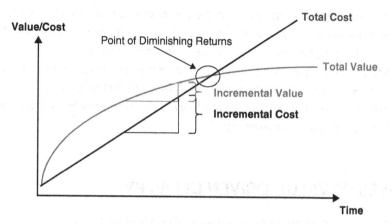

FIGURE 13.2   Typical Agile project cost-value curve

What Are the Advantages of Value-Driven Delivery?

There are several important advantages of a value-driven delivery approach:

1. **Financial return**

   ■ Prioritizing the value to be delivered and using an incremental development approach to deliver portions of the overall solution as quickly as possible can significantly increase the financial return of a project.

2. **Time-to-market**

   ■ With a classical plan-driven approach there can be a very large, missed opportunity to deliver at least some portion of the solution early to achieve faster overall time-to-market.

3. **Risk**

   ■ Another major problem is in risk. In a classical plan-driven project, you don't know if the project will really be successful until the very end and by that time it may be too late to do anything about it or the cost and difficulty of making corrections may be enormous.

   ■ An Agile, value-driven approach avoids this problem by delivering at least some partial value as early as possible which provides important early feedback as to whether the project is on the right track or not.

4. **Flexibility and adaptivity**

   ■ A classical plan-driven approach that places an excessive emphasis on meeting cost and schedule goals can create a very inflexible approach. The reason for that, of course, is that if you want to control the cost and schedule of the project, you must control the scope; and, in order to control the scope, you have to lock in the requirements upfront with some level of change control and you also need to minimize uncertainty by creating assumptions about any areas of ambiguity. That kind of approach just doesn't work well in rapidly changing environments with high levels of uncertainty.

- Locking in the requirements upfront with some level of change control makes it difficult to resolve uncertainties in the requirements as the project progresses.
- When you make arbitrary assumptions to try to minimize uncertainty, many times those assumptions turn out to be wrong.

PRINCIPLES OF VALUE-DRIVEN DELIVERY

The following are some important principles of value-driven delivery:[3]

1. Goal

The first principle is that having a clearly defined goal for a project is extremely important, and the goal should be defined in terms of the desired outcome of the project in terms of the business value that the project is trying to create, not in terms of how it will be achieved. The following are some characteristics of well-defined goals:

- Goals should be specific.
- They should be measurable and achievable.
- They should be realistic and timely.

A useful acronym to remember in this regard is "SMART" which stands for:

- Specific
- Measurable
- Achievable
- Realistic
- Timely.

2. Uncertainty

- We need to accept and acknowledge that some level of uncertainty is a fact of life in all projects, and it cannot be ignored.
- We need to recognize that you can't make all uncertainty go away.
- We need to develop an approach that is appropriate to the level of uncertainty.

3. Trade-offs

There is a tendency with Agile to rush too quickly into a solution which may not be optimal, and a systematic approach is needed to evaluate alternatives. Software consultant Nicolas Gouy has identified a number of problems that teams typically have in evaluating alternative solutions:

- The first is difficulty comparing solutions—an objective way is needed to evaluate the relative merit of alternative solutions and it must be done quickly and efficiently. The best judge of that is the customer, so it is essential for the customer to be engaged in that process.

- Next is difficulty prioritizing features which includes evaluating the contribution of the feature to the overall goal of the project against the cost and effort to develop the feature. A common problem in this area is to try to directly compare two different features on the basis of "what" they do and that can be very difficult to do. You have to do the comparison at a higher level on the basis of their contribution to the overall goals of the project (the "Why" and the "How"). The "Why" part helps to keep focus on the goals of the project.

- Next is becoming irrational—emotions, cognitive illusions, conscious and unconscious biases, fallacies, or fear of regret can cause mistakes in analyzing trade-offs. It's important to be aware of these factors and mitigate their impact on the analysis.

- Finally, another problem is fighting for the best solution. There is definitely a point of diminishing returns where attempting to arrive at a perfect solution is not worth the effort. In order to identify that point, good practice is to use an incremental and iterative approach to developing the solution. Start with something simple that is designed to meet the primary need and enhance it only as needed to provide additional value.

4. Speed

The final point, of course, is that speed is important. All of this has to be done quickly and efficiently to arrive at a solution. Taking time to evaluate alternative solutions is important but we can't let it drift into "analysis paralysis."

Focus on Customer Needs Rather Than Solutions

The most important thing to consider is that, many times, customers will want to tell you what they want in terms of a solution but that should be avoided in most cases. It's good practice to always focus on clearly understanding customer needs before jumping too quickly into the solution to those needs.

- Sometimes a little digging may be needed to uncover the real customer need. A technique that might be useful for that is called "The Five Whys" which was discussed earlier.

- It consists of starting with what you believe to be the customer need and asking "Why" up to five times to get to the real customer need. Here's an example:[4]

A customer is asking for the option to add tags to different records:

1. **Why tags?** They can use the tags to organize their data.
2. **Why organize?** They can find a particular piece of information at a later date.

3. **Why do they need to find it?** The customer needs to refer to historical data to solve current problems.

4. **Why does the customer need historical data?** It prevents them from having to start from scratch.

5. **Why do they not want to start from scratch?** To save time.

"Now that you have this information—that the customer actually wants to save time—you will understand that simply building a tags feature isn't enough. To really save time, the customer needs to be able to easily search or filter using this tag as well."[5]

The Pareto Rule

A very important concept to keep in mind in doing customer-value prioritization is the Pareto Rule which says that many things have an unequal distribution:

- The Pareto principle was first developed in 1906 by Italian economist Vilfredo Pareto, who, in the course of researching his ideas, made an interesting observation: 80% of the land in Italy, he discovered, was owned by just 20% of the people.

- Exploring this relationship in other countries, he found that the situation was the same all over Europe. Over time, he became aware that this 80/20 split was not limited to landowners—or even to human affairs. In fact, he found, 20% of the pea pods in his garden produced 80% of the peas he harvested!

Forty years after Pareto published his ideas, business theorist Joseph Juran stumbled across the 80/20 rule, and wondered if it could be applied to business situations. Could it be that 80% of business problems were generated by just 20% of the related causes? Of course, the answer was a resounding "yes."

In a typical product, a relatively small number of features will result in the majority of the usage and impact of the product. And adding additional features beyond that point can quickly reach a point of diminishing returns and just add complexity to the product without significantly adding value.

There are many examples of this in the real world. Take some of the more commonly used products, like Microsoft Word and Excel ,those products have thousands of features; but, in actual practice, most users only use a very small subset of the features that the product offers.[6]

> A limitation of the Pareto Analysis technique is that it only looks at impact and does not include cost (or effort) required.

CUSTOMER-VALUE PRIORITIZATION OVERVIEW

In this section, we're going to begin discussing a number of different models and best practices for doing customer-value prioritization. Customer value may not be easy to assess, and many people underestimate the difficulty and complexity of this task:

- "Value" can be difficult to assess and quantify and it can also be very subjective.

- Customer ideas about quality are often confused and difficult to see clearly.

- Different stakeholders may have different views of value and it might be difficult to reach consensus on an overall solution.

It's definitely not as easy as asking a customer what they want and getting a simple answer. It is very easy to underestimate the difficulty and complexity associated with identifying and prioritizing customer values and it can require a systematic approach to do that. In the rest of this section, we're going to provide a brief overview of some potential models and best practices for doing that.

Levels of Prioritization

There are at least several levels of prioritization that you might find in an Agile project:

- **Strategic:** The first is strategic. Many times Agile development is not bound to well-defined projects. It is more of a continuous development effort that is ongoing and is continuously re-planned as it progresses. This provides a much more flexible and dynamic way of planning and prioritizing efforts. An example might be a company where the senior executives get together quarterly to plan and prioritize initiatives. This effort will typically be done at a high level based on themes and epics. A theme is a higher-level strategic objective that might cut across a number of projects like improving employee morale and an epic might be a major high-level project goal that consists of a number of different features such as developing an online order tracking tool.

- **Project:** The next level is at the project level. The primary goal at this level is to determine what is included in the scope of a project and what is not included. That's a somewhat different process in an Agile environment because the scope of the project is typically much more fluid and may not be rigidly defined. As a result, the product backlog is prioritized from top to bottom, the items are developed in priority order, and it is understood that at some point the project might reach a point of diminishing returns and, at that point, the remaining items might not be developed at all.

- **Tactical:** The final level is a tactical level associated with planning and prioritizing the order of development of stories and features within a project.

In an Agile environment, all of these levels of planning and prioritization interact with each other in a dynamic way.

In a classical plan-driven environment, you might find a very different environment. You might have much more longer-range plans that are much more static and not well-integrated. For example, at a strategic level, you might find multi-year long-range strategic plans that aren't updated frequently and lower-level plans at project or program levels that may or may not be well integrated.

The whole planning and prioritization process in an Agile environment is typically much more dynamic and, as a result needs to be much more integrated. Changes at a strategic level can propagate much more easily down to lower levels to rapidly shift directions if needed to meet business needs.

Factors to Consider in Prioritization

Dr. Dan Rawsthorne has identified five factors in prioritizing User Stories:[7]

1. **Business value and return on investment:** The first and most important is, of course, business value and return on investment. This factor is, of course, important at all levels of prioritization. However, business value alone does not tell the whole story.

2. **Architectural significance:** Another factor is that some stories have architectural significance and should be addressed early for that reason.

3. **Dependencies:** A related factor is that some stories depend on other stories which needs to be considered in prioritizing the work to be done.

4. **People availability:** Another very important factor is the availability of people to do the work. Some stories may require specialized resources which need to be planned and the total number of resources required to implement a feature is also a very important factor to consider.

5. **Placeholders:** Dr. Rawsthorne has also identified the idea of "placeholders." The idea behind this is that you often want to reserve some capacity for items that have not been fully identified yet. Bug fixes might be one example.

MoSCoW Prioritization

A very simple model for customer value-prioritization is the MoSCoW model. The different levels of priority include:

- **Must have:** These provide the Minimum Usable SubseT (MUST) of requirements that the project guarantees to deliver.

- **Should have:** A "Should Have" may be differentiated from a "Could Have" by reviewing the degree of pain caused by it not being met, in terms of business value or numbers of people affected.

- **Could have:**
  - wanted or desirable but less important
  - less impact if left out (compared with a Should Have).
- **Won't have:** These are requirements the project team has agreed it will not deliver.

VALUE-DRIVEN DELIVERY TOOLS

This section provides an overview of some basic ideas and tools that can be used to support a value-driven delivery approach. Much more advanced ideas and tools are available but are beyond the scope of this book.

Minimum Viable Product

One idea that is particularly important in customer-value prioritization is the idea of the "Minimum Viable Product" which is defined by Eric Ries as follows:[8]

- The Minimum Viable Product is "that version of a new product which allows a team to collect the maximum amount of validated learning about customers with the least effort."
- Eric Ries did a lot of research into new startup companies and found that a large number of startup companies failed. The pattern that he saw was that companies developed elaborate ideas for products and launched an extensive effort to develop those products and then discovered after all of that effort that the product wasn't what the market wanted at all.
- This problem is a direct result of not recognizing the level of uncertainty associated with the acceptance of a new product in the marketplace.
- We have to recognize that there is a lot that we don't know and if we assume that we know everything that there is to know about a new product, we will very frequently be wrong.

A much better approach is to start with something very simple that can be created with minimum investment to get feedback and inputs from customers before making a full-blown investment in developing a complete product. That is what the idea of the Minimum Viable Product is all about.

Minimum Marketable Feature

Another idea that is important in this area is the concept of a "Minimum Marketable Feature" which is defined as "The smallest set of functionality that must be realized in order for the customer to perceive value."[9] What that means is that each feature does not necessarily need to be fully developed to get customer input and feedback. It is very similar to the idea of a Minimum Viable Product—each feature can be developed incrementally from a very basic idea of a minimum marketable feature.

There are several essential characteristics of a Minimum Marketable Feature:

- **Feature:** A "feature" is something that is identifiable and perceived, of itself, to have value by the user.

- **Marketable:** "Marketable" means that it provides significant value to the customer. Value may include revenue generation, cost savings, competitive differentiation, brand-name projection, or enhanced customer loyalty.

- **Minimum:** "Minimum" means the smallest set of functionality that must be realized in order for the customer to perceive value.

SUMMARY OF KEY POINTS

Value-Driven Delivery Overview

1. Delivering value to the customer should always be the most important goal.

   - Meeting cost and schedule goals certainly has some value, but it may not be the most important value in a particular project.

   - The value of meeting cost and schedule goals needs to be balanced against delivering a solution that provides an acceptable level of business value to the customer.

   - If a project doesn't meet stakeholder expectations, it can't be considered successful even if it met cost and schedule goals.

   A major difference between a classical plan-driven project and an Agile project is that in a classical plan-driven project, no value is typically delivered until the very end of the project. In an Agile project, the requirements are prioritized, and value is delivered much more incrementally as the project is in progress.

2. Value-driven delivery has a number of important advantages:

   - **Financial return:** Prioritizing the value to be delivered and using an incremental development approach to deliver portions of the overall solution as quickly as possible can significantly increase the financial return of a project.

   - **Time-to-market:** With a classical plan-driven approach there can be a very large, missed opportunity to deliver at least some portion of the solution early to achieve faster overall time-to-market.

- **Risk:** Another major problem is in risk. In a classical plan-driven project, you don't know if the project will really be successful until the very end and by that time it may be too late to do anything about it or the cost and difficulty of making corrections may be enormous.

- **Flexibility and adaptivity:** A classical plan-driven approach that places an excessive emphasis on meeting cost and schedule goals can create a very inflexible approach. That kind of approach just doesn't work well in rapidly changing environments with high levels of uncertainty.

Principles of Value-Driven Delivery

The following are some important principles of value-driven delivery:

- **Goal:** A clearly defined goal for a project is extremely important, and the goal should be defined in terms of the desired outcome of the project in terms of the business value that the project is trying to create, not in terms of how it will be achieved.

- **Uncertainty:** We need to accept and acknowledge that some level of uncertainty is a fact of life in all projects, and we need to develop an approach that is appropriate to the level of uncertainty.

- **Trade-offs:** There is a tendency with Agile to rush too quickly into a solution which may not be optimal, and a systematic approach is needed to evaluate alternatives.

- **Speed:** All of this must be done quickly and efficiently to arrive at a solution. Taking time to evaluate alternative solutions is important but we can't let it drift into "analysis paralysis."

- **Focus on customer needs rather than solutions:** Many times, customers will want to tell you what they want in terms of a solution but that should be avoided in most cases. It's a good practice to always focus on clearly understanding customer needs before jumping too quickly into the solution to those needs.

- **The Pareto Rule:** A very important concept to keep in mind in doing customer-value prioritization is the Pareto Rule which says that many things have an unequal distribution.

- **MoSCoW model:** This is a very simple model for customer value-prioritization.

Customer Value Prioritization Overview

There are multiple levels to consider in prioritizing value:

- **Strategic:** Many times, Agile development is not bound to well-defined projects. It is more of a continuous development effort that is ongoing and is continuously re-planned as it progresses, and an integrated planning approach may be used across a number of projects to reach an important overall goal.

- **Project:** The next level is at the project level. The primary goal at this level is to determine what is included in the scope of a project and what is not included.

- **Tactical:** The final level is a tactical level associated with planning and prioritizing the order of development of stories and features within a project.

Factors to Consider in Prioritization

There are five major factors to consider in customer-value prioritization:

- Business value and return on investment
- Architectural significance
- Dependencies
- People availability
- Placeholders

Value-Driven Delivery Tools

- "Minimum Viable Product" is "that version of a new product which allows a team to collect the maximum amount of validated learning about customers with the least effort."
- "Minimum Marketable Feature" is defined as "The smallest set of functionality that must be realized in order for the customer to perceive value."

DISCUSSION TOPICS

Value-Driven Delivery Overview

1. How is value-driven delivery in an Agile environment different than classical plan-driven delivery?

2. What advantages does value-driven delivery provide?

Customer Value Prioritization Overview

1. What are the different levels of value-driven prioritization and how would they interact with each other in a typical Agile environment? How would it be different in a typical plan-driven environment?

2. What is the MoSCoW model and how would it be used in customer value prioritization?

Principles of Value-Driven Delivery

1. How would you go about doing value-driven prioritization in a typical project? What principles would you consider?

2. What is the significance of the Pareto Rule in value-driven prioritization?

Value-Driven Delivery Tools

1. Why is the idea of a Minimum Viable Product and Minimum Marketable Feature important and what role does it play in doing value-driven delivery analysis?

NOTES

1. Peter Drucker, quoted in InfoQ, http://www.infoq.com/resource/minibooks/agile-guts/en/pdf/AgilewithGuts-final.pdf.
2. Henry Ford, quoted in InfoQ, http://www.infoq.com/resource/minibooks/agile-guts/en/pdf/AgilewithGuts-final.pdf.
3. http://www.infoq.com/resource/minibooks/agile-guts/en/pdf/AgilewithGuts-final.pdf.
4. "How Can the Five Why's Technique Make You a Better Product Manager?," https://shippingtmrw.substack.com/p/how-can-the-five-whys-technique-make.
5. Ibid.
6. BetterExplained.com, "Understanding the Pareto Principle," https://betterexplained.com/articles/understanding-the-pareto-principle-the-8020-rule/.
7. Dan Rawsthorne, "5 Prioritization Factors," http://blog.3back.com/scrum-tips/5-prioritization-factors/.
8. Eric Ries, quoted in LeanStack.com, "What Is a Minimum Viable Product?," http://ask.leanstack.com/en/articles/902991-what-is-a-minimum-viable-product-mvp#:~:text=When%20Eric%20Ries%2C%20author%20of,customers%20with%20the%20least%20effort.
9. Solutions.com, "Minimum Marketable Features,"https://www.solutionsiq.com/agile-glossary/minimum-marketable-features/.

14 Adaptive Planning

IN CHAPTER 4, WE TALKED BRIEFLY ABOUT AGILE PLANNING PRACTICES. This chapter will go into much more detail on that subject because the approach for doing planning is one of the most important subjects for any Agile Project Manager.

Many people seem to think of planning as an "all-or-nothing" proposition—either you develop a highly detailed plan, or you do no planning at all. I don't believe that to be the case. The right solution is to fit the planning approach to the nature of the project. There are two major problems associated with why people have this difficulty with planning:

1. **Dealing with uncertainty:** Many people seem to have difficulty dealing with an uncertain environment—they want things to be crystal-clear, black-and-white and, in an uncertain environment, they think that it is a waste of time to do any planning at all.

2. **Unrealistic expectations:** A related factor is that many people develop unrealistic expectations about planning: If they develop a well-thought-out plan, they expect that it should work every time but Murphy's Law often contradicts that belief.

Adaptive planning is the general term for a planning approach that evolves as a project is in progress. It is particularly well suited to an environment with a higher level of uncertainty that makes it difficult to plan well-defined and detailed requirements for a project prior to its start. There are several important and related points to cover in this area:

- **Rolling-wave planning:** It's important to understand an overview of what rolling-wave planning is and why it makes sense and then we will discuss some best practices for doing rolling-wave planning.

- **Progressive elaboration:** Next, we're going to discuss the concept of progressive elaboration and a number of best practices related to that area.

- **Multi-level planning:** Finally, we're going to discuss multi-level planning" and how the typical levels of planning or implemented in an Agile project. That is a key element of how an Agile project implements progressive elaboration.

ROLLING-WAVE PLANNING

A lot of people have the mindset that if you develop a plan, you have to stick with it without deviation or at least control any changes to the plan. That way of thinking is deeply rooted in classical plan-driven project management thinking because there is so much emphasis on developing highly detailed upfront plans, as well as managing and controlling changes to manage the scope, costs, and schedule of the project.

- That mode of operation might work in a highly predictable environment, but it doesn't work well in environments with high levels of uncertainty; and, in today's world, it's rare to find any project that doesn't have at least some uncertainty associated with it.

- If you shift the emphasis of the project management approach from an emphasis on control of project scope, costs, and schedules to put more emphasis on providing value and encouraging creativity and innovation, a much more adaptive approach is needed.

We can learn a lot about an adaptive approach to planning from the military. A lot of people think of the military as a very highly planned and rigidly controlled organization but that is not necessarily the case. Here is a quote from General Dwight D. Eisenhower that I particularly like:

> In preparing for battle I have always found that plans are useless, but planning is indispensable.[1]

What he is saying is that there is a lot of value in doing planning, but you should never consider a plan to be absolutely firm and not subject to change. The military trains heavily in many different tactics but on the battlefield, adaptivity is important because you can't completely anticipate what the enemy is going to do.

- A similar thing is true of managing projects in a highly uncertain project environment. In that kind of environment, more information is typically known as the project is further along in progress.

- So, it often makes sense to defer planning decisions until more information is available to support those decisions. Attempting to make too many planning decisions without sufficient information to support those decisions will often be wrong and result in wasted effort.

Overview of Rolling-Wave Planning

Rolling-wave planning is a technique that enables you to plan for a project as it unfolds. This technique, then, requires you to plan iteratively. The planning technique is essential in Scrum or other

Agile approaches. Essentially, when you use rolling-wave planning, plan until you have visibility, implement, and then re-plan.

- Rolling-wave planning is used when you just don't have enough clarity to plan the entire project in detail prior to the start of the project. This lack of clarity could come from various factors, such as emerging requirements and/or technological uncertainty about how the solution will be implemented. Rolling-wave planning is particularly useful in projects with high uncertainty.

- Rolling-wave planning is the process of planning for a project in waves as the project becomes clearer and unfolds. It acknowledges the fact that we can see more clearly what is in close proximity, but if looking further ahead, our vision becomes less clear.

> The core concept behind Rolling-wave Planning is: you cannot plan in detail what you do not know.[2]

The approach that I think makes best sense for doing rolling-wave planning is something like this:

1. **Objectively evaluate the level of uncertainty in the project and identify the knowns and unknowns as best you can:**

   - The word "objectively" is very important here—it is very important to take a mature approach to acknowledge that "we don't know what we don't know."
   - It is foolish to try to develop a detailed plan based on unreliable and incomplete information, but it is also foolish to ignore and not take advantage of information that we do know. Starting with a blank sheet of paper rarely makes sense.

2. **Next, develop an initial plan based on the knowns and an approach for further defining the unknowns as the project progresses:**

   - The approach for doing this might vary significantly from one project to the next depending on the level of uncertainty in the project:
     - At one extreme, you might start out with just some high-level objectives and a vision statement and further elaborate more detailed plans as the project progresses. For example, the planning might start by defining large chunks of work as epics, forecasting the work with roadmaps (refer to Table 15.4 for Product Roadmap elements), and elaborating the work as the project progresses.[3]
     - On the other hand, in a more predictable and less uncertain environment, you might start out with a much more detailed upfront plan with fewer unknowns to elaborate as the project progresses.

3. **Then, as the project is in progress, more detailed planning can be done as more information becomes available to support those planning decisions:**

Effective assessment and management of uncertainty are the most critical elements of doing rolling-wave planning.

Comparison of Planning Approaches

All projects start out with some level of uncertainty but there is a major difference in the approach to managing uncertainty between a classical plan-driven approach and an Agile or adaptive approach.

Classical Plan-Driven Projects

Figure 14.1 shows what is commonly called "the cone of uncertainty" and how a classical plan-driven project would typically be planned.

- In a classical plan-driven project, a key aim is to provide predictability over project costs and schedules.

- As a result, the planning approach attempts to remove uncertainty by more clearly defining the goals and requirements for the project and more clearly outlining the process required to produce the results as well as the level of effort and time required.

A typical Waterfall or plan-driven project attempts to reduce the level of uncertainty associated with the project to a very low level before the project starts.

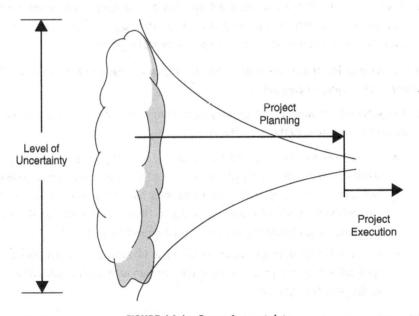

FIGURE 14.1   Cone of uncertainty

- That is a reasonable approach with some projects but it's not reasonable to force that kind of approach on a project that has very high levels of uncertainty.

- Attempting to do that on a project with very high levels of uncertainty might force you to make many assumptions about the requirements for the project based on very sketchy and incomplete information.

A basic problem with the plan-driven approach is that sometimes Project Managers make assumptions like that, lock them into a requirements document to define the project, and then make it difficult to change those assumptions through a formal method of change control later.

- That creates an "illusion of control." On the surface, it looks like the project is very well controlled, but a project can only be as well controlled as the requirements are certain.

- As a result, you might start the project with what appears to be a very detailed plan that is designed to control the project and after large numbers of change requests later, discover that it wasn't so well controlled after all.

Adaptive or Agile Projects

Figure 14.2 shows the general way that an Agile rolling-wave planning approach progressively reduces the uncertainty in a project as the project progresses, rather than attempting to remove all the uncertainty upfront. By breaking up the project into releases and iterations, a lot of the detailed

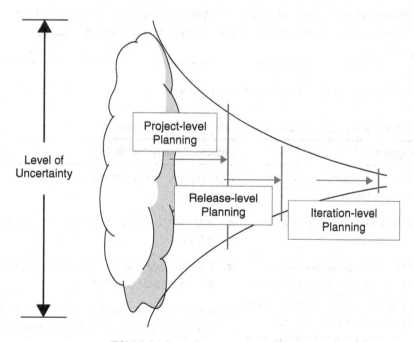

FIGURE 14.2 Agile cone of uncertainty

planning can be deferred until that release or iteration is ready for development and then only the detailed planning associated with that release of iteration needs to be done, rather than attempting to do a detailed plan for the entire project upfront.

What typically happens is project planning is done once at the beginning of the project and establishes the Vision, Product Backlog, Product Road Map, and potentially also some cost and schedule estimates at whatever level of detail is necessary. That forms the high-level plan for the project which is then further elaborated in more detail at the lower levels as the project progresses.

- For example, at the release planning level, the general goals for that release would be defined and the high-level epics to be included in that release would also be defined. Then, at the iteration or sprint level, the stories to be included in that sprint would be finalized and estimated to fit within the capacity of that sprint.

- As more detailed information is planned at the lower levels, there may be some re-planning needed at the higher levels. For example, if the release planning finds that it is going to take significantly longer to complete a release than planned, that information should probably be bubbled up to the project-level plan and used to adjust the project level plan as necessary.

> There is no need to make decisions any further in advance than required to support whatever actions are required at that point in time. Deferring a decision until "the last responsible moment" means delaying the decision to the point where delaying it any further might impact the outcome of whatever actions that decision is designed to support.

Table 14.1 shows a summary of these two approaches:

TABLE 14.1   Comparison of plan-driven and Agile planning approaches

Plan-driven approach	Adaptive/Agile approach
Plan-driven projects attempt to plan the entire project upfront	Agile projects use a "rolling-wave" planning approach
■ Attempting to plan too far in advance involves speculation	■ Only essential, high-level planning is done upfront
■ Many times, that speculation will be wrong and will require wasted effort in re-planning. For example, the work may be defined in firm phases and mile-stones when these can change over time	■ More detailed planning is deferred until "the last responsible moment" ■ Better decisions can be made when more is known about the project (Just-in-Time planning)

A key point about this is that this is not an all-or-nothing decision between very extensive and detailed upfront planning and no upfront planning at all.

- We should use good common sense to adapt the level of planning to the nature of the project.

- Adaptive planning means fitting the planning approach to the nature of the project.

PROGRESSIVE ELABORATION AND MULTILEVEL PLANNING

Progressive Elaboration

Here's a definition of progressive elaboration.

> Progressive elaboration is a continuous iterative process of refining and further detailing the product characteristics based on more detailed information and insight that become available as the project progresses.[4]

Figure 14.3 shows the relationship of progressive elaboration to rolling-wave planning in a classical plan-driven environment. Progressive elaboration is generally considered to be a broader term that encompasses both rolling-wave planning and some form of learning from implementing the plan.

In a classical plan-driven project management context, progressive elaboration usually means further refining the project plan to further define the scope, costs, and schedule of the project by evaluating the estimates of work against actual progress to further refine the estimates for completing the project. It assumes that the solution and the requirements are relatively stable, and the primary variable is the estimated work to complete the project.

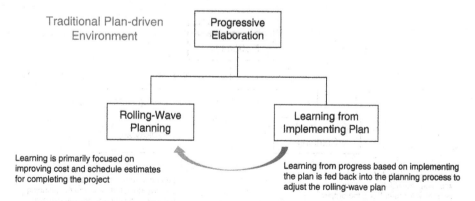

FIGURE 14.3   Relationship of progressive elaboration to rolling-wave planning in a classical plan-driven environment

In an Agile environment, progressive elaboration takes on a broader meaning that places a much heavier emphasis on prototyping the solution or on developing a minimum viable product to get additional feedback that is incorporated into the rolling-wave planning process.

- An Agile environment does not assume that the solution and the requirements are stable; so, in addition to learning about the estimated work to be done, there is a significant amount of additional learning to be done about the solution itself.

- In an Agile environment, all that learning is fed back into the rolling-wave planning process, as shown in Figure 14.4.

Progressive elaboration is an important concept in planning a strategy for the definition of requirements in an Agile project; however, it is not new to Agile. Progressive elaboration has been a classical plan-driven project management practice for a long time; however, Agile emphasizes its use much more.

- In a classical plan-driven project management environment, it is typically used on a limited basis to adjust the projected final schedule and cost of the project as the project is in progress, based on progress to date. It normally is not used to make significant changes in the requirements for the project.

- In an Agile project, it is used on a much broader basis to further define and elaborate the requirements that the project is designed to fulfill as the project is in progress.

Multilevel Planning

In a typical Agile project, rolling-wave planning is normally done at different levels, as illustrated in Figure 14.5; however, each of these levels is inter-related—in other words, you don't do project-level planning at the beginning of the project once and never do it again.

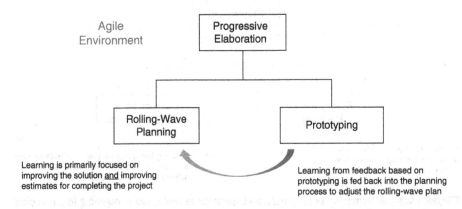

FIGURE 14.4 Relationship of progressive elaboration to rolling-wave planning in an Agile environment

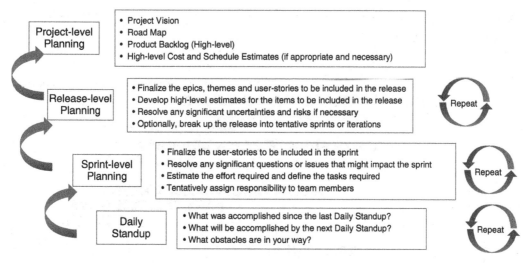

FIGURE 14.5 Multilevel planning

Project-Level Planning

Typically, project planning is done once at the beginning of the project and establishes the Vision, Product Backlog, Product Road Map, and potentially also some cost and schedule estimates at whatever level of detail is necessary. That forms the high-level plan for the project which is then further elaborated into more detail at the lower levels as the project progresses. Figure 14.6 shows an example of a Product Roadmap provided by Aha!.

Release-Level Planning

At the release planning level, the general goals for that release would be defined and the high-level epics, stories, and themes to be included in that release would also be defined.

- At the release level, the level of effort to complete the release will typically be estimated in order to pin down the projected release date. If the projected release date turns out to be unacceptable, trade-offs may be needed, such as removing items from the release to bring the date in.

- Release-level planning is naturally more detailed than project-level planning and is repeated for each release. Any significant new information that results from release-level planning should be fed back to adjust the project-level plan if necessary.

Sprint-Level Planning

Then, at the sprint level, the stories to be included in that sprint would be finalized and estimated to fit within the capacity of that sprint, and any significant issues or questions that might impact successful completion of the sprint are resolved.

FIGURE 14.6   Product Roadmap example

- The team would then identify the tasks required to implement the stories and tentatively assign responsibilities for the tasks among the team members.

- Sprint-level planning is naturally much more detailed than project-level and release-level planning and is repeated for each sprint.

- Any significant new information that results from sprint-level planning should be fed back to adjust the release-level plan and project-level plan, if necessary.

  Sprint-level planning uses a capacity-based planning process as discussed in Chapter 1.[5]

Daily Standup Planning

The final level of Agile planning happens during the sprint at the Daily Standup. During that meeting, each individual reviews what they're working on, what progress has been made, and what obstacles stand in their way, and the plan for completing the sprint is adjusted as necessary. The Daily Standup meeting is naturally repeated daily as the sprint is in progress.

Multilevel Planning Summary

Keep in mind that the overall level of planning may also be very different in an Agile environment. Instead of having a detailed project plan such as you might find in a classical plan-driven environment, a project plan (if one exists at all) in an Agile environment is likely to be much more loosely defined and is expected to go through much more significant changes as the project progresses as a result of progressive elaboration.

> A key point is that in this multi-level planning model, all the levels of planning are inter-related and inter-dependent. If a significant change is discovered at a lower level, it should be bubbled up to adjust the higher-level plans as necessary.
>
> In a similar way, if re-planning is done at a higher level, the lower levels may need to be re-planned to be consistent with the higher-level plan.

SUMMARY OF KEY POINTS

Adaptive Planning

1. Adaptive planning is the general term for a planning approach that evolves as a project is in progress. It is particularly well suited to an environment with a higher level of uncertainty that makes it difficult to plan well-defined and detailed requirements for a project prior to the start of the project.

2. Many people seem to think of planning as an "all-or-nothing" proposition—either you develop a highly detailed plan, or you do no planning at all. There are two major problems associated with why people have this difficulty with planning:

   - **Dealing with uncertainty:** Many people seem to have difficulty dealing with an uncertain environment—they want things to be crystal-clear, black-and-white and, in an uncertain environment, they think that it is a waste of time to do any planning at all.

   - **Unrealistic expectations:** A related factor is that many people develop unrealistic expectations about planning:

     - If they develop a well-thought-out plan, they expect that it should work every time.
     - Many people are also unrealistic that everything about a project will go perfectly all the time, but Murphy's Law often contradicts that belief.

3. Don't get bogged down in planning requirements too far in advance, requirements should only be defined to the extent needed to support any decisions or actions at that point in time.

Progressive Elaboration

1. Defer planning decisions until "the last responsible moment" which is defined as the point in time where delaying decisions further will have an impact on the project.

2. In a classical plan-driven project management context, progressive elaboration usually means further refining the project plan to further define the scope, costs, and schedule of the project by evaluating the estimates of work against actual progress to further refine the estimates for completing the project.

   ■ It assumes that the solution and the requirements are relatively stable, and the primary variable is the estimated work to complete the project.

3. In an Agile environment, it takes on a broader meaning that places a much heavier emphasis on prototyping the solution to get additional feedback that is incorporated into the rolling-wave planning process.

   ■ An Agile environment does not assume that the solution and the requirements are stable; so, in addition to learning about the estimated work to be done, there is a significant amount of additional learning to be done about the solution itself.

   ■ In an Agile environment, all that learning is fed back into the rolling-wave planning process.

Multilevel Planning

4. In a typical Agile project, rolling-wave planning is typically done at different levels; however, each of these levels is inter-related:

   ■ **Project-level planning:** At the project level, there may be a need to understand the overall feasibility and scope of the project to evaluate the resources, costs, and schedule required; however, that can typically be done based on a very high-level understanding of the goals and high-level requirements without details.

   ■ **Release-level planning:** At the release level, there may be a need for a little more accuracy in the estimated time and effort to complete a release and that may call for understanding the requirements in a little more detail, but that also may not require a significant amount of detail.

   ■ **Sprint-level planning:** Finally, at the sprint level, there are typically two needs that should be satisfied prior to the start of a sprint:

     ■ The team needs to have a reasonable estimate of the requirements in terms of story points to determine if they can fit into the capacity for a sprint.

     ■ The team should not take any requirements into a sprint that have major uncertainties or issues associated with them that might block or delay the development effort. However, even at the sprint level, many details of how the requirements will be implemented can be worked out while the sprint is in progress.

   ■ **Daily sprint-level planning:** At the daily level in a sprint there is a need to further elaborate details of requirements to define implementation and to evaluate alternative approaches.

DISCUSSION TOPICS

Adaptive Planning

1. How is an adaptive planning approach different between an Agile and a classical plan-driven environment?

2. What are the most important factors to consider in choosing an appropriate planning approach for a project?

Progressive Elaboration

1. What does it mean to defer planning decisions until "the last responsible moment"?

2. What is the impact of making decisions in a project earlier than required?

3. Can you give an example of how that principle might be applied in a typical project?

Multilevel Planning

1. How is the idea of "rolling-wave planning" related to the idea of "multilevel planning"?

2. What are the different levels of planning in an Agile project environment?

3. How are the different levels of planning inter-related to each other?

NOTES

1. Dwight D. Eisenhower, quoted in BrainyQuote.com, http://www.brainyquote.com/quotes/quotes/d/dwightdei164720.html.
2. Mosaic White Paper, "Rolling-Wave Planning," http://www.mosaicprojects.com.au/WhitePapers/WP1060_Rolling_Wave.pdf.
3. Winston Gonzalez, email comments to the author on book review.
4. Project Management Lexicon, "What is Progressive Elaboration?," http://www.projectmanagementlexicon.com/progressive-elaboration/.
5. Winston Gonzalez, email comments to author on book review.

15 Agile Planning Practices and Tools

IN THIS CHAPTER, WE WILL DISCUSS SOME AGILE PLANNING PRACTICES AND TOOLS that can be used to implement an adaptive planning strategy that we have just discussed. These include:

- Product/Project Vision
- Product Roadmaps
- Exploratory 360 Assessment
- Agile Functional Decomposition
- Agile Project Charter

PRODUCT/PROJECT VISION

A very important benefit of an Agile approach is increased focus on business value. A good approach is to start with a vision statement that clearly defines the business value that the solution will provide. A vision statement should be short and succinct. The vision defines the "Why" of the project. This is the higher purpose, or the reason for the project's existence.

What Is a Product/Project Vision?

Once a high-level vision statement has been defined, it is a good technique to use functional decomposition to break down that vision statement into the functionality that will be needed to achieve that overall vision. Functional decomposition becomes particularly important on large projects where there could be hundreds of user stories. It provides a hierarchical approach for organizing requirements and is an essential technique for prioritizing requirements.

Without an effective approach for functional decomposition that aligns with an overall value statement, it is very easy to "get lost in the weeds" of individual requirements or user stories and lose sight of the big picture of the value you're trying to deliver.

> A vision statement is a "brief statement of the desired future state that would be achieved by developing and deploying a product or project. A good vision should be simple to state and provide a coherent direction to the people who are asked to realize it."[1] A vision statement can be at the level of an entire company or at the level of a particular product, project, or program

Table 15.1 shows an outline for a typical Agile vision statement.[2] This is just a general model and should be modified as necessary to fit the situation.

Product/Project Vision Examples

Table 15.2 is a specific example of an online application intended for grocery shoppers.

TABLE 15.1　Agile vision outline

For (target customer)	It's very important to clearly define the target customer who is going to derive the most benefit from the product or project
Why (statement of the need or opportunity)	Next is to specifically identify the customer need
What (product/project name)	Then define what the product, solution, or service is
So that (key benefit, compelling reason to buy)	Define how it solves the problem
Unlike (primary competitive alternative)	Identify primary competitive products or approaches
Our product (statement of primary differentiation	Identify how this product/project is different from other competitive or alternative approaches

TABLE 15.2　Agile vision example: grocery shopping

For	Grocery shoppers in <xyz> region, city, or country
Why	Shoppers want an easy way to buy grocery items
What	The Grocery Mall is a web-based, online grocery shopping solution
So that	Customers can buy items online
Unlike	Solutions that require physically going to a conventional store
Our product	Provides a totally web-based shopping experience with delivery options for in-store pickup or home delivery

TABLE 15.3   Agile vision example: online training course

For	Project Managers with a classical plan-driven project management background
Why	Want to develop new knowledge and skills to continue developing their careers in today's Agile environment
Our PMI-ACP Training Curriculum	Is a complete online training curriculum
That provides a very comprehensive base of knowledge related to Agile Project Management	Based on real-world experience and is very well-designed for learning
Unlike	Other PMI-ACP training programs
Our Product	Goes beyond simply passing the exam and provides a base of knowledge and skills geared towards a high-impact, real-world role Also provides a deeper understanding of the principles behind the information which should result in a much higher level of learning

Table 15.3 is another example of an online training application intended for project managers:

Tips for Creating a Compelling Vision

Roman Pichler has defined "Eight Tips for Creating a Compelling Product Vision":[3]

1. **Describe the motivation behind the product:** "The product vision is the overarching goal you are aiming for, the reason for creating the product."
2. **Look beyond the product:** "Be clear on the difference between the product vision and the product. The product vision should describe the motivation for creating the product, not how it is implemented."
3. **Distinguish between the vision and the product strategy:** "Your product vision should not be a plan that shows how to reach your goal."
4. **Employ a shared vision:** "You can come up with the most beautiful vision for your product. But it's useless if the people involved in making the product a success don't buy into it."
5. **Choose a motivating vision:** "If you are working on something exciting that you really care about, you don't have to be pushed. 'The vision pulls you,' said Steve Jobs. Your vision should therefore motivate people, connect them to the product, and inspire them."
6. **Think big:** "Make your product vision broad and ambitious so that it engages people, and it can facilitate a change in the strategy."

7. **Keep your vision short and sweet:** "As your vision is the ultimate reason for creating the product, it should be easy to communicate and to understand."

8. **Use the vision to guide your decisions:** "Use the vision to guide your product decisions and to focus everyone on the ultimate reason for creating the product."

PRODUCT ROADMAPS

"A product roadmap is a high-level plan that describes how the product is likely to grow. It allows you to express where you want to take your product, and why it's worthwhile investing in it."[4] Table 15.4 shows the typical elements that would be included in a product roadmap.

What Are the Benefits of a Product Roadmap?

The following is a summary of the benefits of a Product Roadmap by Roman Pichler.[5]

1. **Communications:** The Product Road Map is an excellent tool to help communicate how the product or project will be developed.

2. **Alignment:** It helps align the product or project and the company strategy.

3. **Coordination:** It helps coordinate the development, marketing, and sales activities with important events and other activities and stakeholders that might be outside of the direct project team.

4. **Portfolio management:** From a portfolio management perspective, it helps synchronize the development efforts of different products and projects.

5. **Integration:** And, finally, it provides a way of integrating more tactical project details with higher-level strategic initiatives as well as with related products and projects.

Tips for Creating a Product Roadmap

The following are some tips from Roman Pichler for creating an effective Product Roadmap.[6]

1. **Do the prep work:** "Describe and validate the product strategy—the path to realize your vision—before you create your roadmap and decide how the strategy is best implemented."

TABLE 15.4 Product Roadmap elements

Date	The planned release date or timeframe
Name	The name of the new release
Goal	The reason for creating the new release
Features	The high-level features or epics necessary to meet the goal
Metrics	The metrics to determine if the goal has been met

2. **Tell a convincing and realistic story:** "Your product roadmap should tell a coherent story about the likely growth of your product. Each release should build on the previous one and move you closer toward your vision."

3. **Have the courage to say "NO":** "While you want to get buy-in from the key stakeholders, you should not say yes to every idea and request."

4. **Keep it simple:** "Resist the temptation to add too many details to your roadmap. Keep your roadmap simple and easy to understand. Focus on your goals and capture what really matters; leave out the rest."

5. **Get buy-in:** "Your roadmap is worthless if the people required to develop, market, and sell the product don't buy into it. The best way to create agreement is to involve the key stakeholders in creating the roadmap."

6. **Choose the right timeframe:** "Choose a realistic timeframe for your roadmap—a timeframe where you can anticipate the growth of your product without resorting to speculation."

7. **Prioritize date vs. goal:** "When building the roadmap, ask yourself if meeting a date or fully achieving a goal is more important for the success of your product."

8. **Determine the right innovation cadence:** "To determine how often you should launch a new product version, consider how ambitious your goals are, how difficult is it to build the product, and how often your users and customers can take advantage of new product features without feeling overwhelmed or confused."

9. **Goals come first, features second:** "The features should be the key product capabilities or themes required to reach the goal."

10. **Select helpful metrics and KPIs:** "Once you have selected a goal, ask yourself how you will know that you have met it successfully and determine the appropriate metrics or key progress indicators (KPIs). Your metrics hence depend on you goal."

EXPLORATORY 360 ASSESSMENT

Some people may have the misunderstanding that there is no upfront planning in an Agile project, but that would be incorrect. You would never see an Agile project that started out by writing code with no upfront planning at all.

- The upfront planning in an Agile project is intended to be streamlined and efficient to address the most significant issues and decisions that might have an impact on the project without getting bogged down in planning details.

- An "Exploratory 360 Assessment" is a planning practice that fits that model.

The idea of an Exploratory 360 Assessment was first created by Alistair Cockburn in conjunction with the Crystal Clear methodology in the early 1990s. An Exploratory 360 Assessment is a way of quickly identifying some knowns and unknowns associated with a project.

At the start of a new project, usually during the chartering activity, the team needs to establish that the project is both meaningful and they can deliver it using the intended technology. According to Alistair Cockburn, it would include the following:[7]

1. Business Value Sampling

- "Business value sampling consists of capturing, with key stakeholders, what the system should do for its users and their organization(s).
- This should result in the names of the key use cases for the system, along with the focal roles the system should serve, the personalities and functions it should present to the world."

2. Requirements sampling

- "Requirements sampling consists of low-precision use cases that show what the system must do, and with what other people and systems it will have to interact.
- Often that drafting exercise turns up interfaces between organizations or technology systems that had not formerly been identified."

3. Domain model

- "In larger and more complex projects, it may be useful to create a domain model of the system concurrently or from the use case drafts.
- This sample serves to highlight the key concepts the developers will be working with, the core of the business, the programming and discursive vocabulary.
- It also helps the team to estimate the size and difficulty of the problem at hand."

Cockburn offers some additional items that might be included in an Exploratory 360 Assessment:[8]

1. Technology plans

"Technology plans to evaluate potential uncertainties associated with any technology that is needed for the project.

This may result in some experiments or spikes to analyze and resolve any uncertainties associated with the technology required for the project."

2. Project plan

"The team might create a coarse-grained project plan to make sure the project is delivering suitable business value for suitable expense in a suitable time period."

3. Team makeup

"The assessment might also look at the team makeup to determine if the size and makeup of the team are appropriate for the project."

4. Process or methodology

"And, finally, it may be worthwhile to discuss the process or methodology to be used in the project to ensure that the process to be used is appropriate to the project and the level of uncertainty and risks involved."

The key point is that an Exploratory 360 Assessment is a way of doing a high-level feasibility study of a project and to identify the major risks and issues that need to be addressed for the project to be successful. I want to emphasize that these items are just a checklist of *potential* items to consider in developing a plan for a new project; however, these are probably the major ones that need to be considered in most projects.

AGILE FUNCTIONAL DECOMPOSITION

Functional decomposition was previously discussed in Chapter 1. This chapter provides further detail and examples. The tools defined so far may be all that is needed for a high-level plan. Optionally, it may be worthwhile:

- to use functional decomposition to break down the functionality into high-level epics and stories
- to develop a high-level estimate of the effort and schedule required
- to summarize the high-level plan in a Project Charter document.

These are tools that you can choose to use or not use depending on the nature of the project, the level of uncertainty, and the level of planning that is appropriate.

The first of these is functional decomposition. Here's a definition of Functional Decomposition from Investopedia:

> A method of business analysis that dissects a complex business process to show its individual elements. Functional decomposition is used to facilitate the understanding and management of large and/or complex processes and can be used to help solve problems.[9]

Relationship of Functional Decomposition to Agile

You might ask, "Why is this relevant to Agile?" Agile is about maximizing the business value that a project produces. On small, simple projects, you might be able to easily discover what that business value is based on direct face-to-face discussion with the Product Owner and individual stakeholders. On large enterprise-level projects, that may not be so easy to do.

Functional decomposition also helps to ensure that all the functionality in the project is well aligned with the overall business goals. That is particularly important in large complex projects.

- Using functional decomposition to organize stories into epics and themes makes it possible to keep all the stories well aligned with producing the higher-level business value that the project is intended to produce.

- It makes it a lot easier to effectively manage the product backlog. When you prioritize and move the items in the product backlog, it can be done much more easily at a higher level rather than being forced to reorganize the whole product backlog at an individual user story level.

■ It also provides a capability for traceability—you can look at the top-level functionality and then look down into the functional decomposition to verify that the lower-level functionality is complete and sufficient to support the higher-level functionality.

In an Agile environment, functional decomposition would typically break down the functionality required into themes, epics, and User Stories.

> In Agile development, an epic represents a series of user stories that share a broader strategic objective. When several epics themselves share a common goal, they are grouped together under a still-broader business objective, called a theme. Another important distinction is that a user story can be completed within the timeframe of an Agile sprint. An epic will typically require development work covering several sprints.[10]

There are a number of potential benefits of doing a functional decomposition of a project or product:

■ It provides a way of scoping and defining the overall work to be done and what's in scope and what's not in scope.

■ It breaks up the project into smaller manageable "chunks" that are easier to estimate and manage.

■ It helps to analyze alignment, dependencies, and inconsistencies among functions.

The level of functional decomposition to be done as part of the upfront planning in an Agile project will depend on the nature of the planning process and the level of predictability required by the project.

■ At one extreme, it may not be necessary to do functional decomposition at all as part of the upfront planning process. If the project is small and not very complex and predictability is not important, the functional decomposition might be deferred until later in the project when it becomes more essential to support the work to be done at that time.

■ At the other extreme, if a higher level of predictability is required in the project, it might be necessary to do more functional decomposition upfront to break down the project into smaller chunks of functionality to enable an estimate to be developed of the level of effort required.

■ Of course, there are lots of alternatives between those extremes. A typical practice on many Agile projects would be to break down the functionality required into epics.

Functional Decomposition Examples

Online Banking Application Example

Here's an example of how functional decomposition might be used in a large, online consumer banking application.

1. **Break up the project into epics:**

   ■ As a banking customer of <xyz> bank, I want to be able to easily make an on-line deposit of a check into my bank account through my smart phone so that I can save the time required to send a check for deposit through the mail and I can have the money immediately credited to my checking account as soon as the deposit is completed electronically.

2. **Decompose the epics into user stories:**

   ■ **Scan a check image:** As an electronic banking customer, I want to be able to scan an image of the front and back of a check into my smart phone so that it can be deposited electronically.

   ■ **Enter deposit information:** As an electronic banking customer, I want to be able to enter the deposit information associated with an electronic deposit so that the correct amount will be deposited into the correct bank account when the electronic deposit is processed.

   ■ **Submit deposit:** As an electronic banking customer, I want to be able to electronically submit a scanned check and deposit information to the bank for deposit so that I can save the time associated with sending deposits by mail.

   ■ **Receive confirmation:** As an electronic banking customer, I want to be able to receive confirmation of a completed electronic deposit so that I will know that the deposit was successfully processed.

Figure 15.1 shows a simple example of how functional decomposition might be used to define a process for handling orders at a fast-food restaurant. Note that this is similar to a project management work breakdown structure (WBS). The difference is that this is decomposed on a functional basis. A WBS is a more general kind of decomposition that may not be functionally broken down. For example, a WBS might break down an item of work into the item and the subtasks needed to complete that item of work.

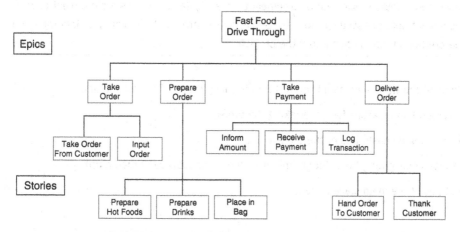

FIGURE 15.1   Functional decomposition example

User Story Mapping

Jeff Patton has developed a nice process called "User Story Mapping" that is essentially the same thing as functional decomposition.[11]

PROJECT CHARTER

A Project Charter plays a very significant role in a classical plan-driven project:

- "A Project Charter (PC) is a document that states a project exists and provides the project manager with written authority to begin work."

- According to PMBOK®, a project should not be allowed to start without a project charter and the project charter essentially defines a contract between the customer and the project team

  If it is used at all in an Agile project, it would likely serve a very different purpose:

- In a classical plan-driven project, a Project Charter is essentially a contract between the customer and the project team.

- In an Agile project, there is much more of a partnership relationship with the customer and the customer should be intimately involved in the project as it progresses with lots of communication, so there is less of a need for a formal Project Charter. However, depending on the nature of the project and the relationship with the customer, a Project Charter might serve a useful purpose to document a shared understanding of the project with the customer and important stakeholders.

> The general rule in producing any kind of documentation in an Agile environment is that we shouldn't produce documentation for the sake of producing documentation. Any document should provide value to someone. In an Agile project, a simple project charter document might be valuable to managers and executives to have a simple definition and description of information about the project.

Some of the typical elements you will find in many Project Charters include:

- background and reasons for undertaking the project
- objectives and constraints of the project
- directions concerning the solution (might include product/project road map)
- identities of the main stakeholders.

Some additional typical elements you will find in many Project Charters include:

- in-scope and out-of-scope items

- risks identified early on

- target project goals and benefits

- key deliverables (might include functional decomposition)

- optional cost and schedule estimates.

Of course, I want to emphasize that a Project Charter is not normally required in an Agile environment but it can add value by documenting the goals and purpose of the project, the scope, and some of the risks and constraints associated with a project. By the same reasoning, judgment is needed as to what should be included in the Project Charter if one is used at all, but the general rule is to keep it short and succinct and rely much more heavily on other forms of direct communications between the customer and the project team.

Appendix C provides an example of a Project Charter template.

SUMMARY OF KEY POINTS

Product/Project Vision

1. What is a vision statement?

- A vision statement is a brief statement of the desired future state that would be achieved by developing and deploying a product. A good vision should be simple to state and provide a coherent direction to the people who are asked to realize it.

- A vision statement can be at the level of an entire company or at the level of a particular product, project, or program.

2. Important tips for creating a compelling vision:

- **Describe the motivation behind the product:** The product vision is the overarching goal you are aiming for, the reason for creating the product.

- **Look beyond the product:** Be clear on the difference between the product vision and the product and don't confuse the two. The vision should be a high-level statement of the desired future state that would be achieved by developing and deploying a product or project. It is not a detailed product description.

- **Distinguish between the vision and the product strategy:** Your product vision should not be a plan that shows how to reach your goal.

- **Employ a shared vision:** You can come up with the most beautiful vision for your product. But it's useless if the people involved in making the product a success don't buy into it.

- **Choose a motivating vision:** If you are working on something exciting that you really care about, you don't have to be pushed.

- **Think big:** Make your product vision broad and ambitious so that it engages people, and it can facilitate a change in the strategy.

- **Keep your vision short and sweet:** As your vision is the ultimate reason for creating the product, it should be easy to communicate and to understand.

- **Use the vision to guide your decisions:** Use the vision to guide your product decisions and to focus everyone on the ultimate reason for creating the product.

Product/Project Roadmaps

1. What is a Product/Project Roadmap?

    - A Product/Project Roadmap is a high-level timeline that describes how the product is likely to grow. It allows you to express where you want to take your product, and why it's worthwhile investing in it. A roadmap typically highlights the major epics or features that may be included on a timeline, illustrating one or more release cycles. Sometimes roadmaps also include milestones for important deliveries such as tentative completion of a given Minimum Viable Product (MVP).[12]

2. The following are important tips for creating a Product/Project Roadmap:

    - **Do the prep work:** Describe and validate the product strategy—the path to realize your vision—before you create your roadmap and decide how the strategy is best implemented.

    - **Tell a convincing and realistic story:** Your product roadmap should tell a coherent story about the likely growth of your product. Each release should build on the previous one and move you closer toward your vision.

    - **Have the courage to say "NO":** While you want to get buy-in from the key stakeholders, you should not say yes to every idea and request.

    - **Keep it simple:** Resist the temptation to add too many details to your roadmap. Keep your roadmap simple and easy to understand. Focus on your goals and capture what really matters; leave out the rest.

    - **Get buy-in:** Your roadmap is worthless if the people required to develop, market, and sell the product don't buy into it. The best way to create agreement is to involve the key stakeholders in creating the roadmap.

    - **Choose the right timeframe:** Choose a realistic timeframe for your roadmap—a timeframe where you can anticipate the growth of your product without resorting to speculation.

- **Prioritize date versus goal:** When building the roadmap, ask yourself if meeting a date or fully achieving a goal is more important for the success of your product.

- **Determine the right innovative cadence:** To determine how often you should launch a new product version, consider how ambitious your goals are, how difficult is it to build the product, and how often your users and customers can take advantage of new product features without feeling overwhelmed or confused.

- **Goals come first, features second:** The features should be the key product capabilities or themes required to reach the goal.

- **Select helpful metrics and KPIs:** Once you have selected a goal, ask yourself how you will know that you have met it successfully and determine the appropriate metrics or key progress indicators (KPIs). Your metrics hence depend on your goal.

Exploratory 360 Assessment

- An Exploratory 360 Assessment is a way of quickly identifying some knowns and unknowns associated with a project.

- The key point is that an Exploratory 360 Assessment is a way of doing a high-level feasibility study of a project and identifying the major risks and issues that need to be addressed for the project to be successful.

Agile Functional Decomposition

1. Functional decomposition is a method of business analysis that dissects a complex business process to show its individual elements.

2. Functional decomposition is used to facilitate the understanding and management of large and/or complex processes and can be used to help solve problems.

3. Using functional decomposition to organize stories into epics and themes:

   - makes it possible to keep all the stories well aligned with producing the higher-level business value that the project is intended to produce

   - makes it a lot easier to effectively manage the product backlog.

4. Some of the benefits of functional decomposition are:

   - It provides a way of scoping and defining the overall work to be done and what's in scope and what's not in scope.

   - It breaks up the project into smaller manageable "chunks" that are easier to estimate and manage.

   - It helps to analyze alignment, dependencies, and inconsistencies among functions.

Project Charter

1. A Project Charter plays a very significant role in a classical plan-driven project. "A Project Charter (PC) is a document that states a project exists and provides the project manager with written authority to begin work."

2. In an Agile environment, the role of a Project Charter is not required; however, it can add value to a project. If it is used, the purpose it serves is more informational in nature. In essence, it summarizes the planning that has been done to define the project for the benefit of anyone concerned.

3. Some typical elements you will find in many Project Charters include:

   - background and reasons for undertaking the project

   - objectives and constraints of the project

   - directions concerning the solution (might include Product/Project Roadmap)

   - identities of the main stakeholders.

4. Some additional elements that you might find in a Project Charter include:

   - in-scope and out-of-scope items

   - risks identified early on

   - target project goals and benefits

   - key deliverables (might include functional decomposition)

   - optional cost and schedule estimates.

DISCUSSION TOPICS

Product/Project Vision

1. What is the role of a product/project vision in an Agile project?

2. What benefits does it provide?

3. What are some characteristics of a good product/project vision?

Product/Project Roadmaps

1. What are the key elements of a Product/Project Roadmap?

2. What benefits does it provide?

3. What are some characteristics of a good product/project roadmap?

Exploratory 360 Assessment

1. What is the purpose of an Exploratory 360 Assessment and how would it be used in a typical project?

2. What items do you think would be reasonable to include in an Exploratory 360 Assessment?

Agile Functional Decomposition

1. What is the benefit of using functional decomposition in an Agile project?

2. How would functional decomposition be used in a typical Agile project?

Project Charter

1. What role would a Project Charter play in an Agile project?

2. How is it different than how a Project Charter is used in a classical plan-driven project?

3. How would you determine if it would be useful to use a Project Charter or not?

NOTES

1. Geoffrey A. Moore, *Crossing the Chasm: Marketing and Selling High-Tech Products to Mainstream Customers* (New York: Harper Business Books, 1999).
2. Ibid.
3. Roman Pichler, "Eight Tips for Creating a Compelling Agile Product Vision," http://www.romanpichler.com/blog/tips-for-writing-compelling-product-vision/.
4. Roman Pichler, "Working with a Product Roadmap," http://www.romanpichler.com/blog/agile-product-roadmap/.
5. Ibid.
6. Ibid.
7. Alistair Cockburn, "Exploratory 360 Assessment," http://alistair.cockburn.us/Exploratory+360.
8. Ibid.
9. Investopedia, "What Is Functional Decomposition?", http://www.investopedia.com/terms/f/functional-decomposition.asp.
10. ProductPlan.com, "What Is an Epic?", https://www.productplan.com/glossary/epic/.
11. Jeff Patton, "User Story Mapping," jpattonassociates.com, https://www.jpattonassociates.com/story-mapping/.
12. Winston Gonzalez, email comments to author on book review.

16 Agile Stakeholder Management and Agile Contracts

STAKEHOLDER MANAGEMENT IS PROBABLY ONE OF THE MOST IMPORTANT ASPECTS of managing any project successfully. A project is not successful unless it meets stakeholder expectations:

- A classical plan-driven project sometimes assumes that meeting the defined requirements within the approved budget and schedule defines success; however, there have been many projects that have met their cost and schedule goals but failed to deliver an acceptable level of business value to the client. It's dangerous to assume that requirements documents accurately reflect the needs of all important stakeholders.

- Agile projects sometimes assume that the Product Owner totally represents all stakeholder interests and that is not necessarily the case either:

  - The role of the Product Owner is designed to make the project execution more efficient by putting a representative of the business users directly in contact with the project team, but it is very dangerous to assume that eliminates the need to talk directly to stakeholders outside of the project team.

  - As an example, there have been many Agile projects that have gotten totally consumed with developing an important application and completely neglected the need to engage others involved in the release process to successfully release and support the application once it is complete.

  - In large, complex projects, it is particularly challenging for a single Product Owner to represent all of the potential stakeholders of the project.

WHAT IS A STAKEHOLDER?

Before we go any further, let's review what a stakeholder is:

> - A stakeholder is either an individual, group or organization who is impacted by the outcome of a project.
> - Stakeholders have an interest in the success of the project and can be within or outside the organization that is sponsoring the project.[1]

It's important to recognize that there can be both positive stakeholders and negative stakeholders:

> - A positive stakeholder sees the project's positive side and benefits by its success. These stakeholders help the project management team to successfully complete the project.
> - On the other hand, a negative stakeholder sees the negative outcome of the project and may be negatively impacted by the project or its outcome. This type of stakeholder is less likely to help the project be completed successfully; however, that does not eliminate the need for managing these stakeholders.[2]

There are also internal and external stakeholders.

Internal Stakeholders

Internal stakeholders are internal to the organization that is doing the project. Examples include:

- the Business Sponsor who is paying for the project
- an internal customer or client (if the project arose due to an internal need of an organization)
- the project team who is responsible for the project deliverables needs to be kept satisfied that their contributions are being respected and they have been given an appropriate level of empowerment and support to be successful
- management of related organizations—this could include functional managers who are responsible for managing the people assigned to the project or managers of support organizations who are responsible for supporting the project deliverables.

External Stakeholders

External stakeholders are external to the organization that is developing the project. Examples might include:

- an external customer or client (if the project arose due to a contract)

- an end user of a product that is being sold to customers

- a supplier or subcontractor on whom the project is dependent

- the government that may have regulations that need to be satisfied.[3]

WHY IS STAKEHOLDER MANAGEMENT IMPORTANT?

Many times, there is a lot of internal focus on making the development process as efficient as possible and that's important, but that's not enough.

Having an efficient development process is not sufficient, in itself, to create a successful project. For example, there was a very large Agile implementation in the Boston area with a major financial services company. The teams were up and running and thought they were doing Agile very effectively, but the senior management of the company was very dissatisfied because they felt they had lost touch with what was going on in the projects.

A key point is that stakeholder management should not be taken for granted. In large, complex projects, a defined and planned approach is needed to manage stakeholder expectations and, ideally, the stakeholders should buy into that approach and agree with it.

Stakeholder Management Can Be Difficult

Stakeholder management can be a difficult thing to do for a number of reasons:

- **Sometimes you need to deliver bad news that stakeholders weren't expecting**. Hopefully, setting stakeholders' expectations upfront, educating them that there is some level of uncertainty in the project, and keeping them informed openly and transparently will help mitigate these surprises of bad news.

- **Sometimes you have to say "No."** Stakeholders can be very demanding and even unreasonable in their expectations. Keeping the most important stakeholders closely engaged in the project so that they are aware of the real-world constraints and issues that the project has to deal with should help them develop an understanding of why you sometimes have to say "No."

- **Different stakeholders may not be in complete agreement with each other.** Thus, it may require some amount of consensus-building to get them to agree on the important issues.

- **The stakeholders in a project may change over time.** Or their expectations might change as well and multiple stakeholders may not agree with each other.

What Can Go Wrong?

Here are a few examples of what can go wrong.

Case Study: US Financial Services Company No. 1

■ The company's senior executives saw the Agile process as having significant benefits to make IT development go faster; however, they saw it as an IT development process only and didn't see the benefits of investing beyond that level. As a result, the implementation of Agile was somewhat limited, superficial, and mechanical and didn't take full advantage of the benefits it can provide.

■ In this kind of situation, it's important to set everyone's expectations early about the level of commitment required to make Agile successful to fully realize the benefits. In this particular situation, the senior executives saw Agile as a "silver bullet" and saw only the benefits without fully realizing the level of commitment to make it work.

Case Study: US Financial Services Company No. 2

■ The project teams were fully involved with implementing Agile at the team level and producing results but failed to keep senior management informed of what they were doing. The impact was that senior management of the company was disappointed in the Agile implementation because they felt that they had lost touch with what was going on in the projects.

■ In this kind of situation, it's important to keep all levels of management engaged and keep them informed. You want your teams to be empowered and self-organizing but they shouldn't become so independent that outside management is left out of the loop.

Case Study: Far East Manufacturing Company

■ The project teams failed to engage the appropriate IT production staff in the development of a major production application.

■ The impact was that when it came time to release the application to production, it could not be released, and the project was significantly delayed while the necessary coordination with the release process was established.

■ In large enterprise-level projects, it is particularly important to keep people who are outside of the project engaged in order to successfully release software. In this particular situation, if the need for this coordination had been realized earlier, the delay in releasing the project to production would not have been necessary.

Common Stakeholder Management Mistakes

Here are some common stakeholder management mistakes:

1. **Identifying and prioritizing the wrong stakeholders:** If you misread the importance of a stakeholder to a project, then it is impossible to objectively and accurately evaluate their opinions on how a project should progress.

2. **Being unrealistic with your key stakeholders:** Project managers are notorious for overpromising on a project and under-delivering.

3. **Failing to develop a stakeholder communication plan:** Take the time to discover what each stakeholder is interested in, and what challenges you face with them.[4]

STAKEHOLDER MANAGEMENT PROCESS

Figure 16.1 shows a general stakeholder management process. Let us now go through the stages of this process in detail.

Identify and Analyze Stakeholders

The first steps are to identify and analyze stakeholders. The following are some questions to consider when doing this:

- Who are the stakeholders who will be impacted by the project?

- How will they be affected by the work?

- Will they be openly supportive, negative or ambivalent?

- What are their expectations and how can these be managed?

- Who and/or what influences the stakeholders' view of the project?

- What are their expectations for communications? For example, how would they like to receive communications, what are they most interested in, and how frequently would they want communications?

- Who would be the best person to engage with the stakeholder?[5]

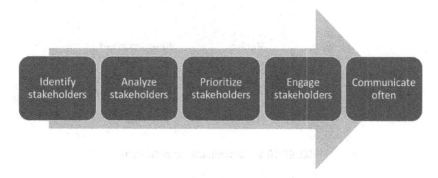

FIGURE 16.1   Stakeholder management process

Prioritize Stakeholders

Figure 16.2 shows a commonly used technique for stakeholder analysis and prioritization that breaks up stakeholders into four quadrants based on two factors:[6]

- their relative power or influence over the project
- their level of interest and influence in the project.

  Here's a summary of each quadrant:

- **Low influence and low interest quadrant:** the strategy for stakeholder management for this quadrant might be just to monitor the people in this quadrant to see if their power or interest level changes that might cause them to need additional attention.

- **High power or influence but low interest quadrant:** These are people who can have a lot of potential impact on the project if they are aroused but they have a low level of interest. The strategy with the people in this quadrant would be to keep them satisfied but since they have a low level of interest in the project, you don't want to bore them with excessive communications.

- **High interest with low power or influence:** These people need to be kept informed to ensure that they remain satisfied.

- **High interest with high power:** This is obviously the most important quadrant that needs a lot of attention and the people in this quadrant need to be managed very closely.

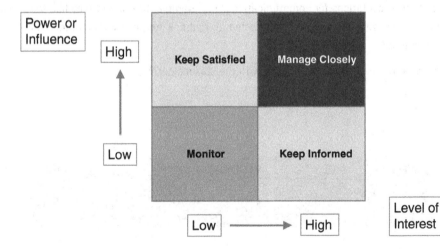

FIGURE 16.2   Stakeholder prioritization

WHAT'S DIFFERENT ABOUT AGILE STAKEHOLDER MANAGEMENT?

One of the biggest differences in an Agile environment is that in a classical plan-driven project, there is more of a contractual-style relationship between the customer and the project team. There is normally some amount of negotiation that goes on to arrive at a well-defined set of requirements that the customer approves, and the project team commits to delivering those requirements within a given time and budget.

An Agile environment requires more of a collaborative relationship based on a spirit of trust and partnership between the customer and the project team. Instead of having a well-defined set of requirements and a contract to deliver those requirements, the customer and the project team agree on the general scope and vision for the project and perhaps at least some critical requirements and agree to work collaboratively to further define detailed requirements as the project is in progress.

> The key point is that an Agile process won't work without a commitment by the customer to actively engage in the project in a spirit of collaboration, trust, and partnership. If a customer isn't willing or able to do that, it may require more of a hybrid approach or a plan-driven approach to effectively manage the relationship with the customer.

However, I want to emphasize, as I have said many times, that this isn't a binary and mutually exclusive choice between Agile and Waterfall, as many people seem to think. Many times, a hybrid approach between these two extremes that blends an Agile approach and a plan-driven approach in the right proportions to fit the situation makes a lot of sense. Of course, that will also impact how stakeholder management is done and the willingness of the customer to participate actively in a collaborative development process as the project is in progress is a major factor in determining the most appropriate process.

Advantages of an Agile Stakeholder Management Approach

In a classical plan-driven environment, there may be a greater possibility for misunderstandings because the customer is typically not directly involved in the project and the project relies heavily on documentation to set and manage expectations. That approach requires more planning and formality and is prone to misunderstandings.

In an Agile environment, because the customer is much more directly engaged in the project and communication is much more regular, there may be less of a need for formality and planning and less risk of misunderstandings.

Agile Stakeholders Have Rights and Responsibilities

Another big difference regarding stakeholder management in an Agile environment is that stakeholders have both rights and responsibilities, and they are expected to be actively engaged in the project. Here is a summary of some of the more important rights and responsibilities.[7]

Agile Stakeholder Rights

Some of the more important rights of stakeholders are:

- They have a right to receive good-faith estimates.
- They have a right to have their interests understood and respected.
- They should be educated on factors effecting estimates and results.
- They should be kept informed of the team's progress.
- Finally, they should understand the development process and how they fit in.

Agile Stakeholder Responsibilities

On the other hand, they also have responsibilities which include:

- They must remain engaged in the project to provide feedback and inputs.
- They must collaborate with the Product Owner and Project Team.
- They should provide timely decision-making when necessary.
- They should help to develop a spirit of trust and partnership.

Responsibility for Stakeholder Management in an Agile Environment

The lack of a well-defined focus on stakeholder management in an Agile environment can be problematic. In an Agile environment, the Product Owner has primary responsibility for overall stakeholder management; however, since everyone on the team plays some role in interacting with stakeholders in some way or other, everyone on the team has some level of responsibility for stakeholder management.

Eight Tips for Agile Stakeholder Management

I want to present eight tips for effective stakeholder management in an Agile environment from a Tech Republic article that I thought was very good:[8]

1. **Secure sponsorship:** Don't start your Agile project without securing sponsorship from stakeholders' leadership and management.

2. **Spend adequate time on setup:** On an Agile project, you must go slowly to go fast. It's somewhat of a paradox, but it's absolutely true. If you don't prepare people for what to expect, you'll have a lot of problems with stakeholders during execution. When planning the execution of an Agile project, reserve enough time for awareness and education.

3. **Protect the developers' bill of rights:** The rights of all data scientists and other developers must be ferociously guarded. This idea extends from the Extreme Programming days, when the developers' bill of rights was documented and sacrosanct. Among these rights are:

   - the right to have open and immediate communication with the stakeholder;
   - the right to update and own their duration estimates;
   - the right to own their day-to-day schedule.

   Stakeholders can be very demanding and even unreasonable and an Agile Project Manager needs to protect them from that if necessary.

4. **Protect the customers' bill of rights:** Your stakeholders have rights too, and they should be protected just as vigilantly. Among these rights are:

   - the right to choose which user stories take priority;
   - the right to get the most value out of every development day;
   - the right to cancel the project at any time and be left with a working system that's better than when they started.

5. **Give them a job:** The best way for stakeholders to know what's happening is to put them in the middle of development with a real assignment.

   - If stakeholders unplug when they think requirements are done, that can be a problem. They lose contact with the day-to-day issues that crop up and as their expectations stray from reality, their frustration can increase when false expectations aren't met.
   - The best way to avoid this is to give them a real job that contributes to the solution—that keeps them consistently plugged into what's going on.

6. **Make sure they attend every required meeting:** They have a responsibility to attend, and you cannot let them off the hook. It's another way to keep them plugged in so they don't build those false expectations.

7. **Prepare to defend unusual practices:** You must always be prepared to defend unusual Agile behaviors like lightweight documentation, and no firm commitment on scope.

8. **Foster a culture of honesty and trust:** Above all, there must be a culture of trust. In the absence of trust, Agile breaks down quickly. Lack of trust causes developers to sandbag hours, and stakeholders to negotiate with developers on feature completion date.

AGILE CONTRACTS

An Agile contract is one in which there is some kind of commitment for delivering the results of a project using an Agile development approach. Many people will say that having a firm contract with an Agile development approach is inconsistent. It is somewhat inconsistent with a pure Agile development approach to have a firm commitment, but it can definitely be made to work using a hybrid Agile approach.

How Would an Agile Contract Work?

At first glance this may seem to be an impossible situation—how can you have firm contractual requirements to deliver something and still have some level of flexibility and adaptivity to define what you deliver? There are obviously some conflicting goals at work here.

- In a classical plan-driven contract, the deliverables are well defined and the cost and schedule for completing the project may be fixed.

- In an Agile contract, there is some flexibility in either the requirements to be delivered and/or the cost and schedule for delivering those requirements. An Agile contract requires a special kind of stakeholder management:

- There has to be somewhat of a collaborative relationship between the customer and the project team, based on a spirit of trust and partnership to work together to make this a win-win for both parties.

- If it is more of an adversarial relationship, an Agile contracting approach just won't work and more of a classical plan-driven contracting approach may be necessary that is based on typical plan-driven project management with formalized change control.

Types of Agile Contracts

There are several different types of Agile contracts:[9]

1. Time and Materials Contract

Time and materials contracts (commonly called T&M) are widely used because they are relatively easy to implement.

- In a T&M contract, the cost and schedule for completing the project are not fixed. There may be some kind of estimate given to the customer, but it is not a firm estimate, and the customer agrees to absorb whatever the actual costs of the project are.

- This type of contract protects the supplier but doesn't provide any protection for the customer. The customer is fully exposed to the entire risk, while the supplier shares none of that risk. A variant of this type of contract is a capped time and materials contract.

- These are contracts that are focused on time and materials up until a fixed agreed-upon upper bounds (or cap).

- Capped time and materials contracts provide benefit to the supplier early on by fully covering their expense; but also provide benefits to the customer toward the end of the project by providing a limit to the total exposure.

- What typically happens in this situation is that the scope of the contract is somewhat negotiable, and the customer and supplier renegotiate the priorities of some of the requirements at the end in order to stay within the overall cost cap that has been agreed on.

2. Incremental Delivery Model

Another contracting model that works well with an Agile approach is an incremental delivery model:

- Incremental delivery contracts are structured with regular inspection points.

- At each inspection point, the customer makes a decision; they can continue with the development of the product, or they can stop development.

- In stopping development, the customer can push the product into production and save the remaining balance of the contract.

- This style of contract works quite naturally for Agile teams because they simply work in an iterative fashion until the point of inspection.

3. Cost-Targeted Contract

Another type of contract that is less widely used is a cost-targeted contract:

- With cost-targeted contracts, both parties agree on a realistic final price of the product.

- Then, if the supplier comes in under budget, both parties share the benefit of those savings.

- However, if the supplier goes over budget, then both companies pay some penalty.

- The amount of benefit or penalty that a company has to pay is usually in line with the ratio of the two companies.

An Agile Contracting Example

Jeff Sutherland, one of the original founders of Scrum, has come up with a very interesting combination of these approaches that he calls "Money for Nothing, Change for Free."[10]

Change for Free

- The "Change for Free" clause is based on the idea that the customer can make any change they want, provided that the total contract work is not changed.

- This allows new features to be added, provided that an equivalent amount of low priority items are removed from the project.

Money for Nothing

- The "Money for Nothing" clause recognizes the fact that in many projects, the customer always asks for everything that they could possibly need.

- If you prioritize those items and deliver the highest priority items first, at some point you will reach a point of diminishing returns where the cost of developing incremental features exceeds the value that those features provide.

- This clause allows the customer to cancel the contract at that point and save 80% of the cost that would have been spent to complete the remaining items.

- However, the contractor receives a fee of 20% of the cost for early cancellation, which makes it a win-win for both the customer and the supplier.

Figure 16.3 shows how that might work. If the requirements are properly prioritized and the highest value items are developed first, you will see that the incremental value of each incremental feature that is developed will diminish but the incremental cost of developing each feature may be more or less the same.

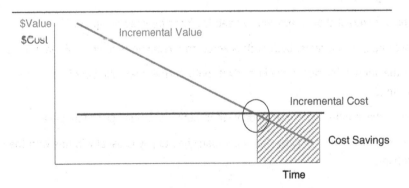

FIGURE 16.3   Incremental costs and incremental values

What frequently happens is that, at some point, you reach a point of diminishing returns where the incremental cost of developing features exceeds the incremental value that those features provide. It is at that point that the customer might decide that there is no need to develop the remaining features and the anticipated cost of developing those features will be saved and shared with the contractor.

SUMMARY OF KEY POINTS

Stakeholder Management

1. Stakeholder management is one of the most important functions of a Project Manager. If a project doesn't meet stakeholder expectations, it can't be considered successful even if it meets its cost and schedule goals.

2. Many projects have failed to meet stakeholder expectations. Some of the major mistakes are:

   - identifying and prioritizing the wrong stakeholders

   - being unrealistic with your stakeholders

   - failing to develop a stakeholder communication plan.

3. Agile stakeholder management relies heavily on a close collaborative relationship with stakeholders where they directly participate in the project:

   - That can simplify the stakeholder management problem somewhat, but that doesn't alleviate the need for doing stakeholder management.

   - The need for stakeholder management is frequently neglected in Agile projects. For example, many times a development team will completely develop a software application and fail to make any contact with the groups who are required to release and support it.

4. The general process for stakeholder management in an Agile environment is essentially the same as a classical plan-driven project but may not need to be as formal. The process requires identifying, analyzing, prioritizing, and engaging stakeholders and then frequent communications to keep them informed.

5. The relationship with stakeholders will be a major factor in determining the project approach.

   - Agile requires a collaborative approach based on a spirit of trust and partnership and it also requires an active commitment by stakeholders to participate in the project.

   - If the stakeholders are not willing to enter into that kind of relationship or make that kind of commitment, an Agile approach won't work and either a hybrid approach or a classical plan-driven approach may be necessary.

   - The key point is that you can't just implement Agile to optimize the efficiency of the project delivery process without considering the impact on stakeholders.

Agile Contracts

1. Agile contracts can be considered a special case of stakeholder management

   - The principles are essentially the same; however, it may be more difficult depending on the terms of the contract.

   - The most important point is that there must be somewhat of a collaborative relationship between the customer and the project team, based on a spirit of trust and partnership to work together to make this a win-win for both parties.

   - If it is more of an adversarial relationship, an Agile contracting approach just won't work and more of a classical plan-driven contracting approach may be necessary, that is based on typical plan-driven project management with formalized change control.

2. There are several types of Agile contracts

   - Time and materials contracts (T&M) are widely used because they are relatively easy to implement.

   - Another contracting model that works well with an Agile approach is an incremental delivery model.

   - Another type of contract that is less widely used is a cost-targeted contract.

3. The "Money for Nothing, Change for Free" approach developed by Jeff Sutherland is a great example of how to make an Agile contract approach a win-win situation for both the customer and the supplier.

DISCUSSION TOPICS

Stakeholder Management

1. Is stakeholder management in an Agile environment more or less important than in a classical plan-driven environment? Why or why not?

2. What's different about stakeholder management in an Agile environment?

Agile Contracts

1. A customer insists on a firm, fixed-price delivery contract. Is it possible to use an Agile contracting approach in this situation? If so, how would it work?

2. How do you go about developing a risk-sharing approach in a contract that is a win-win for both the customer and the supplier?

NOTES

1. Project Manager.com, "What Is a Stakeholder?," https://www.projectmanager.com/blog/what-is-a-stakeholder.
2. PM Study Circle.com, "Types of Stakeholders," https://pmstudycircle.com/2012/03/stakeholders-in-project-management-definition-and-types/.
3. Ibid.
4. Capterra, "Mistakes All Project Managers Make with Their Stakeholders," https://blog.capterra.com/mistakes-all-project-managers-make-with-their-stakeholders/.
5. Association for Project Management, "Identify and Analyze Stakeholders," https://www.apm.org.uk/body-of-knowledge/delivery/integrative-management/stakeholder-management/.
6. Mind Tools, "Stakeholder Analysis – Winning Support for Your Projects," https://www.mindtools.com/pages/article/newPPM_07.htm.
7. Drew Jemilo, "The Stakeholder Management Framework," https://agilealliance.org/wp-content/uploads/2016/01/Stakeholder-Management-by-Drew-Jemilo-Agile2012.pdf.
8. Tech Republic, "8 Foolproof Stakeholder Management Tips for Agile Projects," https://www.techrepublic.com/article/8-foolproof-stakeholder-management-tips-for-agile-projects/.
9. Scrumology.com, "An Overview of Agile Contracts," https://scrumology.com/an-overview-of-agile-contracts/.
10. Scruminc.com, "Money for Nothing, Change for Free," https://www.scruminc.com/agile-contracts-money-for-nothing-and/.

17 Distributed Project Management in Agile

THERE'S A COMMON MISCONCEPTION THAT PROJECT MANAGEMENT IS NOT CONSISTENT WITH AGILE.
A contributing factor to that misconception is that there is normally no one called a "Project Manager" at the team level in an Agile project and the conventional view of project management requires a designated Project Manager. There is actually a lot of project management going on in an Agile project, even though you may not find anyone with the title of "Project Manager."

- It's a different kind of project management with an emphasis on maximizing business value rather than the classical plan-driven project management emphasis of providing predictability on costs and schedules for delivering well-defined requirements.

- The functions that might normally be performed by a single person called a "Project Manager" have been distributed among the different members of the Agile team.

> For many people, there is a strong stereotype that "project management" is something that is only done by a "Project Manager" and that is not necessarily the case anymore.

WHAT IS DISTRIBUTED PROJECT MANAGEMENT?

In an Agile environment, there may be no single person called a "Project Manager" and there is a good reason for that:

- In a classical plan-driven project, there is an emphasis on planning and control. There is also typically somewhat of a contractual relationship with the client to deliver a solution within an approved cost and schedule. In that kind of environment, it is advantageous to have a single point of responsibility for overall planning, control, and communications associated with the project.

- An Agile project has much more of an emphasis on creativity and innovation to maximize the value of the solution. In that kind of environment, empowerment of the project team is essential and allowing enough flexibility and adaptivity to optimize the value of the solution is a very important goal. Too much emphasis on planning and control can stifle the creativity and innovation that are needed.

   Those are some key reasons why a different approach to project management is needed in an Agile environment.

Primary Project Management Emphasis

Table 17.1 compares the emphasis shown between classical plan-driven project management and Agile project management.

TABLE 17.1   Comparison of primary project emphasis

Classical plan-driven project management	Agile project management
Planning and control to achieve cost and schedule goals	Maximize the overall value of the project

Why is it different?

- The primary problem with the classical plan-driven approach is that it doesn't work well in an uncertain environment where it is difficult, if not impossible, to predict the requirements in detail prior to the start of the project.

- Too much emphasis on planning and control can stifle the creativity and innovation that are needed to maximize the value of the solution.

> There have been many situations where a project has met its cost and schedule goals but failed to deliver an appropriate level of business value.

Client Relationship

Table 17.2 compares the client relationship between a classical plan-driven project management and an Agile project management.

TABLE 17.2 Comparison of client relationship

Classical plan-driven project management	Agile project management
Contractual relationship	Collaborative, partnership relationship

Why is it different?

■ The contractual relationship in a plan-driven environment is important to control and manage any changes in scope where predictability over costs and schedules is important.

■ A more collaborative relationship is essential in an Agile environment to work jointly to maximize the value of the project in an uncertain and dynamic environment.

> The customer must be amenable to whatever approach is selected and should openly buy off on it. Achieving an open and transparent relationship with a customer that is based on a spirit of trust, partnership, and collaboration can be a difficult thing to do, especially if the customer is not fully committed to participate in that kind of approach.

I also want to emphasize that this is not a binary choice between a contractual relationship and a collaborative relationship, and it is very possible for a hybrid approach to have some contractual commitments but also be collaborative at the same time.

Communications

Table 17.3 compares the type of communications between a classical plan-driven project management and an Agile project management.

TABLE 17.3 Comparison of communications

Classical plan-driven project management	Agile project management
All communications are controlled through the Project Manager	Open and transparent communications directly between the project team and the customer

Why is it different?

In an Agile environment:

■ There is less of a need to control changes to requirements and project scope, and flexibility and adaptivity are needed to meet uncertain customer needs.

■ Developers are expected to talk directly to users to better understand requirements as the project is in progress.

■ The project team is expected to share project status information openly with the client so that the client is actively involved in the management of the project.

- An important goal of the communications approach is to create an environment of shared responsibility with the customer. An Agile project requires a spirit of trust and partnership with the customer and that requires a very different communications approach; however, that does not negate the need to have well-defined mutual responsibilities.

- Another important goal is to avoid the "bottleneck" that happens when all communications must go through a Project Manager.

Overall Responsibility

Table 17.4 compares overall responsibility between a classical plan-driven project management and an Agile project management.

TABLE 17.4   Comparison of overall responsibility

Classical plan-driven project management	Agile project management
Project Manager is responsible for the project	The entire team is responsible for the success of the project; however, the Product Owner has ultimate responsibility for the overall success of the project
Responsibilities for tasks are assigned and clearly defined	The project team is intended to be self-organizing and responsible for planning and managing their own tasks

Why is it different?

In an Agile environment:

- Flexibility and adaptivity require more decentralized and distributed management where everyone on the team takes some level of responsibility for their own work and for the work of the team as a whole.

- An empowered team where everyone on the team feels some responsibility for the success of the project builds morale and is typically more effective and produces better results.

DISTRIBUTED PROJECT MANAGEMENT ROLES

In an Agile environment at the team level, instead of having a single person who is responsible for project management called a Project Manager, the functions that might be performed by a Project Manager have been distributed among other members of the team: the Developer, the Business Analyst, the Project Owner, and the Scrum Master.

Developer Project Management Responsibilities

Primary Responsibilities

The following are some of the tasks that a Developer would normally be expected to perform in an Agile environment that might normally be performed by a Project Manager in a classical plan-driven project management environment.

1. **Developers are expected to take responsibility for planning and managing individual developer tasks which would include:**

   - estimate the amount of work to be done
   - plan and manage the completion of the work
   - take responsibility for the quality of the software developed
   - take overall responsibility for "shepherding" the story through the process all the way to (and including) UAT. The Developer responsible for the story should lead the presentation of the completed story to the Product Owner in UAT (or Sprint Review).
   - track and report progress at Daily Scrums and other forums.

2. **Beyond these individual responsibilities, all Developers are expected to share responsibility for the overall team results which would include:**

   - working effectively as a team member to build a high performance, cohesive team
   - integrating related efforts within the team to build an overall solution.

Other Responsibilities

In an Agile environment, a Developer needs to also assume some other responsibilities that might normally be performed by either a Project Manager or a Business Analyst.

1. **Requirements Definition**

   - Working directly with the Product Owner to clarify and further define the details of how stories should be implemented.
   - Understanding the business purpose of the story and defining and analyzing possible alternative ways of satisfying it.
   - Providing support to the Product Owner and business users in helping to write and refine user stories and to groom the product backlog.

2. **Process Knowledge and Process Improvement**

   - Understanding and effectively participating in all Agile/Scrum processes including Sprint Planning, Daily Standups, Sprint Reviews, and Sprint Retrospectives.
   - Taking a leadership role in presenting completed work to the Product Owner and users for approval at the end of each sprint.
   - Contributing to ongoing team process improvement through Sprint Retrospectives.

> For the Developer, making a transition to Agile means several changes need to be made:
>
> - There is a significantly increased level of responsibility for Developers, and it will not be easy for most Developers to assume this responsibility.
>
> - Some Developers who are used to just writing code may not want to take on this additional responsibility and just may not make the cut to move to an Agile environment.
>
> - At a very minimum it will require Developers to have some training to take on these responsibilities.

In the real world, it is very difficult for some Developers to fully assume all these responsibilities and some compromises may be needed from this ideal model. In these cases, there may be good reason for the Project Manager or a Business Analyst to provide help to the Developer. However, as much as possible, that assistance should be provided in a way that is consistent with Agile. Here are a couple of guidelines for that:

- If a Project Manager is involved at the team level, he/she should be in more of an advisory and support role rather than a management control mode.
- If a Business Analyst is involved at the team level, he/she should not be just an intermediary between the development team and the business users.

Product Owner Project Management Responsibilities

In an Agile environment, the Product Owner takes on many of the roles that might normally be performed by a Project Manager in a classical plan-driven environment. The following is a comparison of the responsibilities of a Product Owner and a Project Manager:

The role of the Product Owner in an Agile project comes closest to the role of a Project Manager, but it really is a hybrid of some *project management* functions and some *product management* functions. Table 17.5 shows how these two roles are different from an overall responsibility perspective.

Overall Responsibility

Table 17.5 compares the Product Owner's responsibility role between a classical plan-driven project management and an Agile project management..

TABLE 17.5 Comparison of Product Owner's responsibility role

Typical Project Manager responsibility	Typical Product Manager (or Product Owner) responsibility
Planning and control to achieve cost and schedule goals	Planning and management to achieve business success

Setting Goals and Requirements

Table 17.6 shows how these two roles are different from the perspective of setting goals and requirements.

TABLE 17.6   Comparison of Product Owner's requirements and role in setting goals

Typical Project Manager responsibility	Typical Product Manager (or Product Owner) responsibility
Execute the project with defined goals and requirements determined by the business sponsor and business users	Some decision-making authority for defining and prioritizing goals and requirements

Financial Management

Table 17.7 shows how these two roles are different from the perspective of financial management to include budgeting, business planning, and cost management.

TABLE 17.7   Comparison of Product Owner's financial management role

Typical Project Manager responsibility	Typical Product Manager (or Product Owner) responsibility
Management of costs within the scope of the project	Management of overall business profitability

Planning Schedules and Product Roadmap

Table 17.8 shows how these two roles are different from the perspective of planning schedules and completing a product/project roadmap.

TABLE 17.8   Comparison of Product Owner's planning role

Typical Project Manager responsibility	Typical Product Manager (or Product Owner) responsibility
Acts as an intermediary between the project team and the business sponsors to establish a schedule	Take a much more active role in setting the schedule by making decisions and prioritizing what needs to be done

Risk Management

Table 17.9 shows how these two roles are different from the perspective of project risk management.

TABLE 17.9   Comparison of Product Owner's risk management role

Typical Project Manager responsibility	Typical Product Manager (or Product Owner) responsibility
Responsible for monitoring and controlling project risks and some decision-making responsibility for resolving risks within the project budget and plan	Similar responsibility but includes more decision-making responsibility for resolving risks

Reviewing and Approving Results

Table 17.10 shows how these two roles are different from the perspective of reviewing and approving results including acceptance testing.

TABLE 17.10   Comparison of Product Owner's approval role

Typical Project Manager responsibility	Typical Product Manager (or Product Owner) responsibility
Responsible for testing the product against defined requirements and facilitating final acceptance testing	Decision-maker on reviewing and approving results and final acceptance testing not limited to meeting defined requirements

Track and Manage Progress Against Goals

Table 17.11 shows how these two roles are different from the perspective of tracking and managing progress against goals.

TABLE 17.11   Comparison of Product Owner's progress management role

Typical Project Manager responsibility	Typical Product Manager (or Product Owner) responsibility
Track and manage progress against goals and report progress	Same, plus some decision-making authority to revise goals and/or reprioritize deliverables as necessary to meet goals

The role of a Product Owner goes beyond the role of a Project Manager in many ways, and in making a transition to Agile, the role of the Product Owner and the significance of the role are often not well-understood. What often happens is that some businessperson who is a subject matter expert in this area is typically given this responsibility with little or no training or experience. In the real world, those businesspeople will often need help in performing that responsibility.

Scrum Master Project Management Responsibilities

The Scrum Master in an Agile environment also takes on some functions that might be performed by a Project Manager in a classical plan-driven environment. In essence, the Product Owner focuses on the "what" (defining what needs to be done) and the Scrum Master focuses on the "how" (working with the team on how the work gets done).[1]

Leading and Facilitating the Team

Table 17.12 shows how these two roles are different from the perspective of leading and facilitating the team.

TABLE 17.12 Comparison of Scrum Master's leading and facilitating role

Typical Project Manager responsibility	Typical Scrum Master responsibility
Assignment, management, and tracking of all tasks and activities assigned to the team	"Servant-leader" role—facilitating the team and coaching the team in the Scrum process
Responsibility for management of all tasks and activities performed by the team	The team is expected to be self-organizing, and the Scrum Master plays more of a facilitation role
Responsible for facilitating team meetings	Similar responsibility

Removal of Obstacles

Table 17.13 shows how these two roles are different from the perspective of removal of obstacles.

TABLE 17.13 Comparison of Scrum Master's obstacle removal role

Typical Project Manager responsibility	Typical Scrum Master responsibility
Responsible for removing obstacles that might be impeding team progress	Similar responsibility

SUMMARY OF KEY POINTS

What Is Distributed Project Management?

1. Even though you may not find anyone called a "Project Manager" at the team level in an Agile project, there is a lot of "project management" going on. Many people may not recognize this as "project management" because:

- It's a different kind of project management, with an emphasis on maximizing business value rather than a more limited emphasis on planning and control to meet cost and schedule goals.

- The project management functions have been distributed among the members of the team instead of being done by a single person called a "Project Manager."

- A broader view of what "project management" is must be adopted in order to recognize this as "project management."

The Developer's Project Management Roles

1. The role of a Developer in an Agile environment goes well beyond simply writing code:

   - A Developer in an Agile environment is expected to perform some functions that might normally be performed by either a Project Manager or a Business Analyst in a classical plan-driven environment.

2. Some of the additional responsibilities normally given to a Developer in an Agile environment include:

   - estimating the amount of work to be done

   - planning and managing the completion of the work

   - taking responsibility for the quality of the software developed

   - taking overall responsibility for "shepherding" the story through the process all the way to (and including) User Acceptance Testing (UAT). The Developer responsible for the story should lead the presentation of the completed story to the Product Owner in UAT (or Sprint Review).

   - tracking and reporting progress at Daily Scrums and other forums.

3. Other responsibilities include:

   - working effectively as a team member to build a high performance, cohesive team

   - integrating related efforts within the team to build an overall solution.

The Product Owner's Project Management Role

1. **The role of a Product Owner** in an Agile environment is most like the role of a Project Manager, but it includes some additional functions that might be considered more similar to "Product Management."

2. **Overall responsibility:** In the area of overall product/project responsibility, the Product Owner's responsibilities go beyond the role of a Project Manager. The Product Owner is responsible for the overall business success of the product or project and not for simply meeting cost and schedule goals.

3. **Goals and requirements:** In the area of setting goals and requirements, the Product Owner's responsibilities also go beyond the role of a Project Manager. The Product Owner has some responsibility for defining and prioritizing goals and requirements that a Project Manager does not normally have.

4. **Financial management:** In the area of financial management, a Product Owner typically has overall responsibility for business value and profitability where a Project Manager is typically only responsible for managing project costs to meet an approved budget.

5. **Planning schedules and Product Roadmap:** In the area of planning schedules, a Product Owner takes an active role in setting the schedule by making decisions and prioritizing what needs to be done, where the authority of a Project Manager for making that kind of decision is typically much more limited.

6. **Risk management:** Both a Product Owner and a Project Manager are responsible for risk management; however, a Product Owner typically has more decision-making authority for resolving risks.

7. **Tracking and managing progress against goals:** Both roles are responsible for tracking and managing progress against goals; however, a Product Owner has more authority for revising goals and/or reprioritizing deliverables to meet the goals.

8. **Review and approval of results:** The Product Owner has more decision-making authority on approval of results where the Project Manager is typically limited to testing the results against defined requirements and facilitating final acceptance testing.

The Scrum Master's Project Management Role

1. The Scrum Master in an Agile project also has a few functions that would be considered similar to a project management role; however, it would not be accurate to think of the general Scrum Master's role as a typical Project Manager's role.

2. The functions of the Scrum Master's role that have some similarity to a project management role include:

   - leading and facilitating the team and team meetings

   - removal of obstacles.

DISCUSSION TOPICS

What Is Distributed Project Management?

1. What are the major factors that call for a different approach to project management in an Agile project? What's different about a distributed approach to project management and why is it necessary?

The Developer's Project Management Role

1. How is the role of a Developer different in an Agile environment?

2. What additional responsibilities are Developers expected to take on that might normally be done by a Project Manager?

3. What are some of the pros and cons for giving this additional responsibility to Developers?

The Product Owner's Project Management Role

1. Whose role has the greatest level of responsibility: a Product Owner or a Project Manager? Why? How do the roles compare?

2. Is the Product Owner's role more like a Product Manager's role or a Project Manager's role? Why?

3. How does the Product Owner's financial responsibility differ from that of a typical Project Manager?

4. How does the Product Owner's responsibility for review and approval of results compare to a typical Project Manager?

The Scrum Master's Project Management Role

1. Is the Scrum Master's role the most similar Agile role to that of a Project Manager? Why or why not?

2. What functions of a Scrum Master's role are most similar to a project management role?

NOTE

1. Winston Gonzalez, email comments to the author on book review.

PART 4

Making Agile Work for a Business

MANY BUSINESSES TODAY ARE STRUGGLING with improving "business agility." Business agility comes when Agile principles are applied throughout the entire organization. In this way, companies can adapt to external changes, speed up the time to market, and decrease costs without losing quality. Achieving overall business agility is not limited to implementing an Agile development and project management process; that is only the foundation for an overall business agility strategy.

There are many precedents for successful implementation of Agile principles and practices at a project team level; however, extending the Agile principles and practices to large-scale enterprise implementations and integrating with a business environment introduce a number of new challenges, which include:

- Large, complex projects that are commonly found at an enterprise level may require some reinterpretation and adaptation of Agile principles and practices. There is also a challenge associated with scaling Agile to projects requiring multiple teams.

- Integrating Agile principles and practices with higher levels of management typically found at an enterprise level, such as project portfolio management and overall business management, can be difficult. However, if an Agile implementation is limited to a development process only and does not address integration with these higher-level processes, it is not likely to be completely effective.

Part 4 is intended to address these topics and provide an understanding of the key considerations that need to be addressed when:

■ scaling an Agile approach to an enterprise level and for large complex projects requiring multiple teams

■ integrating an Agile development approach with a business environment

■ planning and implementing an Agile transformation.

Chapter 18 – Scaling Agile to an Enterprise Level:

Agile and Scrum were originally designed for small, single-team projects and the size of an Agile team is typically limited to 8–10 people. As soon as the scope of a project grows to a level that requires multiple teams, it introduces an additional complication of coordinating the work of these multiple teams. There are two dimensions of coordination that need to be addressed:

■ coordinating and integrating the technical development work of multiple teams

■ providing integrated business planning and direction to multiple teams.

Chapter 19 – Scaling Agile for Multiple Team Projects:

Scaling an Agile approach to an enterprise level requires some reinterpretation of how the Agile principles and practices apply in an enterprise-level environment. It also requires some planning to address how the Agile approach will integrate with higher-level management processes, such as project management and project portfolio management that are found at an enterprise level.

Chapter 20 – Adapting an Agile Approach to Fit a Business:

The company's business environment, management structure, and culture and values are very likely to impact an enterprise-level Agile implementation that may require either some adaptation or enterprise-level Agile transformation.

Chapter 21 – Enterprise-Level Agile Transformations:

Many organizations are not well designed to support an Agile implementation and some amount of enterprise-level transformation may be needed to successfully implement an Agile approach. This chapter provides some guidelines for planning and managing an enterprise-level Agile transformation.

(18) Scaling Agile to an Enterprise Level

AS A RESULT OF THE WIDESPREAD ADOPTION OF AGILE PRACTICES, large corporations are beginning to apply Agile at an enterprise level. This has introduced some new challenges of how to scale Agile principles and practices to an enterprise level and what to do about many of the existing project management office (PMO) practices and other higher-level business management practices that typically are employed in large enterprises to manage portfolios of projects and products.

Scott Ambler and Mark Lines write:

> Our experience is that "core" Agile methods such as Scrum work wonderfully for small project teams addressing straightforward problems in which there is little risk or consequence of failure. However, "out of the box," these methods do not give adequate consideration to the risks associated with delivering solutions on larger enterprise projects, and as a result we're seeing organizations investing a lot of effort creating hybrid methodologies combining techniques from many sources.[1]

Dean Leffingwell identifies two primary challenges involved with scaling Agile to the enterprise level in his book, *Scaling Software Agility: Best Practices for Large Enterprises*:[2]

1. The first challenge is overcoming the challenges inherent in the methodology. Scaling an Agile methodology to an enterprise level requires reinterpreting the values and principles behind the methodology in a much larger context, and typically also requires some adjustments in Agile practices to adjust to that context.
2. The second challenge is overcoming the limitations imposed by the enterprise that will otherwise prevent the successful application of new methods. Implementing an Agile approach at an enterprise level can be like plugging an appliance that requires DC current into an AC outlet; there's a fundamental incompatibility in many cases that will require some adaptation to get an Agile approach to work inside of an organization that wasn't designed to be Agile.

The first challenge will be discussed in this chapter, and the second challenge will be discussed in Chapter 20, "Adapting an Agile Approach to Fit a Business."

ENTERPRISE-LEVEL AGILE CHALLENGES

Beyond the factors previously discussed to adapt an Agile approach to a company's business environment, there are a number of challenges associated with scaling Agile practices to an enterprise level.

Differences in Enterprise-Level Agile Practices

Applying Agile development practices at an enterprise level typically involves some adjustments to those practices.[3] These differences in practices are broken down into:

- customer participation (Table 18.1)
- development team organization (Table 18.2)

TABLE 18.1   Adaptations to Agile practices at the enterprise level in customer participation

Typical small Agile project	Typical enterprise-level Implementation
The customer is integral to the team.	<ul><li>The customer may be remote or may not have the skills or time available to participate directly in the Agile team</li><li>The customer may also consist of a number of different users and stakeholders</li><li>A Business Analyst (BA) on the team many times plays the role of a proxy for the customer, but that BA should ideally be empowered to act on behalf of the customer</li></ul>

TABLE 18.2   Adaptations to Agile practices at the enterprise level in development team organization

Typical small Agile project	Typical enterprise-level Implementation
Developers, Product Owners, and testers are co-located and not separated by time zones and language barriers Ideally, Agile teams consist of peer-level Developers who take responsibility for their own actions with a minimum of direction. It is intended to be a team of equals with no formally designated technical team leader.	It is likely that many team members may be in different geographic locations and different time zones and perhaps even speak different languages<ul><li>For large-scale development teams, this can be very difficult, if not impossible, to achieve.</li><li>Many times, it is necessary to build teams of more junior-level Developers led by a more senior-level tech lead who can provide some level of guidance and direction to the rest of the team</li></ul>

- application architecture (Table 18.3)
- requirements definition and management (Table 18.4)
- project portfolio management (Table 18.5).

Reinterpreting Agile Manifesto Values and Principles

There is also a need to reinterpret the Agile Manifesto values and principles in a different context at the enterprise level. These differences in values and principles are broken down by:

- tools (Table 18.6)
- documentation (Table 18.7)
- collaboration (Table 18.8)
- planning (Table 18.9)
- change control (Table 18.10)
- communications (Table 18.11)
- progress measurement (Table 18.12).

TABLE 18.3 Adaptations to Agile practices at the enterprise level in application architecture

Typical small Agile project	Typical enterprise-level implementation
In a small-scale Agile project, the application architecture is expected to emerge as the project progresses.	■ With larger-scale systems, the costs and difficulty of refactoring the design as the project progresses to accommodate changes in the architecture make it essential in many cases to do more upfront architectural planning and design in the project ■ Large-scale system designs typically require breaking up the design into components, and without having sufficient architecture defined, it becomes impossible to allocate the work to teams[4] ■ In larger enterprise systems, there may be connections to other systems (both internal and external) that need to be considered in the application architecture

TABLE 18.4 Adaptations to Agile practices at the enterprise level in requirements definition and management

Typical small Agile project	Typical enterprise-level implementation
The Agile development effort can take place one story at a time, and the design incrementally evolves over the duration of the project	■ In large enterprise-level implementations, this approach doesn't work very well ■ A more integrated approach may be required to coordinate the development of the stories to ensure that they all really work together to produce releasable functionality that fulfills the business need

TABLE 18.5 Adaptations to Agile practices at the enterprise level in project/product portfolio management

Typical small Agile project	Typical enterprise-level implementation
Typical Agile projects do not provide a mechanism for higher-level integration to fulfill typical corporate needs for portfolio management of a large set of Agile projects or products	■ It can be very difficult to integrate a number of Agile projects or products into a typical enterprise-level project portfolio management approach; however, some level of integration and management is necessary to make portfolio management decisions ■ This will many times require adopting a hybrid approach to provide the necessary balance of predictability, control, and agility

TABLE 18.6 Agile Manifesto and enterprise-level implementation: tools

Typical small Agile project values	Typical enterprise-level implementation
Individuals and interactions over processes and tools	There is a greater need for tools at an enterprise level: ■ There is more of a need for a defined process to coordinate and synchronize the work of large projects requiring multiple teams as well as coordinating other activities outside the teams ■ Tools can become more important at an enterprise level as the scope and complexity of the effort grow and it is important to have uniform and consistent tools for sharing data among projects throughout the enterprise

TABLE 18.7 Agile Manifesto and enterprise-level implementation: documentation

Typical small Agile project values	Typical enterprise-level implementation
Working software over comprehensive documentation	■ At an enterprise level, solutions tend to be much more complex, and software is only one part of the overall solution. As a result, some form of additional overall coordination is needed to integrate all the components of the overall solution. ■ At an enterprise level, a solution might also include training, business process changes, a support plan, a marketing and rollout plan, and many other requirements beyond just developing software. All of these may increase the need for some kind of documentation.

TABLE 18.8 Agile Manifesto and enterprise-level implementation: collaboration

Typical small Agile project values	Typical enterprise-level implementation
Customer collaboration over contract negotiation	■ At an enterprise level, there is typically a much broader range of customers and stakeholders to consider and managing expectations can be a lot more challenging ■ Some form of Project Charter document may be worthwhile to help manage expectations but it could be defined at a fairly high level

TABLE 18.9 Agile Manifesto and enterprise-level implementation: planning

Typical small Agile project values	Typical enterprise-level implementation
Responding to change over following a plan	As the scope and complexity of an effort at an enterprise level increase, there is typically a need for more planning: ■ to coordinate the efforts of large projects requiring multiple teams ■ to synchronize the efforts of development teams with other activities outside the scope of the development effort ■ to adapt the development effort into higher-level management processes that may be more plan-driven

TABLE 18.10 Agile Manifesto and Enterprise-level implementation: change control

Typical small Agile project values	Typical enterprise-level implementation
Welcome changing requirements, even late in development. Agile processes harness change for the customer's competitive advantage.	■ Change never comes without consequences, and change control can be valuable for configuration management and validating that any new changes are consistent with other previously developed requirements and assumptions ■ If done properly, it does not equate to stifling or *preventing* change. It means ensuring that unnecessary change (as ultimately defined by the Business Owner or Project Sponsor) is rejected, but that necessary change is made to the project with the full awareness of all concerned and that necessary adjustments to designs, plans, timescales, tests, contracts etc. are made with a minimum of wider disruption

TABLE 18.11 Agile Manifesto and enterprise-level implementation: communications

Typical small Agile project values	Typical enterprise-level implementation
The most efficient and effective method of conveying information to and within a development team is face-to-face conversation	■ At an enterprise level, because of the focus on the overall solution, rather than just software, the team is typically wider than simply the people who are developing the software ■ At an enterprise level, the communications strategy must include a broader set of people, such as production operations and support that are important stakeholders in the implementation of the solution

TABLE 18.12 Agile Manifesto and enterprise-level implementation: progress measurement

Typical small Agile project values	Typical enterprise-level implementation
Working software is the primary measure of progress	■ At an enterprise level, success or failure is measured more in terms of delivering *real business value* to the users over simply developing functional software ■ The primary measure of progress should be whatever the Sponsor defines as value (think about user training or process change, for example)

ENTERPRISE-LEVEL OBSTACLES TO OVERCOME

In addition to the challenges previously mentioned, there are a number of obstacles that are commonly associated with an enterprise-level Agile implementation. The most important obstacles to overcome fall into several major areas:

- developing a collaborative and cross-functional approach
- organizational commitment to implementing an Agile approach
- risk and regulatory constraints.

Collaborative and Cross-Functional Approach

One key challenge is developing a collaborative, cross-functional approach that is required for Agile. At an enterprise level, there may be strong resistance to developing that kind of approach because it might require breaking down organizational barriers and involve possibly some significant cultural changes. The challenge to the Agile Project Manager is providing strong leadership to help break down these organizational barriers and implement a more cross-functional management approach. There are several aspects of this challenge.

Development and Quality Assurance (QA)

Quality and reliability standards today for software are higher than ever and new approaches to software testing are essential to meet those goals. An Agile development approach provides a excellent way to meet that challenge. Instead of having an independent QA organization that is separate from the development organization where the QA organization provides direct oversight and control of all the testing process and activities, it is desirable to have a much more integrated and collaborative approach where QA testing is an integral part of the Agile software development team.

There are several significant advantages to having a more integrated Agile testing approach:

1. **A proactive rather than a reactive approach:** Agile testing is a more proactive approach to eliminate defects at the source. A reactive approach to find and fix defects later can require a lot more rework and is much more inefficient.
2. **Immediate feedback to Developers:** An Agile testing approach provides immediate feedback to the development team. If difficult software bugs are left unresolved, they can compound themselves and that can make it much more difficult to find and resolve the defects.
3. **Quality is not someone else's responsibility:** Agile testing makes the quality of the product an integral part of the development process. The development team producing the product owns responsibility for the quality of the product they produce. It is not someone else's responsibility.

Development and Operation

The release of software in an enterprise-level environment needs to be carefully planned and managed to avoid disrupting the production environment. That is normally the responsibility of the IT Operations organization to ensure that any software changes do not cause disruption of the company's critical production processes.

Implementing an Agile software development process to streamline how development is done and make it more efficient has limited impact if at an enterprise level the process for releasing software to production is still cumbersome and inefficient. For that reason, many companies have developed a "DevOps" approach which involves developing a collaborative relationship between the development team and the operations organization to streamline how software is released into production at an enterprise level.

Development and the Business Organization

Another obstacle is associated with developing a collaborative partnership relationship between the business organization and the development organization to work together actively and collaboratively as the project is in progress. To achieve that goal, there is a need to develop a level of trust between these two organizations and a spirit of collaborative partnership rather than an arm's-length contracting approach. That can be very difficult to achieve in many organizations that have not been accustomed to working that way.

This is a very difficult challenge and requires strong leadership from senior executive management to break down these organizational barriers and to implement a more cross-functional management approach.

Organizational Commitment

In many companies, Agile cannot be implemented without some level of cultural change. Implementing it as a development process only and ignoring the need for organizational change will often result in a very limited level of effectiveness. It might be done that way as a first step to show results quickly, but, ultimately, achieving the full benefits of Agile will probably require some level of organizational change.

What Does It Mean to Become Agile?

Many people make the mistake of assuming that you have to force the whole company to be more Agile in order to implement an Agile development process. That is not necessarily the case—a company needs to build its culture around whatever makes sense for the primary business that the company is in. Although becoming more Agile is a good thing in most companies, it must be positioned in the context of the company's overall business strategy. *Just becoming Agile should not necessarily become an end in itself.*

The challenge an Agile Project Manager may face is helping to plan how an Agile development process should be integrated with the company's primary business environment and then helping to lead whatever enterprise-level transformation is needed to achieve that level of integration. Developing the appropriate strategy can be a very challenging role and might involve making compromises between:

- trying to transform the company to become more Agile, and

- adapting an Agile development project management approach to fit with the company's existing business environment and culture.

Although an Agile Project Manager might play a consultative and facilitative role in this, it cannot be done without strong senior management leadership. Once the overall strategy is determined, implementing that strategy can also be very challenging because it can involve a significant amount of change management for whatever organizational and cultural changes may be required.

Training and Coaching Required

Agile requires a lot of skill to implement it successfully and that can require a lot of training at many levels. Not only do the people who have to implement Agile projects need to have a thorough understanding of it, but the business people who need to participate in the process also need to have some understanding of how it works, in order to participate effectively in the process.

In addition to training, some more sophisticated coaching may also be needed to help the teams become effective and to fully integrate the Agile process with the company's business.

Risk and Regulatory Constraints

There may also be factors in the company's business environment that impose constraints on how far you can go with an Agile implementation, that need to be taken into consideration.

- For example, if a company operates in an environment that requires some level of risk and/or regulatory control, it may be necessary to adapt the Agile approach to fit that environment but it's not impossible to do that with the right approach and tools.

- Also, requirements traceability and design control, combined with an effective testing approach, are usually very important criteria for developing an acceptable approach to meet regulatory requirements.

- An Agile project management tool can provide a way of satisfying those constraints—the tool can help demonstrate that the process does indeed provide an acceptable level of control in those areas. The challenge for an Agile Project Manager is in determining how to blend the right level of control and agility to provide the right balance for the company.

ENTERPRISE-LEVEL IMPLEMENTATION CONSIDERATIONS

There are also a number of enterprise-level implementation considerations that impact how Agile projects are effected at an enterprise level. The most important implementation considerations at an enterprise level include:

- architectural planning and direction
- enterprise-level requirements definition and management
- development team integration
- release to production.

Architectural Planning and Direction

Enterprise-level solutions are typically more complex than a small standalone software application, and upfront planning of the design architecture is typically needed for a number of reasons:

- Individual solution complexity:

  - An individual solution often will require multiple teams, and some sort of architectural direction is needed to define how the work among the teams should be organized and coordinated.
  - Because the individual solutions are typically much larger and more complex, there is much greater risk associated with having to redesign and/or refactor the solution after the design is in progress because the level of effort required may be much greater.

- Overall architectural direction:

  - At an enterprise level, there are several types of architectural considerations that should be considered beyond the architecture of individual applications that include overall application architecture standards, technical architecture practices and standards, and the business enterprise architecture strategy.
  - The solution will also need to integrate with other enterprise-level software and conform to whatever standards the organization uses to ensure that it is interoperable with other applications. That will typically require some planning and design reviews early in the project.

What's the Impact on Individual Projects?

Dean Leffingwell has very accurately identified the need for what he calls *intentional architecture* at the enterprise level:

> For small, Agile teams who can define, develop, and deliver a product or application that does not require much coordination with other components, products, or systems, the basic Agile

methods produce excellent results. But what happens when those teams must coordinate their activities as their components integrate into subsystems, which in turn integrate into larger systems? Moreover, re-factoring of these larger-scale systems may not be an option because many hundreds of person-years have been invested and the system is already deployed to tens of thousands of users.

For these systems, the Agile component teams must operate within the context of an intentional architecture, which typically has two characteristics:

(1) it is component-based, each component of which can be developed as independently as possible and yet conform to a set of purposeful, systematically defined interfaces.

(2) it aligns with the team's core competencies, physical locations, and distribution (if this is not the case, it is likely that the teams will realign themselves thereto!).[5]

How Do Teams Implement This?

Of course, this doesn't necessarily mean that a big upfront design approach is needed for every project:

■ Common sense should be used to determine the level of depth that needs to go into upfront architectural planning to reduce the risks and uncertainties involved.

■ For example, if there is significant uncertainty associated with the architecture that would have a high potential risk for the project if the architectural direction is not addressed and resolved early on, this may then necessitate including a special iteration (sometimes called a *spike*) to investigate and resolve that uncertainty.

■ One method that is frequently used is to define and develop a prototype or *slice* of how the ultimate system will be implemented as a proof of concept. That prototype, or proof of concept, can then be used as a reference model by the teams designing and implementing the rest of the system.

Enterprise-Level Requirements Definition and Management

In some cases, there is also a need for more upfront planning of requirements at an enterprise level:

■ Architecture and requirements are intimately related, and it's impossible to define architecture without some idea of what the requirements are. If there is a need to define the architecture prior to development to reduce the risk, it will probably be essential to define more of the requirements upfront in a typical enterprise-level Agile project.

■ The requirements can be a lot more complex, and more upfront analysis of the requirements may be needed to determine the most appropriate solution and the optimum architecture as well as understanding any interdependencies and interrelationships among the requirements. A technique called *functional decomposition* is often used to break down requirements into a logical organization. Functional decomposition is also a useful way of understanding how the requirements are aligned with supporting the business objectives of the system. (See Chapter 15 for more detail on Functional Decomposition.)

■ Typically a larger number of stakeholders are involved in the development of the requirements. For example, a support group will many times have a key role in determining supportability requirements.

There are some significant challenges associated with planning and managing requirements for solutions at an enterprise level.

Dean Leffingwell continues:

Agile's practice of working on a few stories at a time is a wonderful focusing mechanism for the team.

■ But in larger systems, what drives these stories into existence?

■ Who says these are the right stories?

■ Will the summation of all these stories (now in the thousands) actually meet our customer's end-to-end use-case needs?

■ Does our team's Product Owner have clear visibility into stories others are building? Are they likely to affect us? If so, when?

■ And when developing solution sets (large sets of products that must be deployed together and support end-to-end use cases for the user or customers), how do we know that the stories on the table will actually work together to achieve the final objective?

■ Can building an enterprise application, one story at a time, possibly work? Well, perhaps not exactly that way.[6]

However, Michael Hurst, formerly corporate PMO for Harvard Pilgrim Health Care, suggests one clear benefit:

Most Agile teams use their ability to code quickly and efficiently as a requirements discovery process and avoid the overhead of formal specifications. This practice can work effectively because a small team can write and rewrite code at a rate faster than many organizations can attempt to determine and codify their customer requirements anyway![7]

The key thing to consider is that this is not an all-or-nothing decision of having no requirements at all or having large and unwieldy requirements documents.

■ Good common sense should always be used to determine how much upfront planning and what level of detailed definition should go into the requirements for any project.

■ If the requirements planning and definition effort is done using electronic tools rather than traditional Word documents, it can significantly accelerate the development effort once the development is started. For example, defining a structure to the requirements and organizing them as epics and user stories in an electronic tool makes it much easier to plan, organize, and track development and testing tasks against those requirements.

Development Team Integration

Agile was originally designed around small, single-team projects; however, over time, the use of Agile has expanded significantly to include large, complex projects requiring multiple teams. For large projects that require more than one team, some kind of mechanism is needed to coordinate and synchronize the work among individual teams. There are two primary challenges associated with this:

1. The first challenge is integrating the efforts of multiple teams from a development perspective. Development integration can either be very loosely coupled or very tightly coupled.
2. The second challenge is aligning the efforts of all teams with the business objectives of the organization. In many situations, there is a need for both tactical and strategic alignment.

These two challenges are shown in Figure 18.1.

Both of those are very significant challenges, and there is much to learn in both of these areas:

■ Many aspects of the knowledge of how to make Agile work at a team level is relatively mature; however, the knowledge of what needs to be done to scale Agile to an enterprise level is far less mature.

■ This is also an area where it becomes essential to figure out how to integrate Agile principles and practices with classical plan-driven principles and practices in the right proportions to fit the situation.

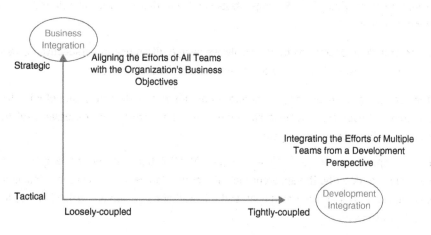

FIGURE 18.1 Integration challenges

An overall approach will need to address both of those challenges. Scrum, on its own, does not provide a mechanism for doing that, but several other approaches can be layered on top of Scrum to achieve that kind of coordination.

Release to Production

The process for releasing mission-critical applications to production is another important factor that increases the complexity of large, complex enterprise-level applications. Dr. Hurst provided some comments on that from his experience with Harvard Pilgrim Health Care, which is one of the major case studies in this book (see Chapter 28).[8]

- We have found in the management of large programs that once the team build passes the acceptance criteria test cases; it is really just the first step in releasing to a complex production environment.

- OK, my component works locally on the development environment (functional testing), and it works on the nearest neighbor environment (contiguous testing), but does it work in the end-to-end production environment for full User Acceptance Testing (UAT)?

- We have found that Kanban process works better than Scrum by the Release Management team since they really need teams to go through a sequence of environments and connections on their way to the full production environment. These stages are highly susceptible to staffing and other resource constraints (test data, test environment updating) that allows only so many things to be in queue at once, but items in queue can be replaced by other things as ones pass through successfully. Perfect production line Kanban.

The whole process of releasing applications to production in an enterprise environment can require a lot of coordination outside of the Agile team, which is a key area of value-added that a Project Manager can provide. The challenges associated with this are typically referred to as DevOps. Dr Hurst continues:[9]

In my experience in operations there has always been a difference in perspective between Dev and Ops, but it's always been more of an impediment than a benefit.

- The common goal should be getting apps deployed as quickly, safely, and efficiently as possible, but each group instead has a more short-term priority not necessarily related to the results the business is looking for.

- Lee Thompson (formerly of E*TRADE, now of DTO Solutions) coined the term "wall of confusion" to describe the apparent inability for development and operations teams to communicate around a common goal, and this wall of confusion is a critical barrier to effective teamwork . . .

Ideally, companies work out a system in which whoever makes the mistake pays the price.

- The reason Ops is so often scared of Dev deploying is that Dev doesn't really care how secure their apps are, how hard they are to deploy, how hard they are to keep running or how many times you have to restart it, because Ops pays the price for those mistakes, not Dev.

- In most organizations the mandate of a developer is merely to produce a piece of software that worked on a workstation—if it worked on your workstation and you can't make it work in production, it's Operations' fault if they can't get that to thousands of machines all around the world . . .

Google is a great example in switching up that process. When they deploy new applications, the developers carry the pagers until the stop going off—only when they stop getting outage alerts does operations take over the operational running of a system.

ENTERPRISE-LEVEL MANAGEMENT PRACTICES

There are different levels of management that typically come into play in large, complex enterprise-level projects. Some of these levels of management are shown in Figure 18.2.

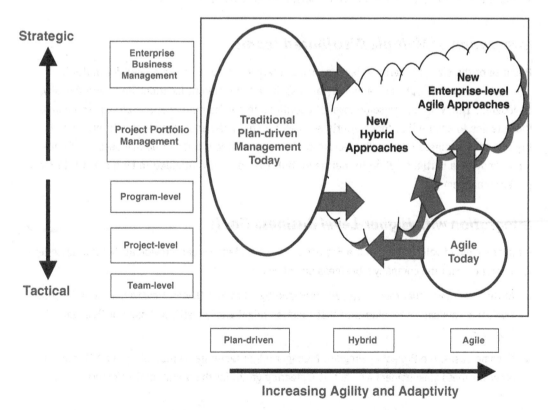

FIGURE 18.2   Typical enterprise levels of management

There are three areas of enterprise-level management practices that are most important for developing an Agile Project Management approach:

- the project/program management approach
- the role of the Project Management Office (PMO)
- the project/product portfolio management approach.

Project/Program Management Approach

At an enterprise level, a number of factors contribute to the need for some level of project/ program management in an Agile project. These factors include:

- integration of multiple distributed teams
- integration with higher-level business goals
- coordination with other organizations and stakeholders
- management of other related activities outside of development.

Integration of Multiple Distributed Teams

Because of the size of enterprise-level projects, there will frequently be a need for multiple teams. In many cases, those teams may be distributed in different locations and cannot easily be co-located. That will require some level of coordination and communications among and across those teams to ensure that those efforts are well synchronized and consistent with producing the intended business results. The Scrum-of-Scrums approach that will be discussed in Chapter 19 may be a partial solution to that need, but it is not really intended to be a substitute for project management.[10]

Integration with Higher-Level Business Goals

A major source of value-added that a Project or Program Manager can provide at the enterprise level is integration with the company's business objectives.

- At the team level, that role is typically provided by a Product Owner. However, at an enterprise level, the workload associated with that function might easily justify a Project or Program Manager.

- In some cases, the Project or Program Manager might facilitate a group of Product Owners or other business stakeholders who act as a steering group for the overall project/program.

Coordination with Other Organizations and Stakeholders

Because large, enterprise-level solutions can have a very broad impact, there is typically a significant need for coordination with other organizations and stakeholders who might be outside of the direct day-to-day project team.

Management of Other Related Activities

At an enterprise level, a number of project-level activities might be outside of the individual development project team, such as planning and managing release to production, business process changes, user training, and support requirements.

The Role of a Project Management Office (PMO)

A Project Management Office (PMO) in a company typically has several major roles which include:

- management consultancy
- project/product portfolio management
- progress tracking and reporting
- project methodology and training.

Management Consultancy

The general role of a PMO is to provide consultancy and training to all management levels on selecting and implementing a project management approach and project methodologies that are well suited to the company's business (see Table 18.13).

TABLE 18.13   General PMO role in management consultancy

Classical plan-driven approach	Agile (value-driven) approach
In a classical plan-driven environment, this role might be focused on defining and implementing plan-driven project management principles, practices, and methodologies needed to fit the company's business and projects	In a more Agile environment, this role might be broader to include defining and implementing an Agile transformation, if necessary, to fit the overall project management approach to the company's business

Project/Product Portfolio Management

In the area of project/product portfolio management, the role of a PMO is primarily to act on behalf of the appropriate business managers to manage the implementation of the company's project/product portfolio management strategy. As a result, the role of the PMO will vary, depending on the level of rigor and control behind the project/product portfolio management approach (see Table 18.14). Project/product portfolio management will be discussed in more detail below.

Progress Tracking and Reporting

Once a project has been initiated, at least some minimal form of tracking and reporting of progress is needed to determine if the projects are, in fact, actually producing the desired business results and return on investment. The actual level of tracking and control should also be commensurate with the level of rigor in the overall project/product portfolio management approach to determine if the projects/products are really fulfilling their goals (see Table 18.15).

Project Methodology, Standards, Training, and Tools

Another typical role of a PMO is to provide a focal point for sponsoring and managing project methodologies and standards used by the organization as well as providing whatever training, tools, and support that may be needed to implement that methodology (see Table 18.16).

TABLE 18.14   General PMO role in project/product portfolio management

Classical plan-driven approach	Agile (value-driven) approach
In a more classical plan-driven project/product management approach, investment decisions to invest in projects will typically require at least a high-level estimate of the costs and schedules of those projects so that they can be intelligently evaluated against other alternatives	In a more Agile environment, a much more dynamic approach with less financial rigor behind it would typically be used

TABLE 18.15   General PMO role in progress tracking and reporting

Classical plan-driven approach	Agile (value-driven) approach
In a more classical plan-driven environment, there is likely to be a much more rigorous and detailed approach to tracking progress against specific goals and milestones, and the PMO will likely play a significant role in consolidating and validating the reporting information	In a more Agile approach, the approach may be less rigorous and might put more responsibility on the individual project teams for reporting their own progress. The PMO might play more of a facilitative role and less of a controlling role.

TABLE 18.16 General PMO Role in project methodology, standards, training, and tools

Classical plan-driven approach	Agile (value-driven) approach
In a classical plan-driven PMO, the methodologies might be fairly rigidly-defined and the PMO might have a responsibility for ensuring compliance with the methodology process requirements, such as completion of required documentation and phase-gate reviews	In a more Agile PMO, there is likely to be a much higher level of flexibility and adaptivity delegated to the project teams in how the methodology is implemented, and the PMO may play more of a consultative and supporting role to provide training and tools to the project teams as needed to help make them successful

Project/Product Portfolio Management

An important enterprise-level management practice relates to how the company does project/product portfolio management. The objective of an enterprise-level project/product portfolio management approach is to plan and maximize the company's overall return on investment from their projects and products by:

■ selecting the right mix of projects and products that is most likely to result in the greatest return

■ managing the efforts associated with developing those projects and products to maximize the return.

They may also have a PMO structure in place to provide overall management of those projects and products and to do some form of resource and capacity planning to support the process.

The approach for doing project/product portfolio management might range from a very formal and rigidly defined process based on planned, quantifiable, and controlled financial metrics in a more classical plan-driven environment to a much more dynamic and less formal process in a more Agile environment

Whatever approach is used for project/product portfolio management, it should be directly related to and very consistent with how the company develops its business strategy. In this section, we will discuss three examples of project/product portfolio management approaches that provide different levels of control versus agility, depending on the level of uncertainty in the company's business, the company's culture, and other factors:

■ traditional quantifiable financial management approach

■ Agile portfolio management approach

■ Lean Startup approach.

The key thing to recognize is that it is not an all-or-nothing decision to have no management at all or totally oppressive over-control with rigorous financial analysis of projects.

Traditional Quantifiable Financial Management Approach

One factor in determining the project/product portfolio management strategy is the level of financial rigor required in making project/product portfolio management decisions.

■ If it is expected that product/product portfolio management decisions will be made on the basis of quantifiable data such as Return on Investment (ROI) or Internal Rate of Return (IRR), it will likely slant the approach toward a more classical plan-driven project/product portfolio management approach.

■ It would be difficult, if not impossible, to base project/product portfolio management decisions on quantifiable financial metrics in a true Agile environment because there typically just isn't that much information available to support that kind of analysis upfront.

Agile Portfolio Management Approach

A more Agile and more dynamic approach is described in the Valpak case study in Chapter 27. Valpak is one of the leading direct marketing companies in North America, The company created high-level epics for each of their major business initiatives and used a high-level team of senior executives to plan and prioritize these high-level initiatives similar to the way you might plan and prioritize product backlog items at a project level. The advantages of this approach are that it is very dynamic and can be shifted easily to adapt to different business conditions and priorities.

How the portfolio management function is implemented might be very different in different companies:

■ In some companies, such as the Valpak case study, where the portfolio management approach is a critical and integral part of the company's business strategy, the senior executives may choose to do it themselves.

■ In other companies where the portfolio management process is not a critical and integral part of the company's business strategy, some portion of this function might be delegated to a PMO.

Lean Startup Approach

An even more Agile approach is the Lean Startup approach that will be described in more detail in Chapter 20.

■ In simple terms, the Lean Startup approach involves making some assumptions and hypotheses about projects and products and testing those assumptions and hypotheses before committing to full-scale development.

■ The Lean Startup approach is very well suited for companies that have a very high level of uncertainty associated with their business initiatives and want to take more of an incremental approach to trying out initiatives to see how they work before making a major commitment to any of them.

The Lean Startup approach is closely associated with the idea of a Minimum Viable Product (MVP). The idea is that many companies make the mistake of launching into significant and expensive product development efforts with lots of risk and uncertainty and without fully verifying the critical assumptions behind the project or product. A better approach is:

■ to start with a Minimum Viable Product to test the market for the product and only expand it as needed to fill the market need;

■ to test and validate any assumptions and hypotheses about the project or product as thoroughly as possible before making a major commitment to the development effort.

SUMMARY OF KEY POINTS

Implementation of an Agile approach at an enterprise level typically requires reinterpreting some of the Agile values and principles in a very different context. It may also require adapting some of the Agile practices to work in a somewhat different environment. The following is a summary of some of the key differences:

1. **Differences in Enterprise-level Agile Practices**

Applying Agile development practices at an enterprise level typically involves some adjustments to those practices particularly with respect to:

■ **customer participation** in a large, complex enterprise-level project can be much more difficult to achieve

■ **development team organization** typically requires provision for coordinating and integrating the work of multiple development teams with each other and keeping all of those teams aligned with the company's overall business direction and priorities

■ **application architecture** may be much more critical and complex and may require much more upfront planning

■ **requirements definition and management** may require a more integrated approach to coordinate the development of all the requirements to develop a well-integrated solution that aligns with the company's business goals

■ **project/product portfolio management:** It can be very difficult to integrate a number of Agile projects or products into a typical enterprise-level project portfolio management approach based on financially quantifiable metrics such as return-on-investment; however, some level of integration and management is necessary to make portfolio management decisions.

2. Enterprise-Level Agile Challenges

There is a need to reinterpret the Agile values and principles at an enterprise level in a different context:

■ Implementing Agile values and principles at an enterprise level requires a focus on overall *solutions*, not just *software*, and implementation of those solutions might typically require coordination with other activities outside of the direct realm of software development.

■ There must be a plan for how to achieve that coordination that goes outside of the boundaries of the immediate software development team.

■ Many times, at an enterprise level there are a number of different stakeholders who have input into the development of a solution. That will require an approach for ensuring that the inputs of those stakeholders are effectively integrated into the project as it progresses.

3. Enterprise-Level Obstacles to Overcome

There are a number of potential obstacles to overcome at an enterprise-level to get an Agile approach to work.

■ One of the biggest challenges is to develop a collaborative and cross-functional approach, and that is very critical to successfully achieving enterprise-level agility.

■ Another major obstacle is that in many companies, Agile cannot be implemented without some level of cultural change. Implementing it as a development process only and ignoring the need for organizational change will often result in a very limited level of effectiveness.

4. Enterprise-Level Implementation Considerations

At an enterprise level,

■ Architectural considerations of how an application interacts with other parts of the architecture can become very significant and may require more upfront planning to ensure that any interdependencies and constraints that must be observed in the architecture are understood and incorporated into the design of the software.

■ The requirements can be much more complex, and more upfront analysis of the requirements may be needed to determine the most appropriate solution and the optimum architecture.

■ The process for releasing projects to production at an enterprise level will typically impose some constraints that must be considered.

5. Enterprise-Level Management Practices

As projects are scaled to an enterprise level, a number of higher-level management considerations beyond the level of individual development teams come into play. There are three areas of enterprise-level management practices that are most important for developing an Agile Project Management approach:

- project/program management approach
- the role of a Project Management Office (PMO)
- project/product portfolio management approach.

DISCUSSION TOPICS

Differences in Agile Practices at an Enterprise-Level

1. What are some of the most important differences in how Agile practices might need to be implemented at an enterprise level? Why?

2. In developing an enterprise-level Agile Project Management strategy, what would be the most important differences in practices to consider? Why?

Enterprise-Level Agile Challenges and Obstacles

1. What do you think is the most significant difference in principles and values that you will encounter in a typical enterprise-level Agile project? Why?

2. What do you think is the most important obstacle to overcome in implementing an enterprise-level Agile transformation? Why?

Enterprise-Level Implementation Considerations

1. What are some of the most important implementation considerations to consider in implementing an Agile approach at an enterprise level? Why?

Enterprise-Level Management Practices

1. How is the role of a Project Management Office (PMO) different in an Agile environment?

2. What is the Lean Startup approach? What value does it provide? Where would it be most useful?

NOTES

1. Scott Ambler and Mark Lines, *Disciplined Agile Delivery: A Practitioner's Guide to Agile Software Delivery in the Enterprise* (Upper Saddle River, NJ: Pearson Education, 2012; Kindle Edition), pp. 349–353.
2. Dean C. Leffingwell, *Scaling Software Agility* (Reading, MA: Addison-Wesley, 2007), p. 87.
3. Ibid., pp. 88–89.
4. Ibid., p. 204.

5. Ibid., p. 190.

6. Ibid. p. 190.

7. Dr. Michael Hurst, email to author, April 20, 2014.

8. Ibid.

9. Blog, "DevOps: What It Is, Why It Exists and Why It's Indispensable," posted by Luke Kanies, August 23, 2011, http://readwrite.com/2011/08/23/devops-what-it-is-why-it-exist#awesm=˜oBYtsF1U5Vw3bm.

10. Mike Cohn, "Advice on Conducting the Scrum of Scrums Meeting," Scrum Alliance, May 7, 2007, http://www.scrumalliance.org/articles/46-advice-on-conducting-the-scrum-of-scrums-meeting.

⟨19⟩ Scaling Agile for Multiple-Team Projects

IN THIS CHAPTER, WE WILL DISCUSS SOME APPROACHES FOR SCALING AGILE FOR MULTIPLE TEAM PROJECTS.

SCRUM-OF-SCRUMS APPROACH

The Scrum-of-Scrums approach is the simplest way to provide coordination for multiple teams; however, it has some significant limitations. The Scrum-of-Scrums approach only provides for a very limited level of technical coordination and that is generally not sufficient for many large, complex projects. Much more sophisticated coordination mechanisms will be discussed later.

With the Scrum-of-Scrums approach, when multiple Scrum teams are engaged in a project, each team does its normal, individual, Daily Standup meeting to discuss items within the scope of that team's own work, and each team sends a representative to the Scrum-of-Scrum meetings to provide a mechanism for coordination and collaboration across different teams, as shown in Figure 19.1.

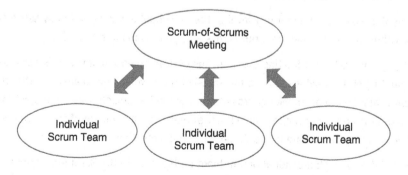

FIGURE 19.1   Scrum-of-Scrums meeting

As explained by Mike Cohn, president of Mountain Goat Software:

The scrum-of-scrums meeting is an important technique in scaling Scrum to large project teams. These meetings allow clusters of teams to discuss their work, focusing especially on areas of overlap and integration . . . Each team would then designate one person to also attend a scrum-of-scrums meeting. The decision of who to send should belong to the team . . .

Being chosen to attend the scrum-of-scrums meeting is not a life sentence. The attendees should change over the course of a typical project. The team should choose its representative based on who will be in the best position to understand and comment on the issues most likely to arise at that time during a project.[1]

The frequency of the meetings should be determined by the teams, depending on the nature of the project and the amount of cross-team communication and coordination required; however, the meetings may not need to be daily. The organization of a Scrum-of-Scrums meeting follows the same general format as a daily Scrum meeting for one of the individual teams. It should be short (typically no more than 15 minutes), and it is focused on the same types of questions as the Daily Standup meeting for individual teams.

Because the scrum-of-scrums meetings may not be daily and because one person is there representing his or her entire team, these three questions need to be rephrased a bit. I also find it beneficial to add a fourth question:

1. What has your team done since we last met?
2. What will your team do before we meet again?
3. Is anything slowing your team down or getting in their way?
4. Are you about to put something in another team's way?[2]

The Scrum-of-Scrums approach is a good mechanism for coordinating the work of multiple teams in a project, but it clearly has its limitations:

- It is highly dependent on the maturity level of the individual teams to be totally self-organizing, not only within each individual team but across all of the teams in the project.

- It is typically limited to coordinating the activities of development teams and is not typically used for coordinating other activities outside the scope of the development teams. For that reason, it is not really intended to be an overall project management approach. Overall, project management is more the domain of the Product Owner(s) than it is the domain of the Scrum team and the people who may participate in the Scrum-of-Scrums meetings.

A representative of each individual Scrum team participates in the Scrum-of-Scrums meetings to represent the interests of each Scrum team and coordinate activities:

- the representative may or may not be the Scrum Master

- the role may be rotated among different team members depending on the need.

LARGE-SCALE SCRUM (LeSS)

The Large-Scale Scrum framework or what is commonly called LeSS was developed by Bas Vodde and Craig Larman in 2002. It's based on the idea that many traditional methodologies for large complex projects are based on the idea of "tailoring down." The normal approach is to take a very large and very complete methodology that is suitable for very large and complex projects and tailor it down by removing the parts you don't need for smaller projects.

LeSS is designed around the opposite approach of starting with a very simple methodology like Scrum and scaling it up as needed for larger and more complex projects. LeSS is a much more comprehensive and more tightly-integrated approach than Scrum-of-Scrums.[3]

Table 19.1 shows a comparison of the Scrum-of-Scrum's approach and the LeSS approach. As you can see from Table 19.1, the Scrum-of-Scrums approach really is only focused on specifying how the teams coordinate work in progress using the Scrum-of-Scrums meeting as a format. In LeSS, there is no formal equivalent of a Scrum-of-Scrums meeting and teams are expected to just talk to each other as necessary to maintain coordination.

The Scrum-of-Scrums process doesn't specify any of the other approaches that LeSS does and leaves that open to determine how the product backlog is managed, how sprint planning and sprint reviews are done, and how work is allocated to teams. LeSS has a recommended approach for each of these:

- It advocates having a single product backlog and advocates that the first part of the sprint planning meeting be joint and all of the sprint review be joint.

- One of the principles behind LeSS is a strong focus on the customer and developing well-integrated solutions and, for that reason, LeSS advocates a feature-based approach to allocating work among teams rather than a component-based approach. The principles behind LeSS are essentially the same as most other Agile methodologies; however, there is a greater level of emphasis on having defined principles.

> LeSS is a much more robust and complete integration mechanism than Scrum-of-Scrums. It focuses on the overall deliverables from a business perspective and not just simply coordination of development efforts.

TABLE 19.1 Comparison of Scrum-of-Scrums and LeSS

	Scrum-of-Scrums	LeSS
Coordination of work	Formal Scrum-of-Scrums meeting	Informal, "just talk"
Product Backlog management	Either single or multiple	Single Product Backlog
Sprint planning	Separate or joint	Joint
Sprint review	Separate or joint	Joint
Allocation of work	By component or feature	By feature

More information on Large-Scale Scrum (LeSS) can be found on the following website:

```
https://less.works/less/framework
```

NEXUS

"Nexus is a framework based on Scrum that is used for scaling Scrum. The framework guides multiple Scrum teams on how to work together deliver results in every sprint. It uses an iterative and incremental approach to scale software development."[4] Table 19.2 presents a summary of how Nexus is different from a standard Scrum process.[5]

TABLE 19.2 Nexus functions

Function	Nexus
Roles	Nexus adds a Nexus Integration Team to provide overall coordination. The Nexus Integration Team consists of: ■ the Product Owner ■ a Scrum Master ■ one or more Nexus team members
Artifacts	All Scrum teams in Nexus use the same, single Product Backlog All Scrum teams in Nexus maintain their own Sprint Backlog A new artifact, the Nexus Sprint Backlog, combines the individual team Sprint Backlogs to create transparency during the sprint
Events	Nexus creates additional events for: ■ Product Backlog refinement ■ Nexus Sprint Planning—includes a representative of each Nexus team to discuss and review the refined product backlog and to select Product Backlog items for each team ■ Nexus Daily Scrum—is like a Scrum-of-Scrums meeting ■ Nexus Sprint Review—is a joint meeting for all Nexus teams to meet with stakeholders ■ Nexus Sprint Retrospective—includes a representative from each Nexus team to identify shared challenges

More information on Nexus can be found on the following website:

`https://www.scrum.org/resources/online-nexus-guide`

SCRUM AT SCALE

Ken Schwaber and Jeff Sutherland are the two original creators of Scrum and there has been some competition in this area between the two on an approach for scaling Scrum. Ken Schwaber created Nexus and Jeff Sutherland created Scrum at Scale.

Scrum at Scale is a more robust framework than Nexus:

- It adds an additional Product Owner Cycle to integrate the Product Owner role.

- It adds an Executive Action Team to provide higher-level executive sponsorship and support to the team.

It puts much stronger emphasis on the need for business integration in addition to development integration. "Scrum@Scale separates these components into two cycles: (1) the Scrum Master cycle (the 'how'), and (2) the Product Owner cycle (the 'what'), intersecting at two components and sharing a third. Taken as a whole, these cycles produce a powerful supporting structure for coordinating the efforts of multiple teams along a single path."[6]

The Product Owner cycle consists of:

- developing the strategic vision

- backlog prioritization

- backlog decomposition and refinement

- release planning

- product and release feedback.

In addition to the Product Owner cycle, Scrum at Scale goes beyond Nexus in adding an Executive Action Team to provide higher-level executive sponsorship and support to the development team.

There are some similarities to Nexus:

- Scrum at Scale uses Scrum-of-Scrum meetings to provide development team coordination

- Scrum at Scale uses a Scaled Retrospective consisting of representatives from each team

- both Nexus and Scrum at Scale attempt to fit the entire organization to a Scrum process.

More information on Scrum-at-Scale can be found on the following website:

`https://www.scruminc.com/jeff-suthlerland-launches-scrum-at-scale-guide/`

SUMMARY OF KEY POINTS

There are a number of methods for scaling Agile for multiple team projects. The most widely used methods have different levels of complexity and sophistication:

1. **Scrum-of-Scrums:** The Scrum-of-Scrums approach is the simplest to implement but it only provides a very limited level of technical coordination. For that reason, it is typically limited to fairly small projects where the business requirements are fairly well defined and not expected to go through very significant changes.

2. **Large-Scale Scrum (LeSS):** The Large-Scale Scrum (LeSS) approach is a much broader and much more sophisticated approach than the Scrum-of-Scrums approach and provides a much tighter level of integration.

   - It goes beyond the basic technical coordination provided by the Scrum-of-Scrums approach and provides for joint coordination of planning and review of the work to be done.

   - That is accomplished by using a single, integrated Product Backlog and jointly holding Sprint Planning and Sprint Review meetings.

3. **Nexus:** The Nexus approach has some similarities to LeSS. It has a single, overall Product Backlog used by all the teams and uses joint sessions for Sprint Planning. However, it goes beyond LeSS by adding:

   - A Nexus Integration team to provide overall integration and coordination of the effort.

   - A Nexus Daily Scrum which is similar to a Scrum-of-Scrums.

   - A Nexus Sprint Retrospective to identify and discuss shared challenges.

4. **Scrum-at-Scale:** Scrum at Scale is a more robust framework than Nexus.

   - It puts much stronger emphasis on the need for business integration in addition to development integration.

   - It puts more emphasis on business integration by adding a Product Owner cycle in parallel with the development cycle.

   - It also adds an Executive Action Team to provide higher-level executive sponsorship and support to the development team.

It is important to note that all these approaches are primarily extensions of Scrum and are limited to development team project integration only.

- They do not provide an overall enterprise-level integration solution.

- They assume that enterprise-level planning and integration are either done separately or are not required for the project that they are applied to.

- That might be suitable for some projects, but very large, complex enterprise-level projects might require a higher-level enterprise integration approach. We will discuss overall enterprise-level frameworks in the following chapters.

DISCUSSION TOPICS

1. What are the limitations of a Scrum-of-Scrums approach?

2. What do you think are the major limitations of all of these approaches and why do you think an enterprise-level framework might be needed?

NOTES

1. Mike Cohn, "Advice on Conducting the Scrum of Scrums Meeting," Scrum Alliance, May 7, 2007, http://www.scrumalliance.org/articles/46-advice-on-conducting-the-scrum-of-scrums-meeting.
2. Ibid.
3. InfoQ, Introduction to LeSS, http://www.infoq.com/presentations/less-large-scale-scrumm.
4. SD Times, Framework and standards are the 'Essence' of agile at scale, https://sdtimes.com/agile/framework-standards-essence-agile-scale/.
5. Nexus Guide, https://scrumorg-website-prod.s3.amazonaws.com/drupal/2018-01/2018-Nexus-Guide-English_0.pdf?nexus-file=https%3A%2F%2Fscrumorg-website-prod.s3.amazonaws.com%2Fdrupal%2F2018-01%2F2018-Nexus-Guide-English_0.pdf.
6. Scrum at Scale Guide, https://www.scrumatscale.com/scrum-at-scale-guide-online/.

(20) Adapting an Agile Approach to Fit a Business

THE MISTAKE MANY COMPANIES SEEM TO MAKE is in failing to see the big picture of how an Agile development process fits into their overall business strategy.

They treat the development process as if it were in a cocoon that can be totally isolated from the higher levels of management that it is part of. There are hundreds of books written about how to optimize every aspect of an Agile development process at a team level inside of that cocoon, but much less is written about how to integrate an Agile development process with a company's overall business environment.

In some cases, they may assume that whatever is good for the development process must be good for the business as a whole and all that is needed is to make the business more Agile like the development process.

Fitting an Agile approach to a business is a significant challenge:

- Many people seem to see Agile as a silver bullet; it seems to be taken for granted that a standard Agile development process will work in all business environments without modification, and the company will somehow adapt their business to work with the Agile development process.

- In many cases, it should work in the other direction—rather than adapting their business environment to fit a development process, the right approach in many cases is to adapt the development process to fit the company's business environment.

- Of course, the exception to that is the case where you know that the business environment itself is dysfunctional, needs to be improved, and/or would benefit from becoming more Agile. That is a situation where it might make sense to adapt the business environment to fit better with the development process. However, many times, more of a compromise or an incremental change is required rather than a radical transformation because of the difficulty of significantly changing the company's overall business environment.

It is not an easy thing to do and requires consideration of:

- the company's primary business environment (either product-based or project-based)

- the company's current management strategy (how the company is organized)

- the company's culture and values.

All those things should be considered and should be in alignment.

THE IMPACT OF DIFFERENT BUSINESS ENVIRONMENTS ON AGILE

One of the biggest factors that causes a misalignment is the difference between a product-based business management model and a more project-based business management model and how those models are funded. Here are a few example scenarios.

Product-Oriented Companies

Agile aligns most easily in companies that are in the business of developing products for external sale. This would include companies whose primary business is selling:

- software products (e.g. Oracle, Microsoft, Intuit)

- services based on software (e.g. Facebook, Twitter, etc.)

- hardware products that contain a substantial amount of software (e.g. Cisco, Garmin).

In this type of company, there is a natural alignment between the principles and practices of an Agile development process and the company's overall business objectives and culture. The business success of this type of company is critically dependent on developing excellent, leading-edge products that dominate their marketplace, and their business success depends on getting those products to market quickly and efficiently with very high levels of quality. Those values are consistent with an Agile development process.

A pure Agile approach is very well suited to these companies. The key thing that differentiates these companies is that the product development process is tightly and directly connected with their business objectives. These companies need to rapidly adapt their products to market needs, which is typically an ongoing and incremental development process. What usually happens in these companies is:

- The company enhances their product(s) continuously to meet market needs and competition.

- Rather than having a single project to complete that has a clearly-defined beginning and end, there is a budget and a team of people dedicated to ongoing improvement and support of the company's product(s).

In this type of environment, managing costs and schedules may not be the highest priority; the costs for the development team are relatively fixed, and responding to change and getting products to market quickly to be competitive are often a much higher priority.

- The budgeted costs are planned to provide a given level of capacity for ongoing development, enhancement, and support of the product(s).

- The primary decision is how to allocate capacity to feature development and support for the product(s), which lends itself very well to a pure Agile development process.

- The focus is more on making product(s) competitive than on managing costs. The key management challenge is: Are we adding the right features fast enough to keep up with what the market wants and to stay ahead of any competition?

At the highest level, a company like Intuit would probably do some product portfolio management to allocate funding to the various products it sells in order to maximize the return on that investment.

A certain amount of investment funding would be allocated to each of the products for ongoing development and support. An Agile development approach can then be used to prioritize the features and capabilities to be developed that would add the most value.

In many of these companies, new technologies such as software-as-a-service (SAAS) and cloud computing have made it possible to get products to market much more rapidly, which has further accelerated the importance of using more Agile development processes in these companies.

Technology-Enabled Businesses

Another type of company where Agile aligns well is a company whose business may not directly involve developing products but where technology plays a very critical and important role in the company's business success. An example would be companies like Amazon.com or Expedia.com—they don't develop products per se; they provide online services to customers.

However, their business success is enabled and critically dependent on Internet technology that underpins the business. Where would Amazon.com or Expedia.com be without having superior Internet technology that powers their business and provides a competitive advantage to them against other potential competitors?

Because the underlying technology plays such a critical role in their business and it's a very competitive and dynamic marketplace, an Agile development process also works well in this type of company.

At the highest level, the company would probably allocate a certain amount of investment funding for the ongoing development and support of the underlying Internet technology that is essential to their business, and the model for allocating that funding to new technology and features would likely be very similar to a product-oriented model.

Project-Oriented Businesses

Where it becomes more difficult to make an Agile development approach work is a project-oriented business, where there is not such a strong natural alignment of an Agile product development process and the factors that drive the company's primary business success. These companies do not sell software as a product or service but might need a significant amount of software application development to support their own internal business operations. They provide this support through their internal IT organization. Examples of companies in that category would be companies that sell products and services that only indirectly use information technology in their business operations for improving operational efficiency, and so on.

For example, Harvard Pilgrim Health Care is a large case study, discussed in Chapter 28. Technology helps Harvard Pilgrim Health Care to more efficiently process huge numbers of healthcare transactions, but they're not a product-oriented company like Intuit or a heavily technology-enabled company like Amazon.com. They do internal company projects to improve operational efficiency and provide higher levels of service to their customers.

The key thing that differentiates these companies is that the software development effort is less directly connected with their business objectives. In these companies, management of costs and schedules is typically much more important. There is a budget and resources established for IT; however, that budget isn't allocated to a particular product. There is likely to be considerable pressure to limit that expenditure as much as possible and to make sure it achieves the desired business objectives. At a high level, what typically happens in this environment is:

■ Business groups propose projects to satisfy their operational needs. They usually justify them based on internal factors such as improved productivity, operational efficiency, and cost reduction rather than direct competition with external products. As a result, projects tend to have a beginning and an end and some well-defined objectives they are expected to accomplish.

■ There is some kind of project/portfolio management approach that prioritizes those potential projects and determines how the IT budget will be best spent to meet the company's overall business needs. That process typically needs at least a rough estimate of costs and schedules for the potential projects to prioritize the effort and to allocate resources.

■ Once resources are allocated to projects and some expectations have been established about the costs and schedule of the projects, some form of tracking and reporting is typically needed to monitor if the projects are on track to fulfill their objectives within the assigned budget.

■ The focus is usually much more on managing costs. Project funding should be spent only to the extent that the investments in projects produce an acceptable return in terms of improved productivity, operational efficiency, reduced costs, and so on. The key management challenge is: Are we investing wisely in projects that will improve our business, and are we getting an acceptable return from those investments?

Hybrid Business Model

Companies in this category don't develop products for external sale, but instead develop internal applications that leverage other revenue-producing products and services. An example of this kind of company is a bank that offers free online services to their customers, such as web-based online checking account access. These services are typically offered to customers at no charge and/or may be heavily subsidized because they are designed to leverage other revenue-producing products or services such as banking services.

This type of company shares some of the characteristics of both of the other two categories:

- Although the company (e.g. the bank) may not sell software directly, the software may have a critical impact on the firm's ability to leverage its primary products or services.

- There is practically always ongoing software development to remain competitive, similar to a company that focuses on software as a product. For example, online bill payment is a common feature that many banks offer, and any bank that does not have that capability might not be very competitive with banks that do.

These companies don't invest in developing projects unless they can produce an acceptable return in terms of leveraged revenue or other factors.

- They typically have some kind of project portfolio management approach where potential projects are prioritized and managed to ensure that the company selects the best mix of projects to align with their business strategy and that the projects generate a reasonable return for their investment.

- The decision process in these companies is similar to the decision process in companies that develop projects for internal use only.

However, in these companies, customer-facing software projects that leverage the company's primary revenue-producing products and services may compete for resources and budget against internal IT projects that are designed to improve operational efficiency.

An integrated approach is needed that ensures that those investments are consistent and complementary to each other.

Adapting an Agile Approach to a Business

Some people will say that Agile only works in product-oriented companies and can't work *at all* in a project-oriented environment. I don't believe that is the case—it can be a lot more difficult to make Agile work in a project-oriented company, it may require some adaptation, and it may take more skill, but it definitely can be done. Some of the challenges in a project-oriented company will center on how projects are funded:

1. Some level of upfront planning and financial analysis might be necessary to support a project portfolio management decision process to determine which projects will provide the most optimum return to the company.

2. Some level of tracking might be needed to validate that the projects are really providing the expected return once they are approved.

In these companies, some compromises might need to be made to make an Agile approach work. If a company insists on a very rigorous level of financial analysis such as Net Present Value (NPV), Return on Investment (ROI), Internal Rate of Return (IRR), etc. to justify and track project investments, it will naturally make it much more difficult to implement a pure Agile development approach. There are two potential compromises that can be made in this kind of environment:

- The company may need to adopt a less rigorous and more dynamic approach to doing financial analysis and tracking of projects. Valpak (which is a case study in Chapter 27) is an example of a company that uses a very dynamic approach for project portfolio management.

- It may also be necessary to adopt more of a hybrid model for projects than a pure Agile approach (see the discussion in Chapter 24 on the Managed Agile Development approach) to provide a plan-driven wrapper around Agile projects. However, that plan-driven wrapper can be as thick or thin as you want it to be—it need not be a very rigorous and highly controlled model.

Even in companies that develop products for external sale, there is a need for some level of overall management, but the management approach for managing investments in an ongoing product development environment is very different than managing investments in internal IT projects. The key point is that in all of these categories, there may be a need to blend some level of classical plan-driven management discipline to manage costs and schedules in order to manage return on investment with an Agile development approach in the right proportions to fit the situation.

TYPICAL LEVELS OF MANAGEMENT

There are several levels of management that are found to some extent in all of these companies, as shown in Figure 18.2 in Chapter 18.

Overall Business Management Level

At the highest level, the overall business management level, the focus is on:

- determining the company's overall business and marketing strategy

- developing a corporate culture and values that are consistent with that strategy

- providing overall leadership to build cross-functional commitment to execution of that strategy.

It's obvious in companies whose primary business is to develop products for external sale that there is a direct relationship between the high-level business management approach and an Agile development approach, and it's very easy to make a strong case for integrating the two. It's also generally easier to implement, because everyone in the organization shares a common focus on making the products that the company produces successful. However, in a company where that is not the case, that can be much more difficult:

- The relationship between the company's overall business goals and an Agile development approach might be much more indirect. There are typically many more organizational challenges to resolve, and it can be more difficult to build the cross-functional synergy that is needed to develop an effective Agile development approach.

- There also may not be a natural alignment between the culture and values of the company and an Agile development approach. For example, in a company whose business success requires a culture with a significant focus on operational excellence, it may be unrealistic and inappropriate to force the company to change its overall business culture to adapt to an Agile development approach.

Overall Management Approach

Many companies are deeply rooted in a classical plan-driven business management approach:

- The very structure of the organization may be a significant obstacle to overcome, and that can put a severe strain on the organization when trying to become more Agile.

- Solving this problem calls for new styles of management that many of these companies are not well prepared to adopt because they typically have hierarchical management structures that are not consistent with an Agile development approach. Jurgen Appelo captured the essence of this problem in his book, *Management 3.0: Leading Agile Developers, Developing Agile Leaders*:[1]

> Some people call it scientific management, whereas others call it command and control. But the basic idea is the same: An organization is designed and managed in a top-down fashion, and power is in the hands of the few . . .
>
> Some people realized that Management 1.0 doesn't work well out of the box, so they created numerous add-on models and services with a semi-scientific status, like the Balanced Scorecard, Six Sigma, Theory of Constraints, and Total Quality Management. Being add-ons to Management 1.0, these Management 2.0 models assume that organizations are managed from the top, and they help those at the top to better "design" their organizations. Sometimes it works; sometimes it doesn't . . .
>
> Management 2.0 is just Management 1.0 with a great number of add-ons to ease the problems of an old system. But the architecture of Management 2.0 is still the same outdated hierarchy . . .

> One important insight is that all organizations are networks. People may draw their organizations as hierarchies, but that doesn't change that they are actually networks. Second, social complexity shows us that management is primarily about people and their relationships, not about departments and profits.
>
> Many of us already knew that "leadership" is just a trendy name for managers doing the right thing and doing things right. But complexity thinking adds a new dimension to our existing vocabulary. It makes us realize that we should see our organizations as living systems, not as machines . . .
>
> We have to replace assumptions of hierarchies with networks because the 21st Century is the Age of Complexity.

That is the essence of what Management 3.0 is about. To fully achieve high performance results with an Agile development approach, we may need to change our thinking about how organizations are managed. Naturally, that is not going to happen overnight, and a transitional approach might be necessary to get there.

Lean Startup Approach

The *Lean Startup* concept is a very Agile approach can be applied at the overall business management level to the development and planning of strategic business initiatives.

- It was developed by Eric Ries and published in his book in 2011.[2]

- The simplest way to describe Lean Startup is an Agile approach for developing and managing a very dynamic and successful overall *business strategy*.

- The word *startup* may be misleading—the approach captures the essence of a very successful, entrepreneurial startup company, but it can be applied to any size company at any stage of maturity. It is not limited to companies that are in the startup phase, and can be used to rejuvenate larger, established companies by infusing the mentality of a successful startup company into the environment.

Eric Ries developed the Lean Startup concept based on lessons learned from his own failed attempts to start a business or launch new products. He characterized the root cause of the failures of There, Inc. and Catalyst Recruiting: "It was working forward from the technology instead of working backward from the business results you're trying to achieve."[3]

The idea behind the Lean Startup approach is that many business ventures or new product launches fail because of some faulty assumptions about what they think the market wants. Instead of recognizing and validating the uncertainty behind those assumptions, companies many times create elaborate business strategies and expensive product development efforts based on some very faulty assumptions that are never fully tested and validated. They don't discover that some of the underlying assumptions behind the business strategy were wrong until an expensive development effort is completed, and by that time it is too late to make a significant change.

The Lean Startup approach starts with a *hypothesis* of what the market wants and, through a series of experiments, attempts to progressively validate and refine that hypothesis. The difference is as follows:

- In a classical plan-driven product development scenario, the business strategy and planning effort and the product development effort are typically sequential. There is an assumption that the business strategy behind the product is complete and is a sound strategy when the development starts, and only details of the requirements need to be resolved.

- In a Lean Startup approach, the business strategy and planning effort are more concurrent with the development effort. When the development effort starts, the business strategy is recognized as only a hypothesis that needs to be tested, validated, and refined as the development effort progresses.

The Lean Startup approach can be a very effective way to develop and validate a business strategy for new products in very uncertain markets. It is most effective when there is a high risk associated with a business strategy being successful because of the uncertainty of market acceptance. Here's how Joe Zulli, CTO at Savings.com, defines Lean Startup:

Lean Startup is a scientific and systematic approach to turning an idea, or vision, into a viable business. One way to think of it is as the scientific method for start-ups.

- Most recent scientific breakthroughs were achieved not as one singular stroke of genius, but rather as a series of smaller experiments to prove or disprove a single hypothesis.

- The beauty of the scientific method is that you always win. Even when your hypothesis is disproven, you still come out better off because you've gained valuable knowledge that will make your next hypothesis that much better.

- Repeat-process enough times, and you will have the breakthrough you were hoping for (assuming you don't give up first). In Lean Startup, we believe the same concept can be applied to business.

- After all, what really is a "vision" other than a collection of hypotheses? If you can break a vision down into a prioritized set of assumptions, then efficiently test those assumptions, you will have much more success, in much less time, for much less money, than if you just build everything whole-hog without testing your assumptions along the way.[4]

The Lean Startup approach forces us to reinterpret several of the basic Agile Manifesto values and principles such as "Working software is the primary measure of progress" in a broader context:

> Above all, Lean Startup is a redefinition of what it means to make progress.
>
> - The Agile Manifesto states that "Working software is the primary measure of progress." Lean Startup, on the other hand, defines the primary measurement of progress as validated learning.
>
> - After all, what good is working software if it doesn't move you forward as a business by making your customers happy? Sure, you may have thousands of lines of beautiful code, but did you really make progress if your company is no better off? Probably not.
>
> - When you follow this new definition of progress, you stop worrying about "velocity" and how much code gets delivered, and start focusing on how you can learn the most the fastest. Sometimes a simple phone interview with half a dozen prospective customers will get you a lot closer to where you need to be as a company than writing half a million lines of code.[5]

Enterprise Product/Project Portfolio Management Level

At the enterprise product/project portfolio management level, the challenge is allocating budget and resources to products and/or projects to optimize the return on investment, as well as tracking performance against planned budget allocations to see if the selected projects are providing the expected return.

In a pure product development company, there may be a mix of different products that need to be managed and optimized. For example, at a high level, Microsoft must determine how much budget to allocate to ongoing development and support of Microsoft Office versus other products such as Microsoft Visual Studio, based on some assumptions as to where it will get the greatest return for that investment.

Once the investment decision is made, the company will want to manage the ROI to determine if the investment in new features and enhancement for Microsoft Office paid off in terms of continued revenue growth in that product.

In either of the other two categories (IT Projects for Internal Use and Hybrid Business Model), it may be more a question of managing a portfolio of projects rather than a portfolio of products, but the management challenges are similar. In those environments, the challenge is how to invest in a portfolio of projects to get the most leverage from that investment either in terms of improving internal operational efficiency or providing new and enhanced capabilities to external customers that drive revenue from the company's primary products and services.

It can be difficult to integrate an Agile development process with an enterprise product/project portfolio management approach, because at the project level, some level of a plan-driven approach may be necessary to at least develop a high-level estimate of costs and schedules to evaluate alternative product/project investments. Also, some form of tracking may be needed to monitor that the investment is yielding an acceptable return.

Achieving that integration may require two things:

1. Putting some kind of plan-driven wrapper around an Agile development approach to provide whatever level of cost and schedule management is desired to support enterprise-level product/project portfolio management decisions.
2. Some tools to roll up data for tracking results across a portfolio of products or projects. Many of these tools also provide capabilities for capacity planning and modeling what-if decisions to evaluate different resource planning scenarios.

In my experience, many companies that implement an Agile development process abandon or neglect this level of portfolio management because it can be very difficult to integrate with an Agile development approach. That typically results in two situations:

1. "Shoot-from-the-hip" judgments about how to invest and manage a portfolio of Agile products and projects, which may not be very effective.
2. Very limited or no capacity planning to allocate resources to projects, and constant juggling of resources among projects, which can destroy the effectiveness of an Agile development process.

Implementing this level of management with an Agile development approach can be somewhat difficult to do, but it is very possible with the right tools and with some level of adaptation of the Agile development approach. Here are a couple of key points:

1. Having the right tools is essential to make this work at the enterprise level. A company that uses very simple paper-based story boards for their Agile teams, without tools that provide the ability to roll up data across projects, will find it very difficult, if not impossible, to do project portfolio management effectively.
2. It may require a hybrid approach that combines an Agile development approach with a plan-driven wrapper to provide some level of predictability and visibility to estimate and manage project costs and schedules:

   - Attempting to do enterprise project portfolio management without some kind of cost and schedule estimates will be difficult, but that is not an all-or-nothing decision.
   - The plan-driven wrapper can be as thick or thin as you want it to be. There are trade-offs to be made when selecting an approach to provide the right balance of agility with predictability and visibility.
   - An example of a hybrid approach that can be tailored to fit a business environment is given in Chapter 24, "Managed Agile Development Framework."

Product Management Level

The product management level is primarily found in companies that develop products for external sale and companies that develop products to leverage other revenue-producing products and services.

It might also be found, to a more limited extent, in companies that develop IT projects for internal use if the internal IT applications are large enough and significant enough to warrant some level of ongoing product management. In those companies, this role might typically be performed by a business analyst rather than a product manager, but that's primarily a change in title only; the skills required are essentially the same.

The product management challenge is to prioritize the features in a given product to maximize the success of the product and, of course, to validate those features with customers to determine if they really have the expected impact.

For companies that do product management of products for external sale or ongoing development of products for internal use, this level is consistent with an Agile development approach. The Product Backlog in an Agile development approach is ideal for managing this kind of effort. However, there still may be a need for more of a hybrid development approach if any higher levels of predictability are required in managing the costs and schedules associated with the effort, and/or to do product/project portfolio management to manage competing investments in multiple projects or products.

Project Management Level

In a typical business environment, the next level down would be management of individual projects. This is where the classical plan-driven project management practices associated with managing scope, costs, and schedules come into play. The level of emphasis on that kind of project management will naturally vary, depending on the relative importance of managing scope, costs, and schedules in the company's overall strategy:

- **This level may not be found at all in product development companies**: In that environment, rather than breaking up an overall product development effort into discrete projects that each have a beginning and an end, the company may treat the effort as more of a continuous, ongoing product development process that may not require a very rigorous project management approach. However, even in product-oriented companies, it might make sense to apply some normal project management discipline to product development projects or programs.

- **Managing costs and schedules to support the company's overall project portfolio management approach requires more focus on project management**: This is especially true in the other two environments where product development is not a major focus. However, even in these environments, a hybrid approach rather than a purely plan-driven, Waterfall-style approach might also be most appropriate.

CORPORATE CULTURE AND VALUES

Cultural obstacles are well known as some of the most difficult issues to overcome in an enterprise-level Agile transformation. However, there is an idea that you should adapt the company's business environment and culture to fit the Agile development process, and that is not necessarily a good idea.

- The overall culture and values of the company should be defined primarily by the success factors that are needed for the markets the business operates in.

- The culture and values associated with implementing an Agile development approach need to be understood in that larger context, but some of the values that are essential for Agile, such as respect for people, are likely to be common to some extent to all business environments.

An important concept that is related to corporate culture is the idea of "Business Architecture":

> Business Architecture reveals how an organization is structured and can clearly demonstrate how elements such as capabilities, processes, organization and information fit together. The relationships among the elements dictate and specify what the organization does, and what it needs to do to meet its common goals.
>
> Business architecture is defined as a blueprint of the enterprise that provides a common understanding of the organization and is used to align strategic objectives and tactical demands. Business architecture is the bridge between the enterprise business model and enterprise strategy on one side, and the business functionality of the enterprise on the other side. [6]

Corporate culture might be thought of as the "glue" that holds the business architecture together.

The Importance of Corporate Culture and Values

A corporate culture and values that are well-defined and aligned with the company's business strategy can be a powerful unifying force in making good companies great. Stephen Covey summed this up very well in Ann Rhoades' book, *Built on Values: Creating an Enviable Culture that Outperforms the Competition:*[7]

> In order to be successful in a volatile world, you must unleash the goodwill and creativity of your people. You must organize your culture in a way that will help your people achieve great things without constant supervision from above. Set this up right, and people will astonish you regularly with their great ideas and ability to take your organization to a higher level . . .
>
> Companies that have significant misalignments between their values and their behavior are all too common, even when the consequences do not make headlines. A company may, for instance, claim to honor the value of cooperation and then set up compensation systems that encourage competition. By their actions and decisions, leaders create a culture, and culture always trumps any strategy you try to implement. To inspire top performance, your organization's strategy needs to be well-aligned with values that are meaningful for your customers and employees.

In a development process that is somewhat prescriptive like the Waterfall process, where people are just expected to follow the rules of the process and there is less room for judgment, the implementation of the process may be less sensitive to the culture and values of a company. However, in an Agile development process, where people are expected to use a lot more judgment and individual initiative to make decisions, and cross-functional synergy is very important, the culture and values of the company take on much more importance and provide:

- an essential framework for guiding people's actions in making decisions

- a unifying force to build cross-functional teamwork around higher-level objectives that supersede individual organizational goals

- a strong motivator for employees engaged in the process who want to see the purpose in what they're doing rather than simply being cogs in the wheel.

There are several possible scenarios:

1. Companies that have a well-defined corporate culture and values that work well for their business and are very consistent with an Agile development process. These companies will naturally have the least difficulty implementing an Agile development approach, and the implementation of that approach will likely make the company stronger.
2. Companies that have a well-defined corporate culture and values that work well for their business but may not be totally compatible with a pure Agile development approach. In these companies, it may not make sense to totally change the company's culture to fit with an Agile development process, but it does require some adaptation of the Agile development process and/or the company's culture to integrate the two.

3. Companies that do not have a well-defined culture and values at all and/or have a dysfunctional culture and values that are not really well aligned with their business strategy and/or are not consistently implemented.

Creating a culture and values for a company where a well-defined culture and values don't exist or changing a company's culture and values that are dysfunctional and not working well can be an extremely difficult thing to do.

Obviously, scenarios #2 and #3 above pose the greatest challenge. In general, the challenge is adapting an Agile approach to fit with the company's existing culture and/or adapting the company's culture to fit an Agile approach.

In any case, it is important to make an assessment of the company's culture and values to identify any obstacles and inconsistencies that need to be overcome to successfully implement an Agile development approach.

Because of the difficulty of changing a company's culture and values, it may make sense to simply implement an Agile development process as a first step without working on higher-level integration with the company's business strategy, culture, and values. But it is important to recognize that the Agile implementation may have limited success without addressing some of these higher-level integration issues.

Value Disciplines

A company's business environment, values, and culture should be determined primarily by the competitive success factors required by the markets it operates in. A mistake some agilists make is that they are so focused on optimizing the development process, that they try to adapt the company's culture to fit what the development process requires as if the whole company revolved around the development process. That may or may not be appropriate, depending on the business environment that the company operates in.

One of my favorite books on this is *The Discipline of Market Leaders* by Treacy and Wiersema.[8] They state:

> . . . no company can succeed today by trying to be all things to all people. It must instead find the unique value that it alone can deliver to a chosen market . . . One point deserves emphasis: Choosing to pursue a value discipline is a central act that shapes every subsequent plan and decision a company makes, coloring the entire organization, from its competencies to its culture. The choice of value discipline, in effect, defines what a company does and therefore what it is . . .
>
> Three distinct value disciplines have been defined by Treacy: The principle is that companies need to at least be sufficient (not deficient) in all three of them but choose one to excel in as its competitive differentiation.

The three value disciplines that Treacy and Wiersema define are as follows:[9]

1. **Operational Excellence:** The first of the three value disciplines is *operational excellence.* Companies that focus on operational excellence succeed or fail by offering products and services more efficiently than their competitors can offer them. Treacy and Wiersema state:

> Companies that pursue this [discipline] are not primarily product or service innovators, nor do they cultivate deep, one-on-one relationships with their customers. Instead, operationally excellent companies provide middle-of-the-market products at the best price with the least inconvenience. Their proposition to customers is simple: low price and hassle-free service. Wal-Mart epitomizes this kind of company, with its no-frills approach to mass-market retailing.

In an operational excellence environment, there is much less room for creativity.

- At McDonald's, there is only one way to cook the hamburgers, and people are required to follow that process in their day-to-day work.
- The company does have a defined way of how employees can suggest and implement improvements to improve the process, but creativity in how the hamburgers are cooked is certainly not encouraged.
- They do the same thing repeatedly at a low cost; that is what their operational excellence is, and it is reflected in their culture.
- McDonald's culture and values emphasize that people may wear uniforms to work and are heavily trained in doing things the same way all the time.

Obviously, there may be more difficulty in implementing an Agile development approach in this type of environment. In this type of company, it may make sense for the company to become more Agile, but it can't lose sight of its primary value discipline of operational excellence.

This is an example of a situation where an Agile development approach might need to be adapted to fit the business environment, and it may not make sense to attempt to completely transform the company's business environment to fit with an Agile development approach.

2. **Product Leadership:** The second of the three value disciplines is *product leadership.* Companies that focus on product leadership succeed or fail primarily by innovating products to meet market needs faster and better than their competitors. According to Treacy and Wiersema:

> The second value discipline we call product leadership. Its practitioners concentrate on offering products that push performance boundaries. Their proposition to customers is an offer of the best product, period. Moreover, product leaders don't build their positions with just one innovation; they continue to innovate year after year, product cycle after product cycle. Intel, for instance, is a product leader in computer chips. Nike is a leader in athletic footwear. For these and other product leaders, competition is not about price; it's about product performance.

A company whose primary value discipline is product leadership needs to create an environment that stimulates creative thought. These companies might be dominated by engineers who might wear jeans and sweatshirts to work. Obviously, this value discipline is very consistent with an Agile development approach.

3. **Customer Intimacy**: The third of the three value disciplines is customer intimacy. Companies that focus on customer intimacy succeed or fail primarily by providing a high level of personalized service to their customers relative to other competitors. Treacy and Wiersema note:

> The third value discipline we have named customer intimacy. Its adherents focus on delivering not what the market wants but what specific customers want. Customer-intimate companies do not pursue one-time transactions; they cultivate relationships. They specialize in satisfying unique needs, which often they, by virtue of their close relationship with—and intimate knowledge of—the customer, recognize. Their proposition to the customer: "We have the best solution for you—and we provide all the support you need to achieve optimum results and/or value from whatever products you buy." Ritz Carlton Hotels is an example of a company that excels at customer intimacy.

In a business environment whose primary focus is on customer intimacy, employees are trained that customer needs and customer service always come first, and their processes need to be flexible to adapt to customer needs.

If a customer has a problem, employees are trained to do whatever it takes to solve the customer's problem with or without a process. As an example, I was traveling with my boss some years ago. He wore a shirt requiring cuff links (French cuffs), and he had forgotten his cuff links. A maintenance man at the hotel we were staying in quickly went to his shop, found two pairs of nuts and bolts, spray painted them black, and turned them into a set of cuff links. Obviously, you wouldn't find a process anywhere to tell someone how to do that—it requires improvisation and employee initiative, and that's what customer intimacy is all about.

The customer intimacy values of adaptability to meet customer needs are very consistent with an Agile development approach; however, you typically find the customer intimacy value combined with other values such as operational excellence in the right proportions.

The idea of value disciplines is that a company can't be deficient in any of the three areas, but the company needs to choose one value discipline as the primary area of focus to excel in. The value discipline that is chosen as the primary competitive differentiator tends to define the whole company and its culture and values.

Many companies have learned to fine-tune their value proposition even further around specific market segments. A very good example of a company that has segmented its customer base is Marriott Hotels:

- Years ago, there was only one kind of Marriott Hotel. It was the typical high-end luxury hotel that Marriott was famous for at one time. However, Marriott recognized that its customers had different values.

- Some customers valued that kind of luxury (liked the covers turned down at night with a piece of chocolate placed on the night table by the bed and were willing to pay a premium price for it). The classical high-end luxury Marriott Hotel fits this value proposition well.

- Other business travelers were more cost-conscious and less concerned about some of these frills. Convenience of getting in and out and staying in a nice comfortable room with the right amenities for business travelers was what was important to them. Marriott Courtyard was developed to fit this need.

- Still other cost-conscious families on a budget wanted a low-cost room and amenities that were designed for a family with children, and Fairfield Inns were created to fill that need.

If Marriott had continued to offer a one-size-fits-all approach, it would have missed the mark in satisfying some of these customers. There would have been a misalignment of their value proposition with what that particular customer segment considered important. This strategy has obviously worked well, as other hotel chains like Hilton have also followed a similar approach.

A misalignment between the primary value discipline a company chooses to pursue and its culture is probably not going to yield optimal results. The same workers in jeans and sweatshirts who were extremely successful in creating new products at a product leadership company are not likely to be as successful cooking hamburgers at McDonald's, where the primary value discipline is operational excellence.

Stephen Covey summed this up very well:

> There is no "right" culture; there is only right fit. Defining the right fit is a process of determining what values are important in your organization's success and committing to them. You must then develop a plan for how people should behave based on those values and put it into practice throughout your organization. The most critical element, of course, is then helping your people adopt those behaviors and live those values, every day.

If you look at the behavior of leaders, you can tell what the real values of a company are. And all too often, those lived values bear almost no resemblance to the stated values. Many leaders want to believe that all they need to do is proclaim a set of values and culture will magically change—but that does nothing to retool the actual values that control day-to-day actions on the front line. Changing those inherent values takes considerably more effort and cannot be accomplished by any leader or set of executives acting alone . . .

If there is one secret shared by companies that create customer-centric cultures, it is that their leaders profoundly understand that people are their biggest asset—and they act on that idea every day. Organizations that value people often don't use it as an advertising slogan. They just do it.[10]

He goes on to identify six principles for creating a values-rich culture:

Principle #1: You can't force culture. You can only create environment. A culture is the culmination of the leadership, values, language, people, processes, rules and other conditions, good or bad, present within the organization . . .

Principle #2: You are on the outside what you are on the inside . . . You can't force people to smile and treat customers well when they feel ill-treated themselves.

Principle #3: Success is doing the right things the right way . . . By defining your values and the behaviors based on them, you also simplify the task of day-to-day decision-making: "Does this make sense in light of our values?" is all you have to ask.

Principle #4: People do exactly what they are incentivized to do . . . Your values will be perceived as hollow and meaningless unless you base compensation and rewards on expressions of behaviors that go along with the values.

Principle #5: Input = Output. Organizations will only get out of something what they are willing to put into it. Values maintenance—what we call continuous improvement—is as important as values creation. In other words, you are never really "done" with culture; you must be always vigilant that no one backslides into old ways.

Principle #6: The environment you want can be built on shared, strategic values and financial responsibility. Conscious action, beginning with determining a set of shared values, can set up a necessary condition for encouraging a culture that will make an organization into a leader in its industry . . . Happy talk values that result in spending huge sums of money on questionable programs are not values that are sustainable in the long run. But neither should you let financial concerns derail the process in its infancy.[11]

SUMMARY OF KEY POINTS

Agile development processes are many times implemented as if they were just a development process that is independent of the company's business environment of which they are a part. There are many books on how to implement and optimize an Agile development process as a stand-alone process but relatively few books that address the big picture of how to adapt an Agile development process to a company's business environment, culture, and values. That is not an easy thing to do because it requires:

- a broad-based understanding of business management principles and practices as well as development process principles and practices (both Agile and plan-driven)

- a deeper understanding of the principles and practices behind various development approaches (plan-driven as well as Agile) to understand how to integrate them in the right proportions to fit a given business environment.

1. The Impact of Different Business Environments

Implementation of an Agile development approach creates some new challenges for businesses that they may have never had to face before.

- If an Agile development approach is not well integrated with the company's business environment, culture, and values, it is not likely to be fully effective.

- Force-fitting a company's business environment and projects to a textbook Agile approach is probably not the best approach in many situations.

- A better approach is to tailor it to fit the company's business environment. This may require blending a combination of principles and practices from a plan-driven management approach, as well as from an Agile approach, to fit the situation.

2. Typical Levels of Management

An Agile development approach does not necessarily require abandoning some of the classical plan-driven management approaches, such as enterprise-level product/project portfolio management; it just requires some adaptation of the Agile development approach to fit with that kind of managed environment or adapting the management environment to fit with Agile.

3. Corporate Culture and Values

An Agile development process is also very sensitive to the company's culture and values. It can be very difficult to implement an Agile development process in a company if there is a misalignment with the company's culture and values or if the company's culture and values are not well defined or are dysfunctional.

DISCUSSION TOPICS

The Impact of Different Business Environments on Agile

1. Why is it more difficult to implement an Agile development approach in a company that has a project-oriented business model? How would you go about resolving this difficulty?

Typical Levels of Management

1. What is the potential impact of having a misalignment of different levels of management in an Agile implementation?

2. What are the potential solutions?

Corporate Culture and Values

1. Why are corporate culture and values so important to successfully implementing an Agile project management approach?

2. Discuss how the concept of value disciplines might apply to some of the companies that you've been associated with. Did the company have a clear idea of what its primary value discipline was? Was it implemented consistently? What were some of the difficulties?

NOTES

1. Jurgen Appelo, *Management 3.0: Leading Agile Developers, Developing Agile Leaders* (Reading, MA: Addison-Wesley, 2011).
2. Eric Ries, *The Lean Startup: How Today's Entrepreneurs Use Continuous Innovation to Create Radically Successful Businesses* (New York: Crown Publishing, 2011).
3. Wade Roush, "Eric Ries, the Face of the Lean Startup Movement, on How a Once-Insane Idea Went Mainstream," *Xconomy* (July 6, 2011).
4. Blog, "Agile/Lean at Savings.com: Interview with Joe Zulli, CTO," posted by Stephanie Stewart, October 4, 2012, http://iamagile.com/2012/10/04/agile-lean-savings-com-joe-zulli-cto/.
5. Ibid.
6. Business Architecture – Definition of Business Architecture, CIO Wiki, https://cio-wiki.org/wiki/Business_Architecture.
7. Ann Rhoades, *Built on Values: Creating an Enviable Culture that Outperforms the Competition* (San Francisco: Jossey-Bass, 2011), p. vi.
8. Michael Treacy and Fred Wiersema, *Discipline of Market Leaders* (Reading, MA: Addison-Wesley, 1995), pp. xiv–xv.
9. Ibid.
10. Stephen Covey, quoted in Rhoades, *Built on Values*, op. cit., pp. ix–x.
11. Ibid., pp. xvi–xviii.

21 Enterprise-Level Agile Transformations

THE WORDS "AGILE TRANSFORMATION" MEAN DIFFERENT THINGS to different companies. There is also an adoption curve associated with Agile, and companies may be at different levels on that adoption curve. An Agile transformation may take a long time to implement, depending on the scope and difficulty of the transformation effort.

No matter what the situation, an Agile transformation typically involves some level of change, and the transformation should be planned. If the company is just starting out on an Agile transformation, it may make sense to pilot the approach on a small scale before attempting to do an enterprise-wide transformation. This chapter provides some guidelines on planning and managing an Agile transformation.

Of course, any transformation of this nature requires a significant amount of leadership and since an Agile transformation may be very broad and cross-functional, it may require a cross-functional leadership team to plan and coordinate the effort.

PLANNING AN AGILE TRANSFORMATION

The following are things to consider in planning an Agile transformation.

Define the Goals You Want to Achieve

I have seen so many companies take a brute force approach to becoming Agile. Doing an Agile transformation just for the sake of becoming more Agile is not necessarily an appropriate goal. What business value do you expect to get from it?

To many companies, being Agile means just going faster. In many cases, that simply means putting pressure on developers to work nights and weekends in "death march" projects to get projects done quickly. Agile isn't necessarily only about becoming faster—if it's done correctly, it can have a dramatic positive impact on the way your whole company operates:

- Transforming the nature of the whole business to make it much more dynamic and adaptive to customer needs.

- Producing higher quality products that are better aligned with customer needs and provide much higher levels of value to the users.

- Dramatically improving the morale and productivity of everyone engaged in the development process because work is done smarter and more efficiently at a planned pace without excessive overtime so that employees are more engaged in the process.

- Developing cross-functional synergy among all organizations and functional areas to break down political boundaries to make the whole company more efficient.

- Developing products faster.

Developing products faster is only *one* benefit of an Agile development approach. It's the one that typically gets the most attention, but an Agile approach is not necessarily faster. A very important aspect of an Agile development process is focusing on business value and being more flexible and adaptive to produce high-value products that are well-aligned with customer needs.

- Sometimes that requires some level of trial-and-error experimentation to iteratively develop a product that is well aligned with customer needs.

- You also have to be receptive to feedback and changes during the development process. Naturally, that can extend the time required to develop a product, but it can result in a much higher quality product in the end.

Before embarking on any Agile transformation, I highly recommend that a company define the goals to be achieved before going too far into an Agile transformation. Developing products faster is only one possible goal. Another consideration is this:

- How Agile do you want to be, and how quickly do you want to get there?

- A pure Agile approach will not be appropriate for all companies, depending on the nature of their business. Even if it is an appropriate strategy, it may not be possible to change the culture of the company overnight to adopt a totally Agile approach, and a step-wise, incremental approach may make more sense.

Becoming Agile Is a Journey, Not a Destination

An Agile transformation can take a long time, and there are many trade-offs to be considered. For example, it may not make sense to tailor an Agile approach to fit a business environment that you know is dysfunctional. We all know that there are limits on how much (and how quickly) you can change a dysfunctional business environment. Many compromises are often necessary, and the initial implementation may be far less than optimal.

Many times, you have to be realistic about what is achievable, take whatever you can get as a starting point, and continuously improve from there. A focus on ongoing, continuous improvement is essential—companies that are looking for a one-shot, quick fix are probably going to be disappointed. When I worked in quality management years ago, one of my favorite expressions was "quality is a journey, not a destination." The same can be said for helping companies become more Agile. It can require a great deal of patience, persistence, and perseverance to implement Agile in any business environment.

Using an Agile approach to plan and manage the transformation itself as an Agile project typically works well:

> In building learning organizations there is no ultimate destination or end state, only a lifelong journey.[1]

- Assess the current situation and develop a vision for where the company wants to wind up after the Agile transformation. What are the problems and limitations in the current situation? What is the scope of the effort required? How will the company be different when the transformation is complete? It is very important to get consensus on what that vision is, as it can be a unifying force for rallying the company around.

- Have conversations to gauge interest of leadership in creating and sustaining the Agile environment. In some organizations, it may be seen as the latest passing fad, only to stop once leadership/administration changes.[2]

- Develop a Product Backlog of all the things that might contribute to achieving that vision. Sometimes it can be difficult to get some traction and momentum established. An incremental and iterative approach works well to break up the transformation into bite-sized chunks that enable getting started quickly and show some quick hits to demonstrate progress.

- Prioritize the items in the Product Backlog and organize the effort into releases and iterations just as you would in an Agile software development project.

- At each step along the way, engage the customer (senior management) and as many key stakeholders as possible in the effort to get feedback and input.

- Use a continuous improvement approach to assess the effectiveness of the effort and take corrective action as needed to make it as effective as possible.

Develop a Culture That Is Conducive to Agile

The cultural transformation required to successfully implement Agile in a software development environment is somewhat similar to the cultural transformation that was needed to implement a Total Quality Management (TQM) approach in the automotive manufacturing industry years ago (see Chapter 11 for more details).

It can require a major culture shift; however, the principles behind TQM can dramatically improve quality and productivity. Agile has the potential to have a similar but even greater impact on a company, but just as with TQM, it may require some rethinking of how the company is managed to create a culture that is more conducive to successfully implementing Agile.

From a management perspective, many of the problems and challenges that need to be overcome are the same as the problems Peter Senge and Dr. Deming noted that needed to be overcome with TQM. These include:[3]

- management by measurement (focusing on short-term metrics and devaluing intangibles)

- managing outcomes (rather than managing the process)

- overemphasis on predictability and control (to manage is to control)

- excessive competitiveness and distrust among people and organizations.

The one most critical objective to accomplish from an organizational culture perspective is the creation of a "Learning Organization" environment. This is something that hasn't changed since the implementation of the TQM movement in the 1990s.

> Learning organizations are skilled at five main activities: systematic problem solving, experimentation with new approaches, learning from their own experience and past history, learning from the experiences and best practices of others, and transferring knowledge quickly and efficiently throughout the organization. Each is accompanied by a distinctive mindset, tool kit, and pattern of behavior. Many companies practice these activities to some degree. But few are consistently successful because they rely on happenstance and isolated examples. By creating systems and processes that support these activities and integrating them into the fabric of daily operations, companies can manage their learning more effectively.[4]

This concept is *absolutely essential* to a successful Agile implementation. Agile is based on the idea of "fail early, fail often," using experimentation and continuous improvement. This concept also hasn't changed significantly since the 1990s. Here's a quote from Peter Senge in 1996 on this:

> When a group of people collectively recognize that nobody has the answer, it transforms the quality of that organization in a remarkable way. And so we teach executives to live with uncertainty, because no matter how smart or successful you are, a fundamental uncertainty will always be present in your life. That fact creates a philosophic communality between people in an organization, which is usually accompanied by an enthusiasm for experimentation. If you are never going to get the answer, all you can do is experiment. When something goes wrong, it's no longer necessary to blame someone for screwing up—mistakes are simply part of the experiment.[5]

Here's another quote on that subject from the same period:

> The big difference between learning and non-learning organizations is not measured in their capacity for error but in their capacity to respond to error. Persistence in error is a sure sign of an organization which, in today's politically correct parlance is "learning-challenged" . . .
>
> But don't be fooled. Dumb organizations are not necessarily staffed by dumb people. In fact, there may be many very smart people in a dumb organization. They are trapped there by a brain-dead organizational apparatus and are often very frustrated. The learning disability usually lies at the organizational level, not with the individual.[6]

Manage Change

Change in any organization is inevitable, and an ability to effectively manage change is an essential element of success for any dynamic and growing business. Migration to a more Agile approach creates new imperatives for change that will put additional pressure on the need for business transformation.

> The pressures of a truly global economy cause today's businesses to increasingly rely on their ability to produce software as a key competitive advantage. Whether it's software for managing manufacturing and customer delivery processes or software improving the efficiency of day-to-day activities, software touches virtually every facet of today's business.
>
> And yet, many CxOs find their software development practices remain little changed from the 1980s. Reliance on prescriptive, plan-based, waterfall-like methods is common despite mountains of evidence that these practices often fail to deliver real value in a timely fashion, and so hamper the company's responsiveness to fast-changing customer requirements and market conditions. And it's not getting easier.[7]

The best companies have learned how to reinvent themselves regularly as necessary to remain competitive and have learned that effective management of change is an essential element of success. However, it is not an easy thing to do and is often unsuccessful.

> To date, major change efforts have helped some organizations adapt significantly to shifting conditions, have improved the competitive standing of others, and have positioned a few for a far better future. But in too many situations the improvements have been disappointing and the carnage has been appalling, with wasted resources and burned-out, scared, or frustrated employees. To some degree, the downside of change is inevitable. Whenever human communities are forced to adjust to shifting conditions, pain is ever present. But a significant amount of the waste and anguish we've witnessed in the past decade is avoidable.[8]

It takes courage to make difficult changes. The company needs to embrace honesty and go to those uncomfortable places to identify what needs to be changed. That's the hardest part about any change initiative. If there are "sacred cows" and things no one wants to discuss or examine, then only symptoms get fixed, not the root causes.[9]

John Kotter has identified eight errors that companies make in managing change:[10]

1. ***Allowing too much complacency.*** By far the biggest mistake people make when trying to change organizations is to plunge ahead without establishing a high enough sense of urgency in fellow managers and employees.

2. ***Failing to create a sufficiently powerful guiding coalition.*** Major change is often said to be impossible unless the head of the organization is an active supporter. What I am talking about here goes far beyond that. In successful transformations, the president, division general manager, or department head plus another five, fifteen, or fifty people with a commitment to improved performance pull together as a team.

3. ***Underestimating the power of vision.*** Urgency and a strong guiding team are necessary but insufficient conditions for major change. Of the remaining elements that are always found in successful transformations, none is more important than a sensible vision.

4. ***Under-communicating the vision by a factor of 10 (or 100 or even 1,000).*** Major change is usually impossible unless most employees are willing to help, often to the point of making short-term sacrifices. But people will not make sacrifices, even if they are unhappy with the status quo, unless they think the potential benefits of change are attractive and unless they really believe that a transformation is possible.

5. ***Permitting obstacles to block the new vision.*** The implementation of any kind of major change requires action from a large number of people. New initiatives fail far too often when employees, even though they embrace a new vision, feel disempowered by huge obstacles in their paths. Occasionally, the roadblocks are only in people's heads, and the challenge is to convince them that no external barriers exist. But in many cases, the blockers are very real.

6. ***Failing to create short-term wins.*** Real transformation takes time. Complex efforts to change strategies or restructure businesses risk losing momentum if there are no short-term goals to meet and celebrate. Most people won't go on the long march unless they see compelling evidence within six to eighteen months that the journey is producing expected results. Without short-term wins, too many employees give up or actively join the resistance.

7. **_Declaring victory too soon._** After a few years of hard work, people can be tempted to declare victory in a major change effort with the first major performance improvement. While celebrating a win is fine, any suggestion that the job is mostly done is generally a terrible mistake. Until changes sink down deeply into the culture, which for an entire company can take three to ten years, new approaches are fragile and subject to regression.

8. **_Neglecting to anchor changes firmly in the corporate culture._** In the final analysis, change sticks only when it becomes "the way we do things around here," when it seeps into the very bloodstream of the work unit or corporate body. Until new behaviors are rooted in social norms and shared values, they are always subject to degradation as soon as the pressures associated with a change effort are removed.[11]

From my experience, there are generally three key elements that are most critical to any successful change management initiative:

1. **Burning platform** (Not creating a sense of urgency—Kotter's error #1): There's got to be a sufficient level of pain associated with the current situation to convince people that the situation is untenable and must be changed.

2. **Vision for the future** (Underestimating the power of vision—Kotter's error #3): There needs to be a clear vision for what the future will look like when the change is complete.

3. **Progress** (Failing to create short-term wins—Kotter's error #6): There are always naysayers and skeptics who will remain on the sidelines waiting to see if the change is likely to be successful before they jump on the bandwagon. That is why it is important to get started and demonstrate progress as quickly as possible.

My experience has shown that if any of these three elements are not in place and a significant change is needed, it will be very difficult to make any progress. For example, if there is no _burning platform_—if people aren't feeling a sufficient amount of pain from the current situation—the motivation to make a change may be insufficient to do anything differently.

Don't Throw the Baby Out with the Bathwater

In migrating to a more Agile approach, it isn't necessary to throw away all the wisdom and knowledge gained from a classical plan-driven approach. In my first book on Agile Project Management,[12] I used a case study based on Sapient to illustrate this point. Sapient developed a very nice hybrid methodology that combines elements of a plan-driven approach with an Agile approach. Sapient recognized that it didn't need to throw away all of its existing traditional processes in order to move to Agile.

Erik Gottesman, author of *Growing Agility in a Large and Distributed Enterprise*, explains it this way:

> Lastly, when transitioning to Agile methods, one should not believe that every practice that is in place before the transition is unnecessary or wrong. Your organization must have been doing some things right to begin with; otherwise, you wouldn't be in business, would you? We point this out because we have seen a disturbing tendency among organizations to think that in order to be Agile, you must unlearn and divorce yourself from all of your old behaviors. This simply isn't true. You must let go of some things. For us, this included evolving our attitude toward fixed-pricing. We also had to get very disciplined about managing WIP and deferring decisions until the last responsible moment in order to avoid speculation and waste.
>
> Conversely, over the course of Sapient's history, we had developed some highly effective processes of our own design for such things as aligning stakeholders around a common vision, eliciting customer requirements, and running Project Management Offices (PMOs). All of this was good and required neither disposal nor significant overhaul in order to integrate with the new techniques introduced.[13]

The key message is to use common sense and intelligence when you apply any methodology to your business. It's always a good idea to clearly identify what's working and what's not. Build on what's working and don't fix what's not broken. Too many teams throw everything up in the air, and it's even more painful to put it back together again.[14]

- It's ironic that the Agile movement started as a revolution against highly prescriptive methodologies like Waterfall, and over time some of the Agile movement has shifted to the point that Agile has become synonymous with particular methodologies like Scrum.

- There's even a view that if you're not doing Scrum rigidly by the book, you're not Agile at all. Scrum is a great methodology (most people call it a framework), but it needs to be applied intelligently in the right context.

- And, as the use of Agile and Scrum has expanded to enterprise-level Agile implementations, Scrum needs to be understood in a much broader context.

Tools Can Be Very Important

Some people in the Agile community have had somewhat of an aversion to tools and only use paper-based stickies on a wall board to plan an Agile effort and track progress. That aversion to using tools probably has its roots in one of the original Agile Manifesto statements:

> Individuals and interactions over processes and tools.[15]

However, relying only on paper-based stickies and no other tools has some serious limitations that become significant when you try to scale Agile projects to a large enterprise environment. Tools become particularly important with distributed teams which is very common in an enterprise-level environment.

Communications and Collaboration among Distributed Teams

When teams are distributed, paper-based tools become extremely limited because they are not easily accessible by everyone on the team. Even for teams that are co-located, tools can play a valuable role in coordinating all the efforts of everyone on the team.

Status Tracking and Reporting

In an enterprise environment, projects typically roll up into some kind of overall portfolio management approach that needs to be tracked and managed, and some form of status reporting is essential for that. This capability can also be essential for achieving integration across different projects.

Requirements Management and Traceability

Tools are very valuable for structuring and organizing requirements, development tasks, test cases, and related information in a project. If there is a need to manage traceability of project requirements, tasks, and test cases, it would be extremely difficult to do that without some kind of tools that are well suited for performing traceability analysis.

If you were using paper-based stickies for User Stories, you would have to re-enter all of that information into some other kind of tool for analyzing and managing traceability. That would be extremely cumbersome and wasteful. It would also be very difficult to keep those sources of information in synch as the project progressed.

The most important point is that tools should support and serve the project team and enable higher productivity, rather than simply controlling and constraining the actions of the people on the team. Erik Gottesman writes:

> I have often argued that Agile methods actually require a higher degree of tooling capability owing to the underlying "need for speed." As feedback cycles in an Agile/iterative process are so small (a continuous integration build, daily stand-ups, sprints, etc.), you need tooling in order to ensure frictionless flow of information to stakeholders; else, you run the risk of "gumming up the works" and petrifying the ability to act at the technical/tactical, project/operational, and organizational/strategic levels.[16]

ADAPTIVE PROJECT GOVERNANCE MODEL

At an enterprise level, some form of project governance model may be needed to manage and sustain an Agile transformation as well as providing a framework for ongoing management of projects going forward. The phrase "project governance" may sound very formal and controlling, but it doesn't need to be that way.

> Governance and Agile methods may seem like an oxymoron—but not so. A means to govern is essential for orderly project functioning. Without governance the advantages of adaptive and evolutionary methods could be overwhelmed by functions bolted together haphazardly and rendered operationally ineffective, expensive to maintain, and disadvantageous to customers and stakeholders . . . In its best form governance empowers action and endows decision-making rights to project management and team leaders.[17]

Having an adaptive project governance framework to provide guidance and direction to an Agile transformation can accomplish several important objectives:

- Engage everyone at the right levels on process decisions to build ownership and consensus on the direction of the effort.

- Align the process from top to bottom with the company's business strategy.

- Make the processes real (not a paper exercise) that people really believe in.

- Provide a mechanism for consistency and ongoing process improvement to sustain the transformation.

The problems this model is intended to address are:

- In many Agile implementations, there is an assumption that the Agile process can be done by the book and there is no need to customize the process to fit the company's business environment. Many times, that is not the case because decisions need to be made to adapt an Agile development process to align with the company's business environment. This governance model provides a framework for making those decisions to help achieve buy-in and consensus on the process across the enterprise.

- Agile heavily emphasizes continuous improvement and learning. That learning often takes place within an Agile team, and there is no vehicle for capturing lessons learned across project teams and then feeding those lessons learned back in the process for other teams to share.

Figure 21.1 shows that an adaptive project governance model is a general model that can be customized as needed to fit a situation based on the scope of the effort and other factors. The model consists of four levels, but those levels can be customized and/or compressed as necessary.

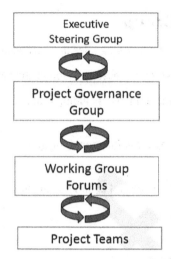

FIGURE 21.1   Multilevel project governance model

The following is a description of suggested roles and responsibilities of each of the levels in this model. Of course, these roles and responsibilities should be customized as necessary to fit a given business environment.

Executive Steering Group

The Executive Steering Group has the following roles and responsibilities in the model:

- Provide high-level strategic direction to align process direction with the company's business strategy.

- Serve as a focal point for building cross-functional integration in the process between all organizations (e.g. IT and the business).

- Make major decisions on project management strategy, such as:

  - Determine appropriate balance of agility and control.
  - Set long-term and short-term goals.
  - Lead any necessary change management initiatives across the organization.
  - Periodically review progress of the effort and alignment of processes with the company's business direction.

- Secure the needed buy-in and support from key stakeholders, without whose visible endorsement the effort may be doomed.

- Provide overall budget and resources for the initiative.[18]

Project Governance Group

The Project Governance group has the following roles and responsibilities in the model:

- Approve the selection and design of processes to support the overall project strategy.

- Review and monitor implementation plans, including training, to ensure effective implementation.

- Serve as a focal point for building cross-functional integration in the process within the development teams.

- Monitor the effectiveness of processes and champion ongoing, continuous improvement efforts.

- Review and approve process changes.

Working Group Forums

The Working Group Forums provide a way for people in similar roles (Scrum Masters, Product Owners, Project Managers, Business Analysts) to share knowledge and best practices, including:

- Provide a forum for reviewing and discussing process effectiveness and leading process improvement initiatives.

- Coordinate activities that span multiple project teams.

- Propose process changes as necessary to the project governance group.

Project Teams

The project teams have the following roles and responsibilities in the model:

- Make decisions to tailor and customize the process as needed to fit project requirements within approved guidelines.

- Escalate decisions to deviate from process requirements as needed for further approval.

- Build strong cross-functional teamwork among project teams.

- Manage the effectiveness of the process implementation and identify opportunities for process improvement.

SUMMARY OF KEY POINTS

An Agile transformation can mean different things to different companies. It can involve very different levels of scope and complexity, depending on the amount of change that may be necessary. The following is a summary of some key points associated with successfully planning and managing an Agile transformation:

1. Define the Goals You Want to Accomplish

Don't make the mistake of force-fitting your business to an off-the-shelf, textbook approach without defining the goals of what you want to get out of it.

- A better approach is to define the goals you want to achieve and tailor a methodology (or combination of methodologies) to fit those goals.

- There are also important trade-offs to consider such as trading off flexibility and adaptability against predictability and control.

2. Becoming Agile Is a Journey, Not a Destination

- Developing an Agile approach can take a lot of skill, long-term commitment, and determination. Companies that are looking for a quick fix that is easy to implement may be disappointed. It generally works well to use an Agile approach during the actual planning and management of the transformation itself:

- Don't try to do everything at once.

> . . . be sensitive to how processes "spider" across the social graph of the project (across roles, teams, divisions, or even organizations). Just as you would risk-adjust your management of software requirements, you must risk-adjust your transformation effort. Get a few relatively simple "quick wins under your belt" while also acknowledging and front-loading riskier behavioral changes by starting collaborative cross-functional conversations early. Demonstrate progress and earn the opportunity to press on to more challenging changes.[19]

- Develop and prioritize a Product Backlog of goals to accomplish.

- Focus on rapid incremental progress to show results quickly.

3. Develop a Culture That Is Conducive to Agile

Organizational culture is well known as one of the biggest obstacles to fully implementing an Agile transformation. It may or may not be possible to change the culture of a company rapidly (or at all) to adapt to an Agile transformation. It is worthwhile to do an assessment of the cultural obstacles that may need to be overcome upfront in the planning process in order to understand the potential impact on a successful Agile transformation.

4. Manage Change

Change management can be a very important aspect of implementing an Agile transformation, and it is essential to be sensitive to the factors that are necessary for achieving successful change in an organization at the enterprise level.

5. Don't Throw the Baby Out with the Bathwater

It isn't necessary to throw away all the wisdom and knowledge companies have gained from implementing plan-driven approaches over the years. Some of that knowledge is still worthwhile—it just may need to be implemented in a very different context when you decide to go to Agile.

6. Tools Can Be Very Important

One of the original values of Agile was individuals and interactions over processes and tools.

- For that reason, some organizations try to implement Agile using very simple paper-based approach by posting user stories on paper stickies on a whiteboard.

- That approach works fine for small projects but doesn't scale well for enterprise-level Agile projects.

- Tools can also have a significant impact on team productivity, especially with distributed teams.

7. Managing an Agile Transformation: Project Governance

It is important to put in place some kind of governance model to effectively manage an Agile transformation and to provide a forum for ongoing, continuous improvement. The model should provide a framework for engaging all the important stakeholders at the right level in the overall management of the process.

DISCUSSION TOPICS

Planning an Agile Transformation

1. What do you think is the most important factor in developing a successful Agile transformation? Why?

2. If you sense that there is some resistance at the enterprise level to implementing an Agile transformation, how would you deal with that resistance?

3. How would you go about adapting an Agile approach to a company that has an existing management structure built around a traditional PMO and a plan-driven approach to project management?

4. What do you think are the three most critical factors in a change management initiative? Why?

Adaptive Project Governance Model

1. Why would a Project Governance model be needed at an enterprise level? What functions would it serve?

NOTES

1. Peter Senge, *The Fifth Discipline: The Art & Practice of the Learning Organization* (New York: Crown Business, 2006).
2. Monica Kay, email comments to author on book review.
3. Senge, *The Fifth Discipline*, op. cit.
4. David A. Garvin. "Building a Learning Organization," *Harvard Business Review* on Knowledge Management (1998), p. 52.
5. Peter Senge, "Systems Thinking," *Executive Excellence* (January 1996), p. 15.
6. Barry Sheehy, Hyler Bracey, and Rick Frazier, *Winning the Race for Value* (New York: American Management Association, 1996), p. 126.
7. Ken Schwaber, "A Playbook for Adopting the Scrum Method of Achieving Software Agility," www.scrumalliance.org/resource_download/66.
8. John Kotter, *Leading Change, With a New Preface by the Author* (New York: Perseus Books Group, 2012; Kindle Edition), pp. 149–151.
9. Liza Wood, email comment to author on book review, May 28, 2014.
10. Kotter, *Leading Change*, op. cit., pp. 149–151.
11. Ibid.
12. Charles Cobb, *Making Sense of Agile Project Management: Balancing Control and Agility* (Hoboken, NJ: John Wiley & Sons, 2011).
13. Erik Gottesman, "Growing Agility in a Large and Distributed Enterprise," Agile Project Leadership Network, Agile Business Conference, London, UK, 2006, p. 5.
14. Liza Wood, email comment to author on book review, May 28, 2014.
15. Agile Manifesto, http://agilemanifesto.org/.
16. Gottesman, "Growing Agility," op. cit.
17. John Goodpasture, *Project Management the Agile Way* (Fort Lauderdale, FL: J. Ross Publishing, 2010), pp. 221–223.
18. Monica Kay, email comments to author on book review.
19. Erik Gottesman, email comments to author on book review.

PART 5

Enterprise-Level Agile Frameworks

PUTTING TOGETHER A COMPLETE TOP-TO-BOTTOM enterprise-level Agile solution can be a very challenging task, especially when some of the pieces are not designed to fit together. To simplify the design of an enterprise-level Agile implementation, it is useful to have some predefined frameworks that can be modified to fit a given business environment, rather than having to start from scratch to design an overall management approach. Three frameworks are discussed in Part 5.

Chapter 22 – Scaled Agile Framework:

The Scaled Agile Framework, developed by Dean Leffingwell, provides a fairly complete Agile approach for an enterprise-level Agile implementation. The Scaled Agile Framework is an excellent approach for companies that want to make a major Agile transformation of their overall business.

Chapter 23 – Disciplined Agile Delivery:

Disciplined Agile Delivery, a project-level framework developed by Scott Ambler, lies in between two approaches also discussed in Part 5: the Scaled Agile Framework and the Managed Agile Development Framework. It is somewhat similar to the Scaled Agile Development framework in terms of adaptivity but more adaptive than the Managed Agile Development Framework.

Chapter 24 – Managed Agile Development Framework:

The Managed Agile Development Framework is a hybrid project-level framework that provides a way of doing a more limited implementation of Agile in a more plan-driven environment. It provides a hybrid of a plan-driven approach layered on top of an Agile development process that can be used to adapt an Agile development process to more of a plan-driven organizational approach.

Chapter 25 – Summary of Enterprise-Level Frameworks:

The three enterprise-level frameworks discussed in Part 5 all provide a way of bridging the gap between an Agile development process based on Scrum and the higher-level management practices found in many typical businesses. This chapter provides an overall summary of these three frameworks.

22 Scaled Agile Framework®

THE SCALED AGILE FRAMEWORK® (SAFE®), developed by Dean Leffingwell, is "a knowledge base of proven, integrated principles, practices, and competencies for achieving business agility using Lean, Agile, and DevOps."[1] Figure 22.1 shows a high-level overview of the latest version of the Scaled Agile Framework. SAFe has evolved over the years from a layered approach to enterprise management to a very robust and well-integrated set of core competencies, principles, and values that can be used to define a complete, top-to-bottom enterprise-level Agile approach.

> Note: The material in this chapter is intended only to provide a high-level overview of the Scaled Agile Framework® (SAFe®). SAFe is very broad and complex, requires considerable judgment and skill, and implementation should not be attempted without further training.

SAFe® COMPETENCY AREAS

SAFe is composed of seven core competency areas:

1. Continuous Learning Culture

SAFe recognizes the importance of having an organizational culture that is consistent with an Agile environment as a foundation for the overall framework:

> The Continuous Learning Culture competency describes a set of values and practices that encourage individuals—and the enterprise as a whole—to continually increase knowledge, competence, performance, and innovation.
>
> This is achieved by becoming a learning organization, committing to relentless improvement, and promoting a culture of innovation.[2]

SAFe 5 for Lean Enterprises

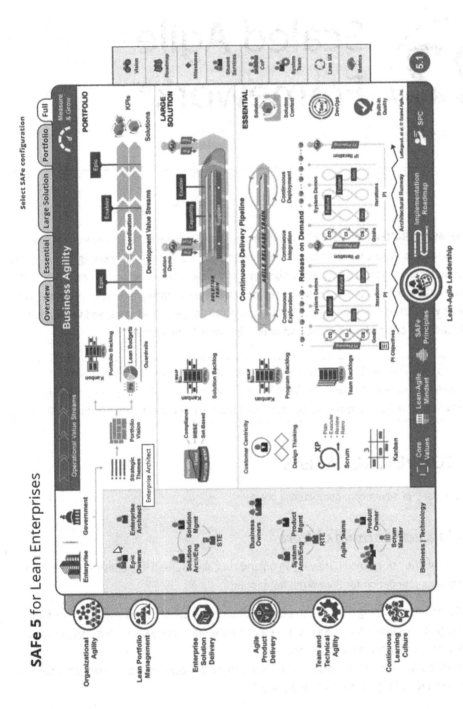

FIGURE 22.1 SAFe 5 for Lean enterprises

2. *Team and Technical Agility*

SAFe uses an approach similar to Scrum including a Product Owner and a Scrum Master for the team-level development approach. However, it is not prescriptive about following a specific Scrum approach and is primarily focused on important team-level principles that are essential to a successful team-level approach.

It also defines an Agile Release Train mechanism for coordinating and integrating the work of multiple development teams. "The Team and Technical Agility competency describes the critical skills and Lean-Agile principles and practices that high-performing Agile teams and Teams of Agile teams use to create high-quality solutions for their customers."[3]

3. *Agile Product Delivery*

SAFe recognizes the need for going beyond typical team-level development practices and focusing on a complete product delivery approach that includes:

- **Customer Centricity and Design Thinking:** Customer centricity puts the customer at the center of every decision and uses design thinking to ensure the solution is desirable, feasible, viable, and sustainable.

- **Develop on Cadence; Release on Demand:** Developing on cadence helps manage the variability inherent in product development. Decoupling the release of value assures customers can get what they need when they need it.

- **DevOps and the Continuous Delivery Pipeline:** DevOps and the Continuous Delivery Pipeline create the foundation that enables Enterprises to release value, in whole or in part, at any time to meet customer and market demand.[4]

4. *Enterprise Solution Delivery*

SAFe recognizes the need for also going beyond a product delivery process and defining a broader process for enterprise solution delivery that includes:

- Requirements analysis

- Business capability definition

- Functional analysis and allocation

- System design and design synthesis

- Design alternatives and trade studies

- Modeling and simulation

- Building and testing components, systems, and systems of systems

- Compliance and verification and validation

- Deployment, monitoring, support, and system updates.

5. *Lean Portfolio Management*

SAFe also recognizes the need for a portfolio management approach that integrates multiple projects and solutions into an overall portfolio management strategy. "The Lean Portfolio Management competency aligns strategy and execution by applying Lean and systems thinking approaches to strategy and investment funding, Agile portfolio operations, and governance."[5]

6. *Organizational Agility*

According to the Scaled Agile Framework:

> In today's digital economy, the only truly sustainable competitive advantage is the speed at which an organization can sense and respond to the needs of its customers. Its strength is its ability to deliver value in the shortest sustainable lead time, to evolve and implement new strategies quickly, and to reorganize to better address emerging opportunities.[6]

The organizational agility competence in SAFe is expressed in three dimensions:

- **Lean-Thinking People and Agile Teams:** Everyone involved in solution delivery is trained in Lean and Agile methods and embraces their values, principles, and practices.

- **Lean Business Operations:** Teams apply Lean principles to understand, map, and continuously improve the processes that deliver and support businesses solutions.

- **Strategy Agility:** The enterprise is Agile enough to continuously sense the market, and quickly change strategy when necessary[7]

7. *Lean Agile Leadership*

According to the Scaled Agile Framework:

> The Lean-Agile Leadership competency describes how Lean-Agile Leaders drive and sustain organizational change and operational excellence by empowering individuals and teams to reach their highest potential. They do this through leading by example; learning and modeling SAFe's Lean-Agile mindset, values, principles, and practices; and leading the change to a new way of working.[8]

SAFe® CORE VALUES

The core values in SAFe are essentially the same as most other Agile approaches and provide alignment to integrate the entire framework.

1. Alignment

SAFe recognizes the need for alignment to integrate all aspects of an enterprise-level Agile approach:

- Alignment is needed to keep pace with fast change, disruptive competitive forces, and geographically distributed teams.

- While empowered, Agile Teams are good (even great), but the responsibility for strategy and alignment cannot rest with the combined opinions of the teams, no matter how good they are.

- Instead, alignment must rely on the Enterprise business objectives . . . Alignment, however, does not imply or encourage top-down command and control. Alignment occurs when everyone is working toward a common direction. Indeed, alignment enables empowerment, autonomy, and decentralized decision-making, allowing those who implement value to make better local decisions.[9]

What that means is having a business architecture including corporate culture where all parts of the organization work together collaboratively and in alignment with the overall business goals of the company.

2. Built-in Quality

SAFe and most other Agile approaches are based on the idea of making quality an integral part of the product development process rather than relying heavily on inspection and testing to find and fix defects later.

- Built-in Quality ensures that every element and every increment of the solution reflect quality standards throughout the development lifecycle. Quality is not "added later."

- Building quality in is a prerequisite of Lean and flow—without it, the organization will likely operate with large batches of unverified, unvalidated work. Excessive rework and slower velocities are likely results.[10]

This principle is essentially the same as the general Agile principle of integrating quality and testing with the development effort so that the development team is responsible for the quality of the product that they produce.

3. *Transparency*

Agile requires a collaborative relationship between the development team and the customer, based on a relationship of trust and partnership.

To achieve that kind of relationship, it is important to be open and transparent in this relationship: To ensure openness—trust is needed.

- Trust exists when the business and development can confidently rely on another to act with integrity, particularly in times of difficulty.

- Without trust no one can build high-performance teams and programs, nor build (or rebuild) the confidence needed to make and meet reasonable commitments. And without trust, working environments are a lot less fun and motivating.[11]

4. *Program Execution*

SAFe recognizes the overall importance of delivering results and business value:

> Of course, none of the rest of SAFe matters if teams can't execute and continuously deliver value.

- Therefore, SAFe places an intense focus on working systems and business outcomes.

- History shows us that while many enterprises start the transformation with individual Agile teams, they often become frustrated as even those teams struggle to deliver more substantial amounts of solution value, reliably and efficiently.[12]

LEAN AGILE MINDSET IN SAFe®

Having the right mindset is important in any Agile implementation and it is particularly important and recognized in SAFe:

- The Lean-Agile Mindset is the combination of beliefs, assumptions, attitudes, and actions of SAFe leaders and practitioners who embrace the concepts of the Agile Manifesto and Lean thinking.

- It's the personal, intellectual, and leadership foundation for adopting and applying SAFe principles and practices. SAFe is firmly grounded in four bodies of knowledge: Lean, Agile, systems thinking, and DevOps... For leaders, it requires a broader and deeper Lean-Agile mindset to drive the organizational change required to adopt Lean and Agile at scale across the entire enterprise.[13]

SAFe® Lean Agile Principles

1. **Take an Economic View:** "Delivering the 'best value and quality for people and society in the shortest sustainable lead time' requires a fundamental understanding of the economics of building systems. Everyday decisions must be made in a proper economic context."[14]

2. **Apply Systems Thinking:** "Deming observed that addressing the challenges in the workplace and the marketplace requires an understanding of the systems within which workers and users operate. Such systems are complex, and they consist of many interrelated components. But optimizing a component does not optimize the system."[15]

3. **Assume Variability, Preserve Options:** "Traditional design and life cycle practices encourage choosing a single design-and-requirements option early in the development process. Unfortunately, if that starting point proves to be the wrong choice, then future adjustments take too long and can lead to a suboptimal design."[16]

4. **Build Incrementally with Fast, Integrated Learning Cycles:** "Developing solutions incrementally in a series of short iterations allows for faster customer feedback and mitigates risk. Subsequent increments build on the previous ones."[17]

5. **Base Milestones on Objective Evaluation of Working Systems:** "Business owners, developers, and customers have a shared responsibility to ensure that investment in new solutions will deliver economic benefit. The sequential, phase-gate development model was designed to meet this challenge, but experience shows that it does not mitigate risk as intended."[18]

6. **Visualize and Limit Work-in-Progress (WIP), Reduce Batch Sizes, and Manage Queue Lengths:** "Lean enterprises strive to achieve a state of continuous flow, where new system capabilities move quickly and visibly from concept to cash."[19]

7. **Apply Cadence, Synchronize with Cross-Domain Planning:** "Cadence creates predictability and provides a rhythm for development. Synchronization causes multiple perspectives to be understood, resolved, and integrated at the same time."[20]

8. **Unlock the Intrinsic Motivation of Knowledge Workers:** "Lean-Agile leaders understand that ideation, innovation, and employee engagement are not generally motivated by individual incentive compensation. Such individual incentives can create internal competition and destroy the cooperation necessary to achieve the larger aim of the system."[21]

9. **Decentralize Decision-Making:** "Achieving fast value delivery requires decentralized decision-making. This reduces delays, improves product development flow, enables faster feedback, and creates more innovative solutions designed by those closest to the local knowledge."[22]

10. **Organize Around Value:** "Many enterprises today are organized around principles developed during the last century. In the name of intended efficiency, most are organized around functional expertise. But in the digital age, the only sustainable competitive advantage is the speed with which an organization can respond to the needs of its customers with new and innovative solutions."[23]

SAFe® ARTIFACTS AND SUPPORTING CAPABILITIES

A SAFe implementation typically includes a number of key artifacts and supporting capabilities:

1. **Vision:** "The Vision is a description of the future state of the Solution under development. It reflects customer and stakeholder needs, as well as the Feature and Capabilities proposed to meet those needs."[24]

2. **Roadmap:** "The Roadmap is a schedule of events and Milestones that communicate planned Solution deliverables over a planning horizon."[25]

3. **Milestones:** "Milestones are used to track progress toward a specific goal or event. There are three types of SAFe milestones: Program Increment (PI), fixed-date, and learning milestones."[26]

4. **Shared Services:** "Shared Services represents the specialty roles, people, and services required for the success of an Agile Release Train (ART) or Solution Train, but that cannot be dedicated full-time."[27]

5. **Communities of Practice:** "Communities of Practice (CoPs) are organized groups of people who have a common interest in a specific technical or business domain. They collaborate regularly to share information, improve their skills, and actively work on advancing the general knowledge of the domain."[28]

6. **System Team:** "The System Team is a specialized Agile Team that assists in building and supporting the Agile development environment, typically including development and mainte-nance of the toolchain that supports the Continuous Delivery Pipeline."[29]

7. **Lean UX:** "Lean User Experience (Lean UX) design is a mindset, culture, and a process that embraces Lean-Agile methods. It implements functionality in minimum viable increments and determines success by measuring results against a benefit hypothesis."[30]

8. **Metrics:** "Metrics are agreed-upon measures used to evaluate how well the organization is progressing toward the portfolio, large solution, ART, and Agile team's business and technical objectives."[31]

SUMMARY OF KEY POINTS

SAFe® Competency Areas

The Scaled Agile Framework (SAFe) is a very complete, robust, and well-integrated framework for implementing an overall Agile approach at an enterprise level.

1. It has evolved over the years from a somewhat prescriptive and layered approach to enter-prise management to a much more general, principles-based approach that recognizes the need to fit the approach to the needs of the organization.

2. It includes a team-level development process similar to Scrum but goes well beyond Scrum in defining an approach for coordinating and integrating the work of multiple teams into an overall Agile Release train.

3. It also recognizes the need for a product development process and an enterprise solution delivery process that goes beyond the typical team-level development found in Scrum and focuses on products and overall solutions.

4. It also defines higher-level, business competencies that are essential for developing an overall business agility approach including

- Lean Portfolio Management
- Organizational Agility
- Lean Agile Leadership.

SAFe® Core Values

SAFe embodies some important core values that are common to most Agile development approaches, but it puts additional emphasis on the importance of these values. The SAFe core values include:

1. **Alignment:** all parts of the organization are well integrated and work together dynamically to achieve an overall business goal.

2. **Built-in Quality:** Quality is not an after-thought. It is integral to the design of the product and the development team is responsible for the quality of the product they produce. It is not some other organization's responsibility to test the product later to ensure that it is a quality product.

3. **Transparency:** Agile requires a collaborative relationship between the development team and the customer based on a relationship of trust and partnership.

4. **Program Execution:** SAFe recognizes the overall importance of delivering results and business value.

Lean Agile Mindset

SAFe requires a mindset shift away from a traditional management mentality to embrace a Lean and Agile mindset that is essential for successful implementation.

Lean Agile Principles

SAFe is based on a number of important principles that are common to most Agile development approaches, but it puts additional emphasis on the importance of these principles. The Lean Agile principles that SAFe is based on include:

1. Take an Economic View.
2. Apply Systems Thinking.

3. Assume Variability; Preserve Options.

4. Build Incrementally with Fast, Integrated Learning Cycles.

5. Base Milestones on Objective Evaluation of Working Systems.

6. Visualize and Limit Work-in-Progress (WIP), Reduce Batch Sizes, and Manage Queue Lengths.

7. Apply Cadence, Synchronize with Cross-domain Planning.

8. Unlock the Intrinsic Value of Knowledge Workers.

9. Decentralize Decision-making.

10. Organize Around Value.

SAFe® Artifacts and Supporting Capabilities

A SAFe implementation normally is based on a number of artifacts and supporting capabilities including:

1. Vision

2. Roadmap

3. Milestones

4. Shared Services

5. Communities of Practice

6. System Team

7. Lean UX

8. Metrics.

DISCUSSION TOPICS

SAFe® Competency Areas

1. How is a SAFe team-level capability similar to Scrum? How does it go beyond Scrum?

2. Why is an appropriate corporate culture important to SAFe? What is the likely impact of not having a well-aligned culture?

3. How does SAFe go beyond a typical development process? Why is that important at an enterprise level?

SAFe® Core Values

1. What do you think is the most important core value in a SAFe implementation? Why?

2. What's different about SAFe Core Values as compared to other Agile development approaches?

Lean Agile Mindset

1. Why is a different mindset needed for implementing SAFe? How is it different from other Agile development approaches? What's the likely impact of not having the right mindset?

SAFe® Lean Agile Principles

1. Do the Lean Agile Principles in SAFe go beyond the principles in the Agile Manifesto? How do they compare to other Agile approaches? Are there any SAFe principles that would not be applicable to any other Agile development process?

SAFe® Artifacts and Supporting Capabilities

1. What is the impact of not having some of the defined SAFe artifacts and capabilities? Why are they important?

NOTES

1. Scaled Agile Framework, © Scaled Agile, Inc., SAFe 5 for Lean Enterprises, https://www.scaledagileframework.com/?_ga=2.62692971.67100464.1647009048-1592209035.1647009048.

2. Scaled Agile Framework, © Scaled Agile, Inc., Continuous Learning Culture, https://www.scaledagileframework.com/continuous-learning-culture/.

3. Scaled Agile Framework, © Scaled Agile, Inc., Team and Technical Agility, https://www.scaledagileframework.com/team-and-technical-agility/.

4. Scaled Agile Framework, © Scaled Agile, Inc., Agile Product Delivery, https://www.scaledagileframework.com/agile-product-delivery/.

5. Scaled Agile Framework, © Scaled Agile, Inc., Lean Portfolio Management, https://www.scaledagileframework.com/lean-portfolio-management/.

6. Scaled Agile Framework, © Scaled Agile, Inc., Organizational Agility, https://www.scaledagileframework.com/organizational-agility/.

7. Ibid.

8. Scaled Agile Framework, © Scaled Agile, Inc., Lean Agile Leadership, https://www.scaledagileframework.com/lean-agile-leadership/.

9. Scaled Agile Framework, © Scaled Agile, Inc., Core Values, https://www.scaledagileframework.com/safe-core-values/.

10. Ibid.

11. Ibid.

12. Ibid.

13. Scaled Agile Framework, © Scaled Agile, Inc., Lean Agile Mindset, https://www.scaledagileframework.com/lean-agile-mindset/.

14. Scaled Agile Framework, © Scaled Agile, Inc., SAFe Lean Agile Principles, https://www.scaledagileframework.com/safe-lean-agile-principles/.

15. Ibid.

16. Ibid.

17. Ibid.

18. Ibid.

19. Ibid.

20. Ibid.

21. Ibid.

22. Ibid.

23. Ibid.

24. Scaled Agile Framework, © Scaled Agile, Inc., Lean Agile Mindset, https://www.scaledagile framework.com/vision/.

25. Scaled Agile Framework, © Scaled Agile, Inc., Roadmap, https://www.scaledagileframework .com/roadmap/.

26. Scaled Agile Framework, © Scaled Agile, Inc., Milestones, https://www.scaledagileframework .com/milestones/.

27. Scaled Agile Framework, © Scaled Agile, Inc., Shared Services, https://www.scaledagileframe work.com/shared-services/.

28. Scaled Agile Framework, © Scaled Agile, Inc., Communities of Practice, https://www.scaled agileframework.com/communities-of-practice/.

29. Scaled Agile Framework, © Scaled Agile, Inc., System Team, https://www.scaledagileframework .com/system-team/.

30. Scaled Agile Framework, © Scaled Agile, Inc., Lean UX, https://www.scaledagileframework .com/lean-ux/.

31. Scaled Agile Framework, © Scaled Agile, Inc., Metrics, https://www.scaledagileframework .com/metrics/.

(23) Disciplined Agile Delivery (DAD®)

THE DISCIPLINED AGILE (DA®) TOOL KIT is an enterprise-level Agile approach from the Project Management Institute (PMI). The PMI describes DA® thus:

> True business agility comes from freedom, not frameworks. Disciplined Agile (DA) is an agnostic, hybrid tool kit that harnesses hundreds of Agile, Lean, and traditional strategies to guide you to the best way of working (WoW) for your team or organization. DA is context sensitive; rather than prescribing a collection of "best practices," it teaches you how to choose and later evolve a fit-for-purpose WoW that is best for you given the situation you face.
>
> The DA tool kit provides straightforward guidance to help organizations streamline their processes in a context-sensitive manner, providing a solid foundation for business agility. It does this by showing how the various business functions such as Finance, Portfolio Management, Solution Delivery (software development), IT Operations, Enterprise Architecture, Vendor Management and many others work together. DA also describes what these business functions should address, provides a range of options for doing so, and describes the trade-offs associated with each option.[1]

It is important to note that DA® is not a framework or methodology and provides choice for enterprises and teams rather than a prescriptive approach. Similar to SAFe®, DA® has evolved over the years from a fairly structured and well-defined approach to a more general set of guidelines and principles that can be used to define a complete enterprise-level Agile approach.

> Note: The material in this chapter is intended only to provide a high-level overview of the Disciplined Agile Delivery (DAD) portion of (DA®), the focus of which is the development of software-based solutions. DA® is very broad and complex, requires considerable judgment and skill and implementation should not be attempted without further training.

DA® LIFE CYCLES

Figure 23.1 shows a general, high-level summary of the DAD life cycle.

DA® supports six different delivery life cycles.[2] These are not really six totally different life cycles, but variations of life cycles that have been optimized for different purposes:

1. **The Agile Life Cycle:** This project life cycle is based on Scrum; however, the activities normally defined in Scrum are extended to recognize additional efforts that would normally be found in a complete enterprise-level solution.[3]

   ■ On the front end of the life cycle, the life cycle adds provision for Product Management and enterprise architecture activities to provide Roadmaps and guidance as well as Portfolio Management to provide the vision and funding for the project.

   ■ On the back end of the project, the life cycle provides additional activities for releasing the solution to production as well as IT Operations and Support which is typically part of DevOps.

2. **The Lean Life Cycle:** The Lean life cycle is based on using Kanban for the development activities rather than Scrum. It is intended for more of a streamlined development effort rather than a full project-oriented development effort that would be more suited to Scrum. This life cycle adds a front end and a back end to the development process similar to Scrum.[4]

3. **The Continuous Delivery: Agile Life Cycle:** The Continuous Delivery life cycle is essentially the same as the Agile life cycle; however, the difference is that the Continuous delivery life cycle is designed around a release to production at the end of each iteration or sprint where releases in the Agile life cycle will typically take more than one iteration or sprint.[5]

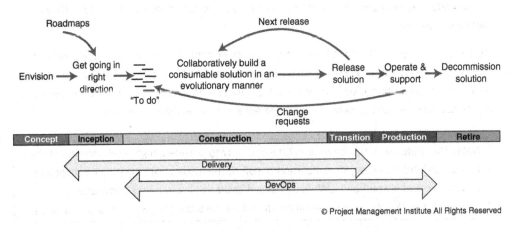

FIGURE 23.1   General Disciplined Agile life cycle.

Source: © Project Management Institute. All Rights reserved.

4. **The Continuous Delivery: Lean Life Cycle:** The Continuous Delivery: Lean life cycle is similar to the Continuous Delivery: Agile life cycle except that the development process is based on a streamlined version of Kanban rather than Scrum.[6]

5. **The Exploratory/Lean Startup Life Cycle:** This life cycle is based on a Lean Startup strategy that uses an iterative approach to rapidly test different hypotheses to better define the solution prior to committing to a full-scale development effort.[7]

6. **Program Life Cycle:** This life cycle is similar to the DA Agile life cycle except that it adds provision for coordinating the work of multiple teams at a program level.[8]

Life Cycle Summary

Figure 23.2 shows the full DA Agile life cycle. The other life cycles follow this general model with some variations.

Table 23.1 shows a summary of the first four life cycles which are primarily variations of each other.

The Exploratory (Lean Startup) life cycle could be used in a situation with a very high level of uncertainty to better understand the solution requirements through an iterative and experimental approach. That life cycle might be used as a front end to one of the other life cycles to reduce the level of uncertainty in the project prior to committing to full-scale development.

The Program life cycle is simply an extension of the standard DA life cycle to support the additional coordination needed in projects large enough to require multiple teams.

DA® ROLES

The foundation of DA® is based on Scrum and some of the roles in DA® are very similar to the roles found in Scrum; however, DA® goes beyond Scrum in more explicitly defining some roles that may not be formally specified in Scrum but are particularly important for large, enterprise-level projects.

Primary DA® Roles

The following are the primary roles in DA®.

Product Owner

The Product Owner role in DA® is essentially the same as the Product Owner role in Scrum although the scope of responsibility in DA® may be greater. "In a system with hundreds or thousands of requirements it is often difficult to get answers to questions regarding the requirements. The product owner is the one individual on the team who speaks as the 'one voice of the customer.'"[9]

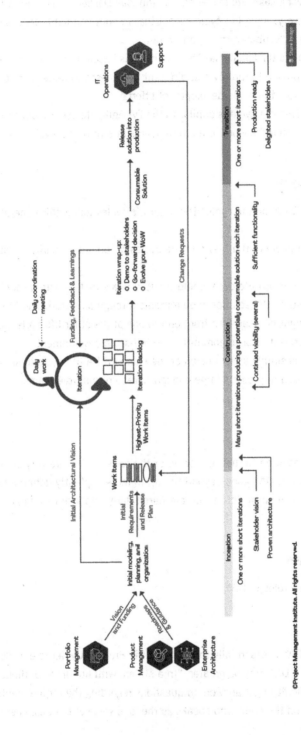

FIGURE 23.2 Full Disciplined Agile life cycle.

Source: © Project Management Institute. All Rights reserved.

TABLE 23.1 Comparison of Disciplined Agile life cycles

Life cycle	Front end	Back end	Development process	Release frequency
Agile	■ Portfolio Mgt ■ Product Mgt ■ Enterprise Architecture	■ Release to Production ■ IT Operations ■ Support	Scrum	Multiple Sprints
Lean	Same	Same	Kanban	Multiple Weeks
Continuous Agile	Same	Same	Scrum	Each Individual Sprint
Continuous Lean	Same	Same	Kanban	Daily

Team Leader

The Team Leader role is essentially the same servant-leader Scrum Master role found in Scrum. However, the idea in Scrum is that all team members are equal, the team is totally self-organizing and there is no need for a formal Team Leader; however, in the real world that is idealistic and not totally realistic.

Defining the role as a Team Lead in DA® is more pragmatic and realistic. In a high-performing team, he/she will probably be more like a Scrum Master; however, in teams at a lower level of maturity, it may be more of a team leadership role.

The PMI state:

An important aspect of self-organizing teams is that the team lead facilitates or guides the team in performing technical management activities instead of taking on these responsibilities him or herself. The team lead is a servant-leader of the team, creating and maintaining the conditions that allow the team to be successful.[10]

Architecture Owner

There is often a need for an architect in Scrum although the role is not formally specified.

DA® just formally recognizes the role since it is particularly critical on large, enterprise-level projects:

Architecture is a key source of project risk, and someone needs to be responsible for ensuring the team mitigates this risk. As a result, DA explicitly includes Agile Modeling's role of architecture owner.

The architecture owner is the person who owns the architecture decisions for the team and who facilitates the creation and evolution of the overall solution design.[11]

Stakeholder

The role of a stakeholder in DA® is essentially the same as the stakeholder role in Scrum; however, DA® explicitly recognizes the importance of a broader range of stakeholders beyond the end-users of the solution. That is particularly important at an enterprise level where a broad range of people might be impacted by the solution.

Team Member

The role of a Team Member in DA® is exactly the same as in Scrum.

Supporting DA® Roles

In addition to the primary roles, DA® recognizes the need for a number of supporting roles. These roles could also be found in a Scrum environment, but DA® makes them explicit because of their importance for enterprise-level projects.

Specialist

An Agile team is typically made up of generalists who provide the most critical skills in the project; however, in any project, there is often a need for specialists for particular skills or tasks, that may not be dedicated full-time to the project team. This is particularly important on large, complex enterprise-level projects.

Domain Expert (or Subject Matter Expert)

According to the PMI:

> The product owner represents a wide range of stakeholders, not just end users, so it isn't reasonable to expect them to be experts in every nuance in your domain, something that is particularly true with complex domains. The product owner will sometimes bring in domain experts to work with the team, for example, a tax expert to explain the details of a requirement or the sponsoring executive to explain the vision for the project.[12]

Technical Experts

The PMI explains:

> Sometimes the team needs the help of technical experts, such as a build master to set up their build scripts, an Agile database administrator to help design and test their database, a user experience (UX) expert to help design a usable interface, or a security expert to provide advice

around writing a secure system. Technical experts are brought in on an as-needed, temporary basis to help the team overcome a difficult problem and to transfer their skills to one or more developers on the team.[13]

Independent Tester

The PMI note: "Although the majority of the testing is done by the people on the DA team themselves, some DA teams are supported by an independent test team working in parallel that validates their work throughout the life cycle."[14]

Integrator

According to the PMI:

> For large DA teams which have been organized into a team of sub-teams, the sub-teams are typically responsible for one or more subsystems or features. The larger the overall team, generally the larger and more complicated the system being built. In these situations, the overall team may require one or more people in the role of integrator responsible for building the entire system from its various subsystems.[15]

DA® MINDSET

The mindset required for DA® consists of principles, promises, and guidelines that are similar to most Agile approaches but they reflect a bit more pragmatic approach, are more explicitly defined, and are particularly appropriate for an enterprise-level environment.

DA® Principles

The following are the principles defined by DA®:[16]

- Delight customers
- Be awesome
- Context counts
- Be pragmatic
- Choice is good
- Optimize flow
- Organize around products and services
- Enterprise awareness.

DA® Promises

DA® promises to do the following:[17]

- Create psychological safety and embrace diversity

- Accelerate value realization

- Collaborate proactively

- Make all work and workflow visible

- Improve predictability

- Keep workloads within capacity

- Improve continuously.

DA® Guidelines

The following are the guidelines defined by DA®:[18]

- Validate our learnings

- Apply design thinking

- Attend to relationships through the value stream

- Create effective environments that foster joy

- Change culture by improving the system

- Create semi-autonomous self-organizing teams

- Adopt measures to improve outcomes

- Leverage and enhance organizational assets.

DA® TOOL KIT

The Disciplined Agile tool kit consists of four layers:[19]

1. **Foundation:** "The Foundation layer provides the conceptual underpinnings of the DA tool kit. This includes the principles, promises, and guidelines of the DA mindset; fundamental concepts from both Agile and Lean; fundamental concepts from serial/traditional approaches; roles and team structures; and the fundamentals of choosing your Way of Working (WoW)."[20]

2. **Disciplined DevOps:** DevOps is the streamlining and integration of the software development process and IT Operations activities associated with the release of the software to a production environment. DAD is part of this layer.

3. **Value Streams:** These are the overall enterprise-level workflows that define the overall business cycle for managing organizational innovation and value realization.

4. **Disciplined Agile Enterprise:** This defines an enterprise-level capability that enables an organization ". . . to sense and swiftly respond to changes in the marketplace. It does this through an organizational culture and structure that facilitates change within the context of the situation it faces."[21]

SUMMARY OF KEY POINTS

DAD® Life Cycles

The Disciplined Agile supports six different life cycle models:

1. The Agile Life Cycle
2. The Lean Life Cycle
3. The Continuous Delivery: Agile Life Cycle
4. The Continuous Delivery: Lean Life Cycle
5. The Exploratory/Lean Startup Life Cycle
6. The Program Life Cycle.

This recognizes the need to choose a life cycle model that is appropriate to the nature of the project.

DA® Roles

The following are the primary roles in DA®:

1. Product Owner
2. Team Leader
3. Architecture Owner
4. Stakeholder
5. Team Member.

In addition, the following supporting roles are recognized in DA® on an as-needed basis:

1. Specialist
2. Domain Expert (or Subject Matter Expert)
3. Technical Experts

4. Independent Tester

5. Integrator.

DA® Mindset (Principles)

The following are the principles defined by DA®:[22]

- Delight customers

- Be awesome

- Context counts

- Be pragmatic

- Choice is good

- Optimize flow

- Organize around products and services

- Enterprise awareness.

DA® Mindset (Promises)

The following are the promises defined by DA®:[23]

- Create psychological safety and embrace diversity

- Accelerate value realization

- Collaborate proactively

- Make all work and workflow visible

- Improve predictability

- Keep workloads within capacity

- Improve continuously.

DA® Mindset (Guidelines)

The following are the guidelines defined by DA®:[24]

- Validate our learnings

- Apply design thinking

- Attend to relationships through the value stream

- Create effective environments that foster joy

- Change culture by improving the system

- Create semi-autonomous self-organizing teams

- Adopt measures to improve outcomes

- Leverage and enhance organizational assets.

DA® Tool Kit

The Disciplined Agile tool kit consists of four layers:[25]

1. Foundation
2. Disciplined DevOps
3. Value Streams
4. Disciplined Agile Enterprise.

DISCUSSION TOPICS

DA® Life Cycles

1. Why do you think that Disciplined Agile supports multiple life cycles? Does that make sense?

2. How would you go about choosing a particular Disciplined Agile life cycle for a project?

3. In general, what is the difference between the Agile variant and a Lean variant of a life cycle?

DA® Roles

1. How are the roles in Disciplined Agile different from a standard Scrum project? Why are they different?

DA® Mindset

1. How are the Disciplined Agile principles different from standard Agile principles? Why are they different?

2. What do you think is the most important of the Disciplined Agile principles? Which principle would be most difficult to follow?

3. Which of the Disciplined Agile promises do you think would be most important? Why?

4. How are the Disciplined Agile guidelines different from guidelines that might be used for an Agile/Scrum project? Why are they different

NOTES

1. PMI Disciplined Agile®, Introduction to Disciplined Agile® (DA™), https://www.pmi.org/disciplined-agile/introduction-to-disciplined-agile.
2. PMI Disciplined Agile, Full Delivery Life Cycles, https://www.pmi.org/disciplined-agile/process/introduction-to-dad/full-delivery-lifecycles-introduction.

3. PMI Disciplined Agile, DAD Life Cycle – Agile (Scrum-based), https://www.pmi.org/disciplined-agile/lifecycle/agile-lifecycle.

4. PMI Disciplined Agile, DAD Life Cycle – Lean, https://www.pmi.org/disciplined-agile/lifecycle/lean-lifecycle.

5. PMI Disciplined Agile, DAD Life Cycle – Continuous Delivery: Agile, https://www.pmi.org/disciplined-agile/lifecycle/dad-lifecycle-continuous-delivery-agile.

6. PMI Disciplined Agile, DAD Life Cycle – Continuous Delivery: Lean, https://www.pmi.org/disciplined-agile/process/introduction-to-dad/full-delivery-lifecycles-introduction.

7. PMI Disciplined Agile, DAD Life Cycle – Exploratory (Lean Startup), https://www.pmi.org/disciplined-agile/lifecycle/exploratory-lifecycle.

8. PMI Disciplined Agile, Program Life Cycle, https://www.pmi.org/disciplined-agile/process/introduction-to-dad/full-delivery-lifecycles-introduction.

9. PMI Disciplined Agile, People First: Roles in DAD, https://www.pmi.org/disciplined-agile/process/introduction-to-dad/people-first-roles-in-dad-introduction.

10. Ibid.

11. Ibid.

12. Ibid.

13. Ibid.

14. Ibid.

15. Ibid.

16. PMI Disciplined Agile, The Disciplined Agile Mindset, https://www.pmi.org/disciplined-agile/mindset.

17. Ibid.

18. Ibid.

19. PMI Disciplined Agile, Introduction to Disciplined Agile (DA™), https://www.pmi.org/disciplined-agile/introduction-to-disciplined-agile.

20. Ibid.

21. Ibid.

22. PMI Disciplined Agile, The Disciplined Agile® Mindset, https://www.pmi.org/disciplined-agile/mindset.

23. Ibid.

24. Ibid.

25. PMI Disciplined Agile, Introduction to Disciplined Agile® (DA™), https://www.pmi.org/disciplined-agile/introduction-to-disciplined-agile.

24 Managed Agile Development Framework

THE MANAGED AGILE DEVELOPMENT FRAMEWORK described in this chapter is different from the Scaled Agile Framework (SAFe®) and the Disciplined Agile® approach. Both of those are designed for full-scale enterprise-level Agile implementations. The Managed Agile Development framework is a project-level framework that is intended to provide a balance of agility combined with some level of predictability and control.

- It is intended for companies that are unable or not ready to move to a more complete top-to-bottom Agile model such as the Scaled Agile Framework.

- It is a hybrid project life cycle model consisting of a blend of an adaptive Agile development approach based on Scrum at the micro-level and a more classical plan-driven approach at the macro-level, as shown in Figure 24.1.

- It can easily be customized to fit a given project and business environment and can be adapted to companies that have more classical plan-driven business and project/portfolio management approaches at a higher level.

- It generally requires no significant transformation of those higher-level processes.

I created this approach initially when I was managing a large government project. In order to meet government contractual requirements, we were required to commit to some plan-driven milestones at the program level, and we were required to report earned-value metrics to measure progress against those milestones. On first glance, it may sound impossible to make an Agile approach work in that environment, but it worked quite well.

Naturally, there are trade-offs between the level of agility and flexibility to adapt to change at the micro-level and the level of predictability and control at the macro-level. It is important that both the client or business sponsor and the development team agree on those trade-offs. The framework

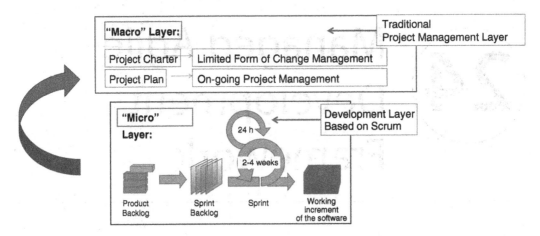

FIGURE 24.1   Managed Agile Development Framework macro-level and micro-level.

Source: http://en.wikipedia.org/wiki/scrum_(development)

provides a mechanism for making those trade-offs by making the macro-level as *thick* or *thin* as you want to fit a given situation.

Instead of having an either-or choice between a fairly rigidly controlled Waterfall approach at one extreme and a completely adaptive approach with very limited or no control over costs and schedules at the other extreme, you can customize an approach that provides the desired balance to fit a broad range of situations. The example process flow shown here is intended for large, complex projects because it is easier to *scale down* a process and to decide to eliminate or minimize activities that are not important to a project than it is to *scale up* and add activities.

MANAGED AGILE DEVELOPMENT OVERVIEW

This framework consists of two layers, as shown in Figure 24.1:

- The macro-level framework is a plan-driven approach, designed to provide a sufficient level of control and predictability for the overall project. It defines the outer envelope (scope and high-level requirements) that the project operates within.

- Within that outer envelope, the micro-level framework utilizes a more flexible and iterative approach based on an Agile Scrum approach designed to be adaptive to user needs.

The combination of these two layers is designed to provide a balance of predictability/control and agility:

- The macro-level process provides a high-level framework for achieving some level of predictability and control in the project.

- The micro-level process provides a more flexible and adaptive approach designed to accelerate the development process and optimize the solution to meet user needs.

These two different levels will need to be synchronized with each other. This framework is intended to be customized and/or scaled up and down to fit particular projects:

- For small, simple projects, the macro-level can be simplified or eliminated.

- Larger, more complex projects may require more emphasis on a plan-driven approach at the macro-level with more detailed requirements.

The Macro-Level

At the macro-level, a Project Charter document is created to define the major deliverables and estimated project milestones based on high-level project objectives. Optionally, a Project Plan could also be created with more detail on the project plan. Once that high-level plan is established, it provides a basis for ongoing management of the project. Changes to those macro-level requirements can be controlled as necessary to manage the overall scope of the project.

The Micro-Level

At the micro-level, requirements are further defined and elaborated in more detail in an iterative approach, with direct participation by the users in an environment that provides more flexibility to adapt to their needs. At a detail level, changes are not finalized until the design has been reviewed and approved by the users at the end of each sprint.

- If the work being done at the micro-level results in a change to the higher-level requirements and plan at the macro-level, those changes are fed back to the macro-level process, and the macro-level plan should be adjusted as necessary for the impact of those changes.

- However, if the process is implemented as it should be, there should be sufficient slack in the macro-level plan to absorb minor changes in requirements at the micro-level so that only significant changes in the micro-level should require replanning at the macro-level.

OBJECTIVES OF MANAGED AGILE DEVELOPMENT

The objectives that this framework is intended to achieve are a combination of the benefits of a classical plan-driven approach with the benefits of a more flexible and adaptive Agile approach.

Plan-Driven Benefits

1. A well-defined process is consistently implemented and is somewhat predictable.
2. High-level milestones provide a framework for managing user expectations and integration with other activities outside the Agile development environment.

3. Scope and cost of projects can be easily and effectively managed.

4. There is improved capacity planning and resource allocation.

5. The process provides a basis for learning and continuous improvement.

Agile Benefits

1. User Satisfaction and Operational Business Results

- Users are engaged more collaboratively in the effort to define requirements.

- There is earlier and more direct feedback on design alternatives from iterations and prototyping.

2. Time-to-Market

- Requirements are ordered and grouped into releases and iterations.

- Efforts can be more concurrent and less sequential where possible.

3. Productivity and Efficiency

- The team is empowered to tailor the process to fit the project.

- The process can be optimized for the project.

- Unnecessary paperwork is reduced (more emphasis on direct team involvement and face-to-face communications).

- The process is a collaborative, cross-functional approach.

Key Differences from a Typical Waterfall Approach

There are several aspects of this framework that are significantly different from a typical Waterfall approach.

1. Partnership with the Business Sponsor and Business Users

The typical Waterfall model is based on a contractual type of relationship between the Business Sponsor/users and the development team. A collaborative, partnership approach is much more suited to an uncertain environment where it is necessary for the customer to take a more active role in working with the project team to define and evaluate the work that is in progress.

- Typically, with the Waterfall model, the business commits to well-defined requirements upfront, and the development organization commits to well-defined schedules, costs, and plans to deliver against those requirements. The Project Manager then takes full responsibility for delivering a solution to meet those agreed documented project requirements.

- In the Managed Agile approach, a high level of understanding of the project scope and high-level requirements exist in the form of user stories at the macro-level. The level of detail in the requirements definition may vary, depending on the nature of the project; however, it is normally not necessary to specify all the details of how a solution will be developed in the project-level planning.

- Factors that might impact the level of detail that goes into the requirements definition in the project-level planning include:

  - the level of uncertainty in the requirements and the level of confidence needed in the cost and schedule estimates
  - the level of risk in the project
  - the need for coordination with other groups outside of the Agile development environment.

Project-level planning should define the project to a sufficient level to develop an estimate of the project costs and schedule to whatever level of accuracy is required (provided, of course, that the desired level of accuracy is realistic and consistent with the level of uncertainty in the requirements).

- The project-level planning in the macro layer simply establishes the outer envelope that the project is expected to operate within.
- It isn't intended to be a rigid contractual agreement. It is essential to have a level of trust and partnership between the business users and the development team to make that work.

Within that envelope established at the macro-level, the Business Sponsor/users and the project team will jointly own responsibility for further defining the details of the solution, as well as optimizing it to make it successful in the micro-level process. This approach will provide more flexibility to adapt to business requirements as the project progresses and provide a much higher level of assurance that the final solution will meet operational business needs with very high levels of quality and user satisfaction.

2. Cross-Functional Team Approach

With a typical Waterfall approach, functional managers might primarily manage the efforts of the functional departments that contribute to the project, such as development and QA. Their efforts are typically sequential, and the Project Manager works to coordinate those efforts into the overall plan. A cross-functional team approach is more suited to maximizing the productivity of the team and encouraging creativity and innovation.

- Implementation of this framework requires more of a cross-functional team approach, where the various functional participants in the team (Developers, Business Analysts, testers, etc.) work concurrently as an integrated team with the business users to jointly define, develop, and test the solution throughout the project.

- The entire project team will also be involved in making collaborative, cross-functional decisions as part of the project-level planning and release-level planning activities, together with the Business Sponsor, Product Owner, and business users.

3. *Rolling-Wave Planning Approach*

With a typical Waterfall methodology, an attempt is made to develop a detailed plan for the entire project at the front end of the project. A rolling-wave planning approach allows for the project plan to be further elaborated as the project is in progress and provides a higher level of flexibility and adaptivity for an uncertain environment.

The Managed Agile Development framework is based on progressive elaboration and rolling-wave planning. Requirements are generally defined only to a high level in the initial project planning, and then requirements are progressively elaborated in more detail as the project progresses. The advantages of that approach are:

- It expedites the upfront project planning process.

- It allows the user to defer decisions about detailed requirements until a point where more information is available to make better decisions. In Agile terminology, this is called the *last responsible moment*, because it is the latest point possible that a decision can be made without having an adverse impact on the delivery of the solution.

- More direct communication with the users minimizes misunderstanding and miscommunication of requirements, and user requirements can be better integrated with the design effort to optimize the design.

- It minimizes the effort required to document requirements along with translation errors that might result from misunderstanding documented requirements. (Note that this approach does not eliminate the need for documented requirements.)

4. *Incremental and Iterative Approach*

With a typical Waterfall methodology, the entire solution is usually designed, developed, and tested sequentially. The disadvantages of that approach are:

- Slow overall development time due to the sequential nature of all activities.

- Feedback on problems and defects may not be discovered immediately.

- Opportunities to get early user feedback and inputs may be limited because users are not involved at all in the development effort to provide feedback, and user acceptance doesn't occur until the very end of the project.

  The advantages of a more incremental and iterative approach are:

- Using a more iterative approach will provide earlier realization of business benefits and greater ROI.

- Using a more iterative approach will provide faster overall development time because more activities can be overlapped and concurrent.

- Defects and problems can be detected and corrected much earlier.

- The user is much more directly engaged in the project as it progresses. This provides direct and immediate feedback and will result in a higher level of assurance that the project will meet user needs.

- Some efforts can be performed concurrently rather than sequentially, based on a risk assessment by the project team. For example, the planning phase for each release can typically be overlapped with the execution phase of the previous release.

FRAMEWORK DESCRIPTION

Project Organization and Work Streams

This approach is dependent on using relatively small teams of approximately 8 to 10 people each to perform the design, development, and testing of the solution.

- For larger projects, the overall project can be broken up into work streams, and each work stream can be assigned to a team with a team leader/Scrum Master.

- Each team will consist of the minimum core team required to design, develop, and test the requirements associated with that particular work stream (Developers, testers, and Business Analysts).

- Specialized resources such as data architects, database developers, and system architects can be centralized as needed to provide support to all work stream teams.

High-Level Process Overview

Each project or work stream will normally be broken into releases and iterations, and each release will typically result in a deliverable product to production, as depicted in Figure 24.2.

- The purpose of breaking the project into releases is to accelerate the delivery of critical features.

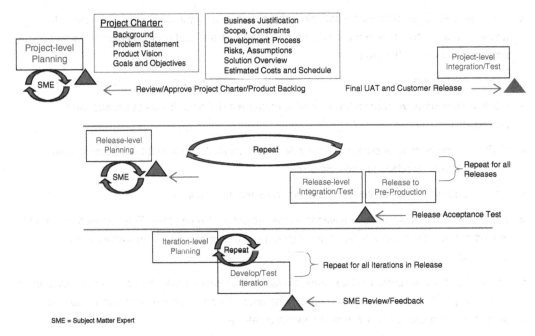

SME = Subject Matter Expert

FIGURE 24.2   Managed Agile Development high-level framework overview

- The features and user stories to be included in each release should be prioritized to deliver the functionality that provides the highest level of value as early as possible.

- That allows critical functionality to be delivered quickly, without waiting for 100% of the total functionality required for the ultimate implementation of the system.

- An important assumption in this is that the deliverables within each release are sufficiently independent of each other that they can be delivered to production in increments.

Project-Level Planning

Project level planning will normally consist of:

- Breaking up the project into work streams if necessary for large projects.

- Organizing the project team and kicking off the project.

- Defining a Project Charter that includes the business objectives to be accomplished, risks, assumptions, and dependencies, and key milestones to be accomplished. See Project Charter Document Template in Appendix C.

- Defining and ordering the Product Backlog required for the project or work stream and developing a high-level estimate of the effort required for each Product Backlog item in terms of story points.

- Tentatively allocating the user stories in the Product Backlog to releases if necessary.

- Developing a high-level plan with resource requirements for completing all of the requirements of that work stream, including identifying any dependencies on shared resources outside of the project team.

- Identifying and resolving any significant issues and uncertainties that must be resolved prior to starting the project and mitigating any significant risks.

Project-level planning will normally be performed once at the beginning of the project, and that plan will be updated as the project progresses. The project-level planning process is shown in Figure 24.3.

Release-Level Planning

Release-level planning will normally consist of tentatively allocating the user stories in the release to iterations, estimating the schedule required to complete each of the iterations required for that release, and resolving any major questions or issues that must be resolved prior to beginning the effort required for that release.

The results of the release-level planning in each project or work stream will normally be fed back into the macro-level process to ensure that the micro-level release-level planning is consistent with macro-level project goals and deliverables.

The release-level planning process is shown in Figure 24.4.

Iteration-Level Process

Each release may be further broken down into iterations. (If a project contains only releases with no iterations, the release-level planning and iteration-level planning will be combined.)

- An iteration is a portion of a release that is segmented from the rest of the release for the purposes of optimizing the delivery of the overall release.

- An iteration normally produces some deliverable functionality that can be demonstrated to the user to show progress and also to get user feedback and inputs; however, an iteration may not be releasable to production.

- The major benefit of segmenting a release into iterations is to provide a mechanism for getting early user feedback and inputs on the results of each iteration.

Fixed-length iterations are preferred because they allow the team to establish a cadence, but iterations may or may not be fixed-length time-boxes. They may be sized to fit the level of effort required to complete the user stories that the team included in that iteration. The length of iterations should be kept relatively short to demonstrate progress and get user feedback quickly.

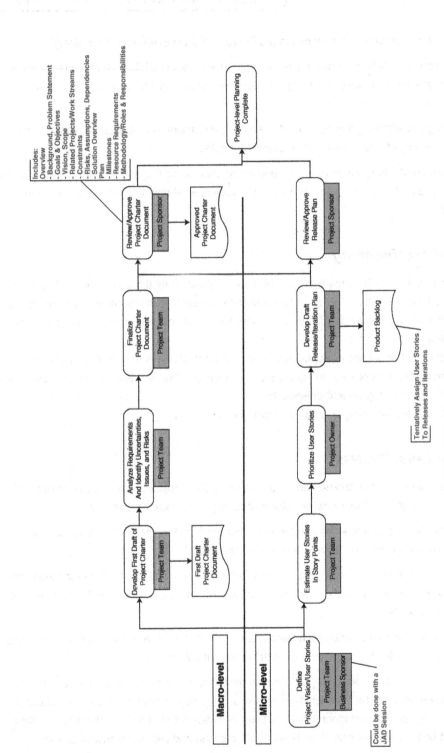

Macro-level

Develop First Draft of Project Charter
— Project Team

First Draft Project Charter Document

Analyze Requirements And Identify Uncertainties, Issues, and Risks
— Project Team

Finalize Project Charter Document
— Project Team

Review/Approve Project Charter Document
— Project Sponsor

Includes:
Overview
 - Background, Problem Statement
 - Goals & Objectives
 - Vision, Scope
 - Related Projects/Work Streams
 - Constraints
 - Risks, Assumptions, Dependencies
 - Solution Overview
Plan
 - Milestones
 - Resource Requirements
 - Methodology/Roles & Responsibilities

Approved Project Charter Document

Project-level Planning Complete

Micro-level

Define Project Vision/User Stories
— Project Team
— Business Sponsor

Could be done with a JAD Session

Estimate User Stories In Story Points
— Project Team

Prioritize User Stories
— Project Owner

Develop Draft Release/Iteration Plan
— Project Team

Review/Approve Release Plan
— Project Sponsor

Product Backlog

Tentatively Assign User Stories To Releases and Iterations

FIGURE 24.3 Project-level planning

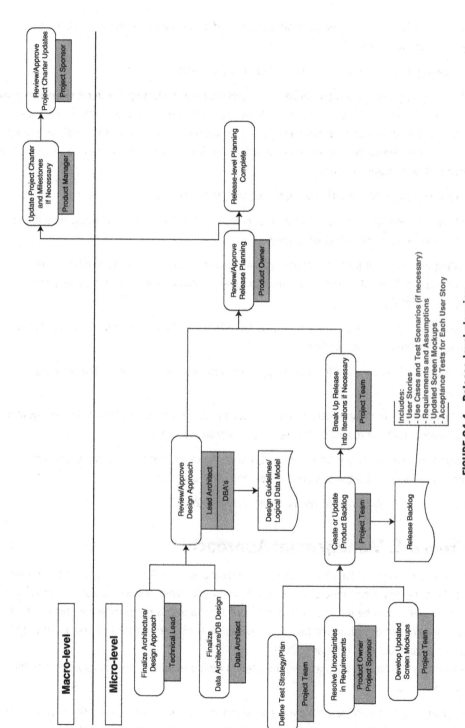

FIGURE 24.4   Release-level planning

The project team will be responsible for planning and performing each iteration. Iteration-level planning will typically consist of:

■ Clearly identifying all user stories to be included in that iteration.

■ Resolving any major questions that need to be resolved prior to starting the iteration and identifying the level of user input required for completing each task during the iteration. If there are significant uncertainties that cannot easily be resolved prior to the start of the iteration, a special iteration or spike should be planned if necessary to resolve those uncertainties prior to beginning the normal development iteration.

■ Defining user acceptance criteria for each user story to be included in the iteration.

■ Identifying the development tasks required for completing that iteration, including testing and allocating the tasks to individual developers.

■ Estimating the effort required for completing each development task in the iteration. The estimate must include all activities required for development and testing of the task (see definition of "done").

The results of each iteration will normally be as fully tested as possible and accepted by the user prior to completion of the iteration:

■ Developers will be responsible for designing, developing, and unit-testing as the software is developed.

■ QA testers who are an integral part of the development team will be responsible for any system testing to verify that the results of the iteration meet requirements.

■ The users will be asked to perform limited user acceptance testing at the completion of the iteration to validate that the results of the iteration meet user needs.

The iteration-level planning process is shown in Figure 24.5.

Requirements Management Approach

The users will be heavily and directly engaged in the development effort as the project progresses to provide direct feedback and inputs. In cases that require a significant amount of user input, a prototyping approach may be used to further define and elaborate user inputs as the design effort progresses. An example of a situation that would require a significant amount of user input might be the design of a graphical user interface (GUI).

A progressive elaboration approach will be used to define requirements in more detail as the project is in progress.

■ At the beginning of the project, a planning session such as a *joint application development* (JAD) session is normally held jointly with the business users, the project team, and all major stakeholders.

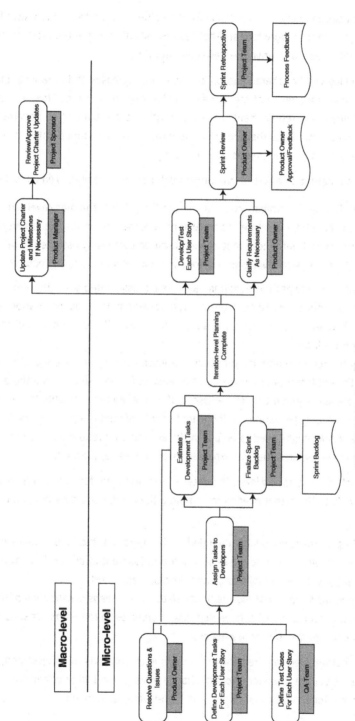

FIGURE 24.5 Iteration-level planning

409

- The result of that session is typically a vision for the solution and a list of high-level features for the solution, usually in user story format. That feature list will become the Product Backlog, which will be used to plan and drive the development effort.

- The Product Backlog will be ordered and broken down into releases and iterations, primarily based on the relative importance of the items in the backlog to the users. However, other factors such as the stability of the requirements associated with the items might be important considerations. (For example, some items that are known to have unstable requirements may be deferred to allow more time for the requirements to stabilize.)

The requirements definition effort for the project will normally be organized as follows:

1. **Project-level:** During the project-level planning of the project, the high-level requirements for the entire project will be identified to the extent necessary to define the requirements at a feature or user-story level. Those requirements will be tentatively assigned to releases.

2. **Release-level:** The release-level planning for each release will primarily focus on:

   - Resolving any outstanding questions and issues associated with the *release backlog* to the extent necessary to start design and development of each iteration in the release. The release backlog is the subset of Product Backlog items included in the current release.

   - Defining the requirements for subsequent releases only to the extent that they might impact the implementation of the current release. For example, during the development of the detailed requirements for release 1, the detailed requirements for releases beyond release 1 will only be defined to the extent that they might impact release 1. As an example, *hooks* might need to be put in the architecture required for release 1 to accommodate features that are expected later in releases 2 and 3.

3. **Iteration-level:** At an iteration level, the detailed requirements for each iteration will be elaborated only to the extent needed to complete the design and development effort for that iteration.

   - Depending on the functionality included in that iteration, the requirements elaboration might be completed prior to the design and development effort, or it might be completed concurrently with the actual design and development effort.

   - *It is important to note that requirements elaboration means clarification of details of how a requirement should be implemented. Major changes to requirements should not be allowed once an iteration is in progress.*

It is assumed that detailed requirements will continue to evolve as the project progresses without formal change control; therefore, formal change management will normally not be done at the micro-level, except for changes that impact the high-level requirements.

- The business user representative (Product Owner) can approve at the micro-level any changes to the detailed project requirements that do not significantly impact the high-level project plan.

- Any change that significantly impacts the high-level scope and direction of the project should be approved by the project sponsor.

Project Scheduling Approach

An implication of breaking the project up into releases and iterations using a rolling-wave planning approach is that, at the onset of the project, only an estimate of the overall project schedule will be known.

- For example, at the beginning of release 1, an estimate of the schedule for completing the release 1 requirements will be made, but normally only a rough estimate of the schedule for subsequent releases and iterations will be available.

- As each release and its iteration are completed, the schedule for subsequent releases and the overall project schedule will be progressively defined and updated.

At the macro-level, an overall project schedule will be developed and maintained throughout the project, but it is understood that because the requirements will be progressively elaborated only at the micro-level within each release and iteration, those estimates of the project schedule are likely to change. Adjustments to the scope of what is included in each release and iteration may be needed to synchronize the deliverables with the macro-level project schedule.

Project Management Approach

The following is a description of the project management approach:

1. The high-level Project Charter document defined at the macro-level will typically define the overall scope and vision for the project and will also include milestones for measuring progress. This high-level Project Charter document defines the *envelope* that the project is expected to operate in; however, it is understood that the scope may change as the project progresses.
2. The high-level Project Charter document will normally be prepared by the Project Manager, with close collaboration with the business user and approved by the project sponsor. Once approved, any significant deviations from these high-level artifacts must be approved by the project sponsor.
3. At the micro-level, a business user representative (Product Owner) and the project team will jointly take responsibility for the execution of the project as long as the project stays within the envelope defined by the high-level Project Charter document. The business user

representative (Product Owner) must be a knowledgeable subject matter expert in the area being developed and must be empowered by the project sponsor to make decisions on how the detailed requirements associated with how the functionality will be implemented. This might be implemented in different ways:

- In the simplest case, the project sponsor, Product Owner, and business user representative could be the same person.
- In a more complex case, you might have a project sponsor who has ultimate approval authority for the project and a Product Owner who has decision-making authority on the details of how the project is implemented.
- You might or might not have business user representatives in addition to the Product Owner to represent different stakeholder needs. (The Product Owner may serve as representing all business needs.)

4. The project team will normally track progress against the high-level milestones at the macro-level and periodically publish status reports to the project sponsor.

    At the macro-level, the major milestones to be tracked might include completion of:

- project-level planning
- release-level planning for each release
- completion of each iteration within the release
- limited UAT for the deliverables for each iteration
- User Acceptance Test (UAT) for the deliverables for each release.
    At the micro-level, the activities to be tracked might include:
- sprint burn-down charts to monitor completion of development tasks and team velocity
- release burn-down charts to monitor completion of development and testing for all project requirements
- test plan progress to monitor progress of completing test cases required by each test plan
- resolution of any bugs.

Communications Approach

Most of the communications within the project team and with business user representatives will rely heavily on direct communications (either face-to-face or remote conferencing). Table 24.1 is a suggested format that can be used for managing project communications.

In large projects requiring multiple work streams, there will naturally be a need for additional weekly meetings to coordinate the efforts of multiple teams.

TABLE 24.1 Format for managing project communications

Meeting	Description	Frequency

Daily Scrums	A Daily Scrum is an Agile practice and is used at the team level within each work stream. The team leader (Scrum Master) responsible for each work stream will facilitate these meetings
	
	■ In a pure Agile environment, the agenda for this meeting is very simple and follows standard Scrum guidelines. Everyone on the team answers three questions:
	
	  ■ What did you accomplish yesterday for the project?
	
	  ■ What are you planning to work on today?
	
	  ■ What obstacles are in your way?
	
	■ In some cases, such as when an offshore development team is involved, there may be a higher need for communications, and going beyond these basic questions may be necessary
	
	■ To keep these meetings short and focused, the rule should normally be not to attempt to resolve issues and questions in this meeting unless they are very simple things to resolve. If anything comes up that requires a significant amount of discussion, another meeting should be scheduled to discuss it outside of the Scrum.
		Daily

Scrum meetings	The micro level will include normal Scrum meetings such as Sprint Planning, Sprint Review, and Sprint Retrospective	
Other project meetings	Other meetings will be scheduled as necessary to resolve specific issues that cannot be resolved in the Daily Scrum meetings.	As needed
Weekly status meeting	This will typically be a weekly meeting to review project status with the project sponsor. This meeting is primarily focused on reviewing macro-level progress with the project sponsor, where the other meetings are more at the micro-level. For small projects, this meeting might not be necessary.	Weekly

ROLES AND RESPONSIBILITIES

Table 24.2 is a suggested description of the major roles and responsibilities of the key participants in this process. Note that some of these roles may be combined in actual practice. These roles and responsibilities are intended to be customized as necessary for a given business and project environment.

TABLE 24.2   Roles and responsibilities of key participants

Role	Responsibility
Business Sponsor	The business sponsor has the ultimate responsibility for the success of the project or program from a business perspective, including: ■ Providing direction on business objectives that the project or program must achieve to maximize the benefits of the project or program to the business ■ Ensuring that the appropriate business personnel are fully engaged in defining and prioritizing requirements for deliverables ■ Reviewing and approving all deliverables prior to implementation ■ Resolution of any issues that cannot be resolved by the project team
Business Process Owner(s)	Business process owners are responsible for the execution of the business processes that are affected by the project. Business process owners are responsible for: ■ Planning and implementing any business process changes that may be required by the solution ■ Ensuring that the solution is consistent with new business processes ■ Planning the cutover of the solution to ensure that any business process changes and other important requirements outside of the development effort, such as user training, are synchronized with the implementation of the solution
Subject Matter Expert (SME)	Subject matter experts provide domain-specific knowledge in an area that is relevant to the project, including: ■ Representing user needs ■ Clarifying project requirements as necessary ■ Reviewing and signoff on business-related project documents

TABLE 24.2 *(Continued)*

Role	Responsibility
Stakeholder	A stakeholder is a person, group, organization, or system that affects or can be affected by an organization's actions. A stakeholder is identified as someone who has a direct or indirect interest in a project. Stakeholders can range from the project sponsor to an end user, and a stakeholder can be any person who can affect or be affected by the products of a project, either during the project or after the project has been completed. Stakeholders are responsible for:

- Representing their area of interest and providing inputs to the project requirements and project planning effort to ensure that there is no unexpected impact to their area of interest

- Ensuring that the solution is consistent with the requirements in their area of interest and validating that the solution has no unexpected impact during the testing and implementation of the solution

- Stakeholders would also include "passive stakeholders" such as regulatory authorities or government who do not directly participate in the project but might need to be managed by the project to keep them satisfied

Role	Responsibility
Project Manager	The Project Manager has overall responsibility for the success or failure of the project and is responsible for:

- Leading the development of an overall Project Charter and developing and implementing the project plan for the project

- Integrating the activities within the scope of the project into the overall project plan and ensuring that they are well aligned with the project's business objectives

- Tracking and reporting the status of all activities within the scope of the project

- Resolving issues or escalating any issues that cannot be resolved within the project manager's responsibility

- Taking full responsibility for the management of their assigned project(s) in accordance with any relevant processes, including:

- Monitoring and guiding each project through to completion, using specific techniques and procedures to establish the framework and structure of the project

(continued)

TABLE 24.2   *(Continued)*

Role	Responsibility
Project Manager	■ Keeping project stakeholders informed and involved in project decisions
	■ Anticipating changes required in project plans and processes and recommending alternative approaches if necessary
	■ Considering impacts to and from other projects and coordinating the planning and implementation of their project(s) with other project teams via project managers as necessary
	■ Taking the initiative to achieve value-added results, within scope of the PM responsibility
	■ Leading the project team in taking accountability for work products and ensuring quality and timely delivery of end results
Scrum Master	The Scrum Master:
	■ Serves the team, not the reverse
	■ Is responsible for maintaining Scrum values and practices
	■ Facilitates most meetings
	■ Removes impediments
	■ Tracks metrics (i.e. burn-down chart)
	■ Communicates with management (status, impediments)
	■ Shields the team from external interferences
Product Owner	The Product Owner:
	■ Creates and maintains the Product Backlog
	■ Prioritizes and sequences the backlog according to business value or ROI
	■ Assists with the elaboration of epics, themes, and features into user stories that are granular enough to be achieved in a single sprint
	■ Conveys the vision and goals at the beginning of every release and sprint
	■ Represents the customer; interfaces and engages with all customer stakeholders
	■ Participates in the daily Scrums, sprint planning meetings, and sprint reviews and retrospectives
	■ Inspects the product progress at the end of every sprint and has complete authority to accept or reject work done

TABLE 24.2   (*Continued*)

Role	Responsibility
Product Owner	■ Can reorder and redefine the Product Backlog at the end of every sprint to reflect additions and changes that do not impact the macro-level scope of the project. (Significant changes that impact the macro-level scope may need to be approved by the project sponsor.) ■ Terminates a sprint if it is determined that a drastic change in direction is required
Business Analyst	The Business Analyst (if required) is the primary resource for the day-to-day efforts of eliciting, analyzing, documenting, and validating the business requirements. The Business Analyst: ■ Analyzes and scopes the project solution and works with Project Managers, Product Owner, and Business Sponsors to clarify the level and complexity of the business analysis effort needed for the project. (Note that on some projects, the Business Analyst may also play the role of the Product Owner.) ■ Selects the appropriate elicitation technique to efficiently identify critical requirements and asks the right questions through the use of interviewing techniques developed specifically for business analysis elicitation ■ Plans an approach for analyzing, categorizing, and managing requirements; determines the level of formality required and considers options for documenting and packaging requirements based on project type, priorities, and risks ■ Analyzes and refines business and functional requirements from a business perspective, identifying any issues and questions that must be resolved and verifying that requirements are testable ■ Builds strong relationships with project stakeholders and conducts effective requirements reviews to improve the quality of requirements deliverables ■ Anticipates issues, thinking proactively and using critical-thinking skills to plan stakeholder elicitation sessions ■ May also perform the role of a Business Systems Analyst ■ Analyzes and refines business and functional requirements from a systems perspective, identifying any issues and questions that must be resolved and verifying that requirements are testable

SUMMARY OF KEY POINTS

Overview of the Managed Agile Development Framework

The Managed Agile Development Framework is a project-level framework that is intended to provide a balance of agility combined with some level of predictability and control. It is a hybrid Agile approach that is intended for companies that prefer a hybrid management approach over a pure Agile approach or are unable or not ready to move to a more complete top-to-bottom Agile model such as the Scaled Agile Framework.

It consists of two layers:

1. **Macro-level:** The macro-level framework is a plan-driven approach, designed to provide a sufficient level of control and predictability for the overall project. It defines the outer envelope (scope and high-level requirements) that the project operates within.
2. **Micro-level:** Within that outer envelope, the micro-level framework utilizes a more flexible and iterative approach based on an Agile Scrum approach designed to be adaptive to user needs.

These two different levels will need to be synchronized with each other. This framework is intended to be customized and/or scaled up and down to fit particular projects.

Objectives of the Managed Agile Development Framework

The objectives that this framework is intended to achieve are a combination of the benefits of a traditional plan-driven approach with the benefits of a more flexible and adaptive Agile approach.

1. **Plan-driven benefits**

   - A well-defined process is consistently implemented and is somewhat predictable.
   - High-level milestones provide a framework for managing user expectations and integration with other activities outside the Agile development environment.
   - Scope and cost of projects can be easily and effectively managed.
   - There is improved capacity planning and resource allocation.
   - The process provides a basis for learning and continuous improvement.

2. **Agile benefits**

   - **User satisfaction and operational business results**

     - Users are engaged more collaboratively in the effort to define requirements.
     - There is earlier and more direct feedback on design alternatives from iterations and prototyping.

   - **Time-to-market**

     - Requirements are ordered and grouped into releases and iterations.
     - Efforts can be more concurrent and less sequential where possible.

- **Productivity and efficiency**

  - Team is empowered to tailor the process to fit the project.
  - The process can be optimized for the project.
  - Unnecessary paperwork is reduced (more emphasis on direct team involvement and face-to-face communications).
  - The process is a collaborative, cross-functional approach.

Key Differences from a Typical Waterfall Approach

1. Partnership with the business sponsor and business users

- The typical Waterfall model is based on a contractual type of relationship between the business sponsor/users and the development team.
- A collaborative, partnership approach is much more suited to an uncertain environment where it is necessary for the customer to take a more active role in working with the project team to define and evaluate the work that is in progress. However, this does not preclude having some level of documented contractual relationship.

2. Cross-functional team approach

- With a typical Waterfall approach, functional managers might primarily manage the efforts of the functional departments that contribute to the project, such as development and QA.
- Their efforts are typically sequential, and the Project Manager works to coordinate those efforts into the overall plan.
- A cross-functional team approach is more suited to maximizing the productivity of the team and encouraging creativity and innovation.

3. Rolling-wave planning approach

- With a typical Waterfall methodology, an attempt is made to develop a detailed plan for the entire project at the front end of the project.
- A rolling-wave planning approach allows for the project plan to be further elaborated as the project is in progress and provides a higher level of flexibility and adaptivity for an uncertain environment.

4. Incremental and iterative approach

- With a typical Waterfall methodology, the entire solution is usually designed, developed, and tested sequentially in one contiguous effort.
- Breaking up the project into incremental iterations with customer feedback at the end of each iteration provides a higher level of confidence that the overall solution will meet customer needs and can also provide for early release of value.

High-Level Process Overview

Each project or work stream will normally be broken into releases and iterations, and each release will typically result in a deliverable product to production. The purpose of breaking the project into releases is to accelerate the delivery of critical features.

1. Project-level planning

Project-level planning will normally consist of:

- Breaking up the project into work streams if necessary for large projects.
- Organizing the project team and kicking off the project.
- Defining a Project Charter that includes the business objectives to be accomplished, risks, assumptions, and dependencies, and key milestones to be accomplished. Defining and ordering the Product Backlog required for the project or work stream and developing a high-level estimate of the effort required for each Product Backlog item in terms of story points.
- Tentatively allocating the user stories in the Product Backlog to releases if necessary.
- Developing a high-level plan with resource requirements for completing all the requirements of that work stream, including identifying any dependencies on shared resources outside of the project team.
- Identifying and resolving any significant issues and uncertainties that must be resolved prior to starting the project and mitigating any significant risks.

2. Release-level planning

Release-level planning will normally consist of:

- Tentatively allocating the user stories in the release to iterations, estimating the schedule required to complete each of the iterations required for that release.
- Resolving any major questions or issues that must be resolved prior to beginning the effort required for that release.

The results of the release-level planning in each project or work stream will normally be fed back into the macro-level process to ensure that the micro-level release-level planning is consistent with macro-level project goals and deliverables.

3. Iteration-level process

Each release may be further broken down into iterations. An iteration normally produces some deliverable functionality that can be demonstrated to the user to show progress and also to get user feedback and inputs; however, an iteration may not be releasable to production. Iteration-level planning will typically consist of:

- Clearly identifying all user stories to be included in that iteration.
- Resolving any major questions that need to be resolved prior to starting the iteration and identifying the level of user input required for completing each task during the iteration.

If there are significant uncertainties that cannot easily be resolved prior to the start of the iteration, a special iteration or spike should be planned if necessary to resolve those uncertainties prior to beginning the normal development iteration.

- Defining user acceptance criteria for each user story to be included in the iteration.
- Identifying the development tasks required for completing that iteration, including testing and allocating the tasks to individual developers.
- Estimating the time required for completing each development task in the iteration in hours. The estimate must include all activities required for development and testing of the task (see definition of "done").

Requirements Management Approach

The users will be heavily and directly engaged in the development effort as the project progresses to provide direct feedback and inputs. A progressive elaboration approach will be used to define requirements in more detail as the project is in progress.

Project Scheduling Approach

- An implication of breaking the project up into releases and iterations using a rolling-wave planning approach is that, at the onset of the project, only an estimate of the overall project schedule will be known.

- As each release and iteration is completed, the schedule for subsequent releases and the overall project schedule will be progressively defined and updated.

Project Management Approach

The following is a description of the project management approach:

1. The high-level Project Charter document defined at the macro-level will typically define the overall scope and vision for the project and will also include milestones for measuring progress. This high-level Project Charter document defines the *envelope* that the project is expected to operate in; however, it is understood that the scope may change as the project progresses.
2. At the micro-level, a business user representative (Product Owner) and the project team will jointly take responsibility for the execution of the project as long as the project stays within the envelope defined by the high-level Project Charter document.
3. The project team will normally track progress against the high-level milestones at the macro-level and periodically publish status reports to the Project Sponsor.

Communications Approach

Most of the communications within the project team and with business user representatives will heavily rely on direct communications (either face-to-face or through remote conferencing).

DISCUSSION TOPICS

Overview of the Managed Agile Development Framework

1. How would you go about adjusting the Managed Agile Development approach to tailor it to provide either more or less planning and control?

2. What are the layers in the Managed Agile Development framework and how are they interrelated?

Objectives of the Managed Agile Development Framework

1. What are the benefits of the Managed Agile Development framework from a plan-driven perspective?

2. What are the benefits of the Managed Agile Development framework from an Agile perspective?

Key Differences from a Typical Waterfall Approach

1. What are the key differences of the Managed Agile Development framework from a typical Waterfall approach?

2. What value do they provide?

High-Level Process Overview

1. What are the levels of planning in the Managed Agile Development approach and how are they integrated into an overall planning approach?

2. What does each level consist of?

Requirements Management Approach

1. How does the general requirements management process work? How far in advance are requirements planned?

2. How would you handle a situation where the requirements are very uncertain?

3. What happens if requirements change as the project is in progress?

Project Scheduling Approach

1. How is project scheduling handled both upfront prior to the start of the project and as the project is in progress?

Project Management Approach

1. What is the project management approach? How are the different levels of the project integrated into an overall project management approach?

Communications Approach

1. What is the project communications approach? How is it different from a typical Waterfall project?

25 Summary of Enterprise-Level Frameworks

THE THREE ENTERPRISE-LEVEL FRAMEWORKS DISCUSSED in previous chapters all provide a way of bridging the gap between a team-level Agile development process based on Scrum and the higher-level management practices found in many typical businesses.

HIGH-LEVEL COMPARISON

Table 25.1 shows a brief comparison of the three frameworks.

TABLE 25.1 Comparison of enterprise-level frameworks

Framework	Pros	Cons
Scaled Agile Framework (SAFe®)	■ More of a complete top-to-bottom Agile approach	■ Requires a more significant transformation to Agile ■ May not be appropriate for many companies
Disciplined Agile Delivery Tool Kit (DA®)	■ Based on a standard Agile development process (Scrum) or lean ■ Provides extensions to the Agile development process for scaling to an enterprise level	■ Not a complete enterprise-level framework but is easily adaptable to existing higher-level management levels

(*continued*)

TABLE 25.1   (*Continued*)

Framework	Pros	Cons
Managed Agile Development Framework	■ Hybrid process with a blend of traditional plan-driven and Agile approaches ■ Minimizes the level of organizational transformation needed in organizations with a traditional management structure (could be done as an interim step in an Agile transformation)	■ Limited to project-level layer only ■ Not a complete enterprise-level framework but is easily adaptable to existing higher-level management levels

HOW THESE FRAMEWORKS HAVE EVOLVED

It's very interesting to look at how these frameworks have evolved since the original version of this book was published in 2015. In 2015, both the Scaled Agile Framework (SAFe®) and Disciplined Agile (DA®) were much more structured and well-defined than they are today.

Rather than defining a somewhat prescriptive approach to implementing the framework, both have shifted to an emphasis on principles and leave it up to the organization to determine how to use those principles to tailor the approach to fit their business.

I think that is the right thing to do but it obviously creates a challenge for anyone implementing one of these frameworks. Instead of implementing a fairly well-defined framework, it requires the organization to use the principles to determine an appropriate approach for their business. Naturally, that requires a higher level of skill.

DISCUSSION TOPICS

DA® and SAFe®

1. How would you go about choosing between Disciplined Agile® and SAFe®? What are the most important considerations? What are the advantages and disadvantages of each?

Managed Agile Framework

1. When and why would you use the Managed Agile Development approach and how would it fit into an overall enterprise solution?

PART 6

Case Studies

IN ANY BOOK OF THIS nature, it's always useful to go beyond theory and concepts and show how companies have actually put these ideas into practice in the real world. Of course, there is no canned, ready-made approach that works for all companies—each of these case studies is different and shows how a different approach may be needed in different situations. Part 6 also includes a chapter on "Not-So-Successful" case studies, which shows some of the problems that can develop in an Agile implementation.

Chapter 26 – "Not-So-Successful" Case Studies:

You can learn just as much or more from situations that don't work very well as you can from more successful implementations.

Agile is a very difficult thing to do and is based on the principle of "Fail early, fail often." For that reason, these case studies should be regarded as learning opportunities rather than failures.

There's another saying that I really like: "If you have never failed at something, you're not trying hard enough." This chapter contains some case studies of companies that were not so successful in implementing Agile.

Chapter 27 – Case Study: Valpak:

Valpak is an example of a company that has done a major transformation of its whole business around a highly Agile approach using the scaled Agile architecture. It illustrates how a company has successfully scaled Agile principles and practices to an enterprise level and how it has begun to thoroughly integrate Agile into many aspects of their business.

Chapter 28 – Case Study: Harvard Pilgrim Health Care:

Harvard Pilgrim Health Care (HPHC) is the only private health plan in the nation to be named #1 for member satisfaction and quality of care for nine consecutive years.

HPHC was faced with a major redesign of its architecture to move to a service-oriented architecture, which involved over 200 different projects that had to be completed within a five-year period without disrupting its quality ranking. It was further complicated by the fact that most of HPHC's development resources were outsourced to another company.

Because of the scope and complexity of this effort, a hybrid Agile approach was needed to provide a blend of agility and control to most effectively manage the overall effort.

Chapter 29 – Case Study: General Dynamics, UK:

The General Dynamics, UK's effort demonstrates how an Agile development approach can be applied to a large and complex government program. Because of the need to manage costs and schedules of the overall program to meet government contracting requirements, a hybrid Agile approach (DSDM) was used.

Chapter 30 – Agile Hardware Development:

Much has been said about applying Agile principles and practices to software development, but comparatively little has been said about Agile hardware development. Joe Justice has been a pioneer in the area of applying Agile to hardware projects and has graciously given me permission to summarize some of his materials in this chapter. This entire chapter is based on Joe Justice's work. It also includes some material on the implementation of this process at Tesla.

Chapter 31 – Non-Software Case Studies:

Many people think of Agile as only applicable to a software development environment. While Agile is widely used in a software development environment, there is no reason why the principles behind Agile cannot be used in non-software projects. This chapter provides a summary of several examples of the use of Agile for non-software projects. These examples don't necessarily use Scrum, but show how to use common-sense Agile principles for everyday projects.

Chapter 32 – Overall Summary:

This chapter provides an overall summary of some of the most important points in the book, plus some recommendations and tips for putting these ideas into practice.

26 "Not-So-Successful" Case Studies

YOU CAN LEARN JUST AS MUCH OR MORE from companies that have attempted to implement an Agile approach and failed or where it has not been completely successful. *Failure* in Agile is seen only as a learning experience and is encouraged—without failure, there probably would not be much learning. Here are some quotes on the subject of failure:

Winston Churchill:

- Success is the ability to go from one failure to another with no loss of enthusiasm.[1]

  Thomas Edison:

- I have not failed 10,000 times. I have not failed once. I have succeeded in proving that those 10,000 ways will not work. When I have eliminated the ways that will not work, I will find the way that will work.[2]

- Show me a thoroughly satisfied man, and I will show you a failure.[3]

  Joan Collins:

- Show me a person who has never made a mistake, and I'll show you someone who has never achieved much.[4]

  Well-known Agile mantra:

- Fail early, fail often.

The examples given in this chapter are companies that have had problems implementing an Agile approach or the implementation has been incomplete. These are real companies; however, naturally, the companies are anonymous.

- Agile can be a very difficult thing to do if a significant amount of cultural change is required.

- It is also a very empirical process, which means sometimes you have to try things to see what works and then make adjustments and corrections.

For that reason, these case studies should be regarded as learning opportunities and not failures. There's another well-known saying that is appropriate here: "If you have never failed at something, you're not trying hard enough." These companies should be applauded for trying to implement Agile and, in many cases, they were ultimately successful after an initial false start. Tables 26.1–26.17 detail the problems and the solutions.

COMPANY A

Background

Company A was a mid-sized IT organization. The company embarked on an Agile implementation and trained most of the IT application development staff of about 80–100 people in Agile practices. An Agile Coach was brought in for over a year to provide coaching to the teams, and the company made some progress on implementing an Agile process at the development team level; however, due to cost-cutting pressure, the Agile Coach was let go, and there was little or no support at the executive level to take the Agile process to the next level.

The Approach

About 80–100 people in the organization were trained in Agile practices, and, at least at a mechanical level, a number of Agile practices such as Daily Standups were being implemented. However, the scope of the effort was very limited to the development organization; the approach was fairly mechanical without an understanding of the principles behind it and without much of an attempt to fit the approach into the company's business environment.

What Went Wrong

Overall Conclusions

This is a great example of an Agile implementation where the company was looking for a "quick hit" and didn't follow through enough to fully develop their Agile approach. Agile is a journey and can require some significant organizational change to fully implement it. It is not an all-or-nothing proposition and sometimes requires a hybrid approach that is designed to fit the business environment. In this case, an Agile Coach was brought in and was successful in developing a foundation of basic Agile practices, but it just didn't go far enough.

TABLE 26.1 Company A: what went wrong: Agile is not just a development process

Problem	Solution
The company's senior executives saw the Agile process as having significant benefits to make IT development go faster; however, they saw it as an IT development process only and didn't see the benefits of investing beyond that level On the surface, the development effort did appear to go faster, but the truth is that people were overworked to make it go faster, and the quality of the products really suffered as a result	Many times companies see Agile as a "silver bullet" that is going to make development go faster; they see it as an opportunity for a quick and easy "win" and not as more than a development process This often results in a partial implementation of Agile that doesn't take full advantage of the benefits it can provide It is often necessary to start out with a limited implementation of Agile to get started with, but it's important to set everyone's expectations early on that it is only a start and much more follow-on effort will be required to fully realize the benefits

TABLE 26.2 Company A: what went wrong: commit resources to teams

Problem	Solution
The company implemented many of the "mechanical" aspects of an Agile development process (e.g. Daily Standups were held), but there was no real change in the way people were assigned to teams People were not dedicated to teams and might be assigned to as many as three to four different teams In some cases, the Developers didn't participate directly in the teams and were represented on the team by their managers This was clearly not consistent with a true Agile approach. Without dedicated people on teams, it was almost impossible to stabilize the velocity of the teams and accurately predict performance	This is, unfortunately, a common practice. Companies implement a few Agile practices and call it Agile, but it really is a very limited implementation of Agile On the surface, it looks "Agile" because some of the Agile rituals like Daily Standups are being followed, but it may be only superficial and not really consistent with the real principles behind Agile. It is very difficult, if not impossible, to make an Agile development process work effectively if the majority of the people on the team are not dedicated to that team and don't even participate directly in it The problem in many cases is that the company sees Agile as a development process that only impacts the IT Development organization; focuses on the mechanical implementation level without understanding the principles behind it; and never follows through with a more complete organizational transformation to make it really work

TABLE 26.3   Company A: what went wrong: change is essential

Problem	Solution
There was no fundamental change in the way the company handled requirements A separate product management group was responsible for producing a business requirements document (BRD) and handing it off to the development team The process was not very collaborative and it was cumbersome. The product management group insisted on control of the requirements and became a "middleman" between the development team and the business users for clarification of requirement details	The company operated in a highly regulated environment within the financial services industry, and the control being exercised by the product management group had been seen as essential to tightly control and manage the product development effort However, control is not an "all-or-nothing" proposition. There are lots of ways to implement an effective level of control over an Agile project without over-controlling it. The Managed Agile Development framework described in this book is an example

COMPANY B

Background

Company B is a mid-sized IT services company. The company has been rapidly developing a software application development business for their clients and the business has experienced significant growth. The company has used a Waterfall process to manage its software development projects and bid fixed-price software development projects. The typical implementation of the process consisted of two steps:

1. There is typically a fixed-price effort to do the planning and design phase to define detailed requirements and design specifications for the project.
2. Following the fixed-price planning and design phase, another fixed-price project is proposed to complete the development and testing phase of the software solution.

Recently, Company B experienced significant problems with one of its largest customers. A critical project with this customer was completely stalled. Company B had accepted a fixed-price contract for delivery of the software based on some incomplete requirements and it wasn't making any significant progress, because Company B's development team was largely idle waiting for requirements to be further defined by the customer.

The customer had made a commitment to complete an initial installation of the software in several months and recognized that it needed to take charge of the situation to get the project moving. The customer replaced their project staff that was directing the project and brought in a new Project Manager and a number of Business Analysts to accelerate the requirements definition process.

The Approach

After the customer replaced their project staff, the customer gave an ultimatum to Company B that the initial installation deadline still had to be met in spite of the earlier delays in completing the requirements. The customer then laid out a plan consisting of a number of sprints that it wanted Company B to meet in order to hit the initial installation date. The following are some of the most significant characteristics of this effort:

- Both the customer and Company B recognized that a more Agile approach was needed to make progress; however, neither Company B nor the customer had any significant experience with implementing an Agile software development approach.

- The methodology that was ultimately implemented was not really Agile or Waterfall—it was really just a brute force effort to get the work done in order to hit the deadline. It was similar to breaking up the overall project into a series of *mini-waterfalls*, each being about two weeks in length that were called *sprints*.

- The customer created a development schedule for completing the project based on what they thought was needed to meet the installation schedule and broke up the functionality into sprints based on their estimates of the level of effort. Company B's development team was not directly involved in those estimates.

- Company B was pressured into making a fixed-price commitment for completing the project by the scheduled delivery date with performance penalties for missing the deadline based only on a very high-level understanding of the requirements.

- The Business Analysts who were brought into the project by the customer to accelerate the requirements definition effort worked with the customer's business stakeholders to create use cases and user stories to document the requirements and there was a very limited amount of direct communication with Company B's development team during that process.

- Prior to the beginning of each sprint, the Business Analysts who represented the customer turned over approved requirements documents in the form of use cases and user stories to Company B to be implemented in the next sprint.

- A limited amount of integration testing and user acceptance testing was included in each sprint due to the time pressures to get the work completed on time, and some time was reserved at the end of the project for doing final integration and testing prior to release.

What Went Wrong

In situations like this when there is a project failure, there is a tendency to take a brute force approach to just put pressure on the situation to make the project work, rather than getting down to the root cause of some of the problems and taking a more systemic approach to address the core

TABLE 26.4 Company B: what went wrong: contracting

Problem	Solution
From a business management perspective, Company B had not adequately recognized the risks and uncertainties associated with software application development projects and had not developed an effective business management approach for managing those risks and uncertainties	More of a collaborative partnership relationship with the customer to develop a mutual understanding of the risks would have been a better approach
Company B wanted the business with this customer very badly and very aggressively over-committed to meet a fixed date with incomplete requirements.	

issues that tend to make any project successful. In this case, there were four systemic issues that needed to be addressed, which are summarized in Table 26.4 in the following four general areas:

■ Project Governance

■ Process

■ People

■ Tools.

In the end, the project turned into a "death march" project to meet a firm delivery date that had been committed to with very incomplete requirements and, that in itself, was very problematic.

Overall Conclusions

This is a great example of a project that was failing, and because of the time urgency of meeting a deadline, a brute force effort was initiated to get the project moving without taking the time to take a more systemic approach to address the core issues that were causing the project to fail.

This is also a perfect illustration of the need for an Agile Project Management approach. In this particular situation, a pure Agile approach would not have provided much confidence of meeting the dates the project had to meet and a hybrid approach was needed that provided some level of predictability and control over the costs and schedule of the effort, blended with a more Agile approach for further elaborating the detailed requirements as the project progressed.

It also indicates the importance of a collaborative spirit of trust and partnership between the customer and the service provider to break down barriers to allow the project to work much more efficiently based on direct communications rather than an arm's-length contractual relationship.

TABLE 26.5 Company B: what went wrong: project governance

Problem	Solution
From a project management perspective, Company B had a PMO that was used for managing other types of projects but the PMO was not heavily involved in software application development projects and had no Project Managers who were trained and experienced in managing software development projects A senior-level Solution Architect and a Technical Director were used to manage the effort for Company B. A manager was assigned to play a supporting role to handle some project administration and reporting.	Company B needed to redefine their management approach for managing software application development projects; however, given the time required by the customer to complete this particular project, it was not possible to do a reset and have a significant impact on the approach for this particular project As a result, Company B had to take a brute force approach to get it done without a well-defined methodology A better solution would have been to take an incremental approach to improving the management process as the project proceeded That approach would have consisted of using an Agile approach to identify and prioritize potential areas for improvement and implementing the most critical actions that could be done without significantly disrupting the project as it was in progress The software development effort was seen primarily as an effort that needed to be led by senior-level Developers—Project Managers were seen primarily as administrators who were more heavily associated with plan-driven, Waterfall-style projects

TABLE 26.6 Company B: what went wrong: project management

Problem	Solution
A senior-level Solution Architect and a Technical Director were used to manage the effort for Company B. A manager was assigned to play a supporting role to handle some project administration and reporting	This is a great example of how the need for effective project management in a software development effort is often overlooked Company B did not fully understand the concept of Agile Project Management and how to better define the project roles to support an Agile Project Management approach

(continued)

TABLE 26.6   (*Continued*)

Problem	Solution
The software development effort was seen primarily as an effort that needed to be led by senior-level Developers—Project Managers were seen primarily as administrators who were more heavily associated with plan-driven, Waterfall-style projects	In order to develop an effective approach, the roles in providing overall project management needed to be much more clearly defined In an Agile project, some of the project management functions are distributed among the members of the team; however, in this particular situation, the team was not at that level of maturity; and, even if it was, that wouldn't necessarily eliminate the need for a defined project management role in managing projects of this nature

TABLE 26.7   Company B: what went wrong: process

Problem	Solution
Because there was no alternative to Company B's Waterfall-style process for software application development, in any situation where that process doesn't work or isn't acceptable to the customer, the fallback was to use no process at all or be at the mercy of a customer-defined process from customers who are not sufficiently experienced to provide that kind of direction Somewhat of an adversarial relationship had developed between Company B and the customer because each side blamed the other for the earlier project failure. That made it difficult for Company B and the customer to develop a joint approach that was optimized to make the project successful	This is a classic case of where it is perceived that it is an "all-or-nothing" choice between a totally planned and controlled Waterfall approach and a totally unplanned and uncontrolled approach A pure Agile process would not have worked in Company B's environment as it would not provide a way of setting and managing customer expectations for the cost and schedule for completing projects A hybrid Agile approach was needed that provided a way of managing customer expectations combined with a sufficient level of flexibility and adaptivity to define the details of requirements as the project was in progress In order to make this kind of approach work, it is essential to develop a collaborative spirit of trust and partnership between Company B and the customer to jointly manage expectations about the project

TABLE 26.8 Company B: what went wrong: people

Problem	Solution
A more Agile software development process was needed but that is heavily dependent on having highly skilled and well-trained people—it also can require a considerable shift in thinking and sometimes there is resistance to that kind of change	In this situation, training of Company B's people was badly needed. However, a standard Agile training course would have had limited effectiveness An ideal solution would be to first better define how a hybrid Agile approach would work and then provide training in the context of that approach; however, it wasn't practical in this particular situation to take that approach
■ Attempting to do a project like this without people who are well trained in implementing the project methodology is not a reliable, repeatable, and scalable approach	■ A more pragmatic approach in this situation would be to incrementally implement process improvements and training as the project was in progress

TABLE 26.9 Company B: what went wrong: tools

Problem	Solution
In a fast-moving effort like this, tools can be essential for coordinating the efforts of the project team as well as tracking and reporting progress	The may not be the most critical aspect of a solution for this situation; however, implementation of tools is one thing that can be done fairly easily and phased in without significantly disrupting the progress of the project and it can have a big impact by providing improved communication and visibility into project progress
Because of the joint nature of the effort between Company B and the customer, the tool needs to be capable of sharing information openly and easily However, that requires a spirit of transparency and openness for Company B and the customer to freely share information about the project	Before a tool can be effectively used to share information freely and openly, there needs to be a collaborative spirit of trust and partnership between the customer and the service provider

COMPANY C

Background

Company C is a company with a small IT organization that has been in the primary mode of supporting existing legacy applications. The company has not had to develop a major new application for a number of years.

- The existing legacy applications evolved gradually and were developed incrementally over a long period of time.

- The company initiated an effort to replace and redesign a large, existing legacy application and decided to use an Agile approach for the development effort although they did not have any significant experience in Agile, especially using Agile for large, complex enterprise-level application development.

The Approach

The company implemented an Agile development process from the "bottom up" within the development organization. The process was limited to a development process only, the business participation in the process was limited primarily to joint application design (JAD) sessions to define the requirements, and the role of the Product Owner was not fully implemented.

What Went Wrong

TABLE 26.10   Company C: what went wrong: Product Owner role

Problem	Solution
The company didn't understand the role of the Product Owner in an Agile Scrum project and used a traditional model for managing requirements	In companies that are in the primary business of developing software products, this role is obvious; however, in a company that uses internal IT applications to manage their business, the strategic importance of those applications may not be appreciated, and the role of the Product Owner (which is really equivalent to a Product Manager) might not be understood
IT was held responsible for the overall success or failure of the project in meeting business objectives, and the business role was limited to providing input to requirements through JAD sessions	
When IT takes primary responsibility for a project of this nature, there is a relatively weak focus on defining the business value that the project should produce without a Product Owner	A Product Owner needed to be appointed and trained to fill that role; however, given the scope and complexity of this development effort, a better alternative might have been to outsource the whole effort rather than doing it internally at all

TABLE 26.11 Company C: what went wrong: project governance

Problem	Solution
The company did not have a clearly defined governance model of how the project would be governed. As a result, the right people at the right levels were not engaged in making the right decisions about the project at the right times	On large enterprise-level projects, there is a need for inputs and decision-making from a number of people at different levels, and those inputs need to be organized
Direction from different people was sometimes in conflict	A good project governance model should engage the right people at the right levels to make the right decisions at the right times about the project. For example, senior executives should be defining measurable business objectives that the project should fulfill and delegating more detailed decisions about how the design of the system will fulfill those objectives to others
Some people were making decisions that they should not have been making. For example, some of the executives in the company were making detailed decisions about such things as UI screen designs	
Some decisions, such as defining the high-level business objectives the project needed to fulfill, were not being clearly defined by anyone at all	The managers who are responsible for the business processes should be defining the business rules of how their processes should work
The project was way behind schedule with no end in sight and senior executives had lost confidence in the project being successful	The users and stakeholders who use the system from day to day should have a key role in defining such things as screen designs to ensure that the system is usable
	Without clearly defining these roles and responsibilities, there may be conflicts among people attempting to give direction, and some direction might be left out
	In this particular case, a clearly defined project governance model needed to be defined and implemented

TABLE 26.12 Company C: what went wrong: development process

Problem	Solution
The company assumed that the development effort could be accelerated by simply breaking up the development process into sprints and using an Agile approach for managing the sprints	For large, complex enterprise-level projects, an Agile process cannot be implemented only at the development level; there has to be some higher-level planning processes associated with it or it is not likely to be successful

(continued)

TABLE 26.12 (*Continued*)

Problem	Solution
However, because the Agile development process was not fully implemented with the business, the project-level planning and release-level planning that should have taken place were neglected The result was that the development team was off-and-running developing code, but there wasn't a clear plan for how that code would be released and what the minimum functional requirements for a production release would be	Project-level planning is necessary to define a roadmap at a high level for how the product will be rolled out in releases Release-level planning may also be needed for how the features will be allocated to releases In many cases, it is also essential to document a high-level plan to define what assumptions have been made about how the project will be rolled out so that there is consensus and buy-in to that plan from all appropriate stakeholders

TABLE 26.13 Company C: what went wrong: Quality Assurance testing

Problem	Solution
There were no formal QA testing resources on the project, and whatever testing was done was done on an ad-hoc basis by Developers and Business Analysts without any formal QA test training This can be a problem for many Agile projects. It was assumed that formal QA was no longer needed, and people on the project, such as Developers and Business Analysts, who did not necessarily have any formal QA training would perform testing The result was that testing was very ad hoc without a plan and without well-defined, repeatable test cases to ensure an adequate level of test coverage was provided	Testing is a science, if it is done properly, and requires people who have some skill in developing well-designed test plans and test cases. An Agile development process does not totally eliminate the need for that A large portion of the testing effort can be done by Developers and Business Analysts, but there is still a role for formal QA testing, especially on large, critical enterprise-level projects. Rather than having a separate QA department perform that function, testers can be integrated into the team, but whoever performs that role on the team should have the appropriate testing skills

TABLE 26.14   Company C: what went wrong: architectural planning

Problem	Solution
In this particular project, there was a significant risk associated with the cutover of the existing legacy system to the new system that needed to be planned There were some significant architectural decisions associated with how the two systems would coexist with each other for some period of time for the transition to be successful The architectural planning associated with that transition was not given a sufficient level of focus, and a solution to this architectural problem was deferred until well into the development process	One of the Agile principles is, "Best architectures and requirements emerge from self-organizing teams" This project illustrates how that principle needs to be reinterpreted at an enterprise level. In this particular case, the risks associated with these architectural decisions were so great that they needed to be addressed and a solution planned early in the project In this situation, none of the Developers on the team had the level of expertise required to do the level of architectural planning necessary Expecting the development team to perform this function without a sufficient level of focus and expertise just isn't realistic A separate work stream might need to be created that is staffed by people with the right level of experience and focus to do the architectural planning in parallel with the primary development effort in the team In many large enterprise projects, particularly ones that require multiple teams, a separate team is responsible for architectural planning and direction

TABLE 26.15   Company C: what went wrong: project management

Problem	Solution
This is an example of a project where a hybrid approach is needed to blend some amount of classical plan project management with an Agile approach. Without some level of project-level planning and management, the project had an insufficient level of overall management	In more mature Agile teams, some of these project management functions might be performed within the team, but they become especially critical on large enterprise-level projects This is a perfect example of the need to fit the methodology to the project rather than force-fitting the project to a pure Agile approach

(*continued*)

TABLE 26.15 *(Continued)*

Problem	Solution
In this particular project, a Project Manager was assigned; however, the company tried to implement a pure Agile approach at the development level without a higher level of planning to perform some classical plan-driven project management functions, such as planning a roll-out strategy, developing milestones, and performing general risk management tasks	It is important to make an assessment of the scope and complexity of the project and develop an approach that is appropriate to the project. In this particular project, a hybrid approach such as the Managed Agile Development approach is probably needed to provide a blend of classical plan-driven project management at the macro-level with a more Agile development process at the micro-layer

TABLE 26.16 Company C: what went wrong: company culture

Problem	Solution
The culture in this particular company was very sales-oriented and also very heavily focused on operational excellence	It's very difficult, if not impossible, to change a company culture like that to make it more compatible with an Agile development approach, and any change in company culture can take a significant amount of time to implement
The individual users had sales goals that they needed to meet, and there was a significant amount of pressure to meet those goals	The best approach is probably to make the senior managers aware of the impact of these cultural differences and make a conscious decision of how to mitigate their impact
The management approach in the company had a strong command-and-control orientation	
As a result, there was a lot of top-down direction to the project without a sufficient level of delegation of responsibility and empowerment of the team	A hybrid approach such as the Managed Agile Development process can be a good way to layer an approach that adapts an Agile development process to a culture like this that is not fully compatible with Agile

TABLE 26.17 Company C: what went wrong: tools

Problem	Solution
An Agile Project Management tool was used on the project, but no one on the project team was fully trained in its use. As a result, the tool was not well utilized, and it was very difficult to plan and organize the project	Tools are essential in most cases to manage large enterprise-level projects, and people on the project team need to be trained in their use to know how to use them effectively

Overall Conclusions

It is very difficult to transform an entire company overnight from a classical Waterfall approach to an Agile approach.

- Many times, it is appropriate to take a bottom-up approach and start with implementing an Agile development process without attempting to transform the higher levels of management in the company; however, when that is done, you shouldn't ignore the higher levels and leave a void in those areas that *isn't filled at all*.

- A hybrid approach, such as the Managed Agile Development process, is many times a good way to integrate an Agile development process with a more traditional higher-level management framework as a first step until those higher levels can be addressed and transformed to a more Agile approach. It's like training wheels on a bike—learning to ride a bicycle for the first time as a young child can be a terrifying experience; children tend to fall over quite often and have many accidents. Training wheels are needed until the child gains a sense of balance and the confidence to ride the bike without them.

   This is also a good illustration of the need for Agile Coaching and training in implementing an Agile transformation. Although the company had some people on the project who were trained in Agile and understood the mechanics of how to apply an Agile process at a development level, there was no one on the team with a sufficient level of training and expertise to implement an Agile process on a large, complex enterprise-level development effort such as this.

DISCUSSION TOPICS

1. What do you think is the most likely problem that would cause an Agile approach to fail?

2. How would you go about preventing such a failure?

NOTES

1. "Miscellaneous Wit & Wisdom," National Churchill Museum, http://www.nationalchurchillmuseum.org/wit-wisdom-quotes.html.
2. Nathan Furr, "How Failure Taught Edison to Repeatedly Innovate," *Forbes* (June 9, 2011), http://www.forbes.com/sites/nathanfurr/2011/06/09/how-failure-taught-edison-to-repeatedly-innovate/.
3. Laurence J. Peter, "Failure," in: *Peter's Quotations: Ideas for Our Time* (New York: Bantam Books, 1977), p. 177.
4. Francesca Rice, "19 Joan Collins Quotes We Wish We'd Said Ourselves," *Marie Claire* (February 4, 2014), http://www.marieclaire.co.uk/blogs/545502/the-joan-collins-quotes-we-wish-we-d-said-ourselves.html. Several unverified sources claim that Albert Einstein said, "Anyone who has never made a mistake has never tried anything new."

㉗ Case Study: Valpak

BACKGROUND

VALPAK, ESTABLISHED IN 1968 AND HEADQUARTERED IN LARGO, FLORIDA, is one of the leading direct marketing companies in North America. In addition, Valpak has one of the largest collections of digital coupons on the Internet with thousands of local products and services, as well as national brands.

Working in partnership with its network of nearly 170 franchisees in the United States and Canada, Valpak helps more than 54,000 businesses a year to achieve their marketing goals.

- Valpak's primary competitors are Money Mailer, Valassis (Red Plum), and Super Coups. Valpak's secondary competition includes newspapers, television, Yellow Pages, and any other forms of advertising.

- All in all, Valpak is trusted by consumers and merchants alike to consistently deliver value. The Blue Envelope® delivers savings and value to nearly 40 million households each month. Annually, Valpak will distribute some 20 billion offers inserted in more than 500 million envelopes.

- Valpak also offers digital solutions with www.Valpak.com®, an online site for printable coupons and coupon codes, which has nearly 70 million offer views each month, as well as apps for smartphone platforms.

VALPAK STAKEHOLDERS

Valpak's IT group builds and supports technology for a wide variety of stakeholders and audiences, including:

- consumers who are focused on saving money with coupons
- Valpak franchisees that need systems to run their business and sales operations

- merchants interested in tracking and maximizing their returns on investment

- traditional internal corporate stakeholders that need to run the core business operations.

   Efforts to serve these distinct audiences required a broad-based approach: To some companies, it may simply mean moving from a Waterfall-style development process to adopting a more Agile development approach. To Valpak, it meant transforming the way the whole company operates.

- That required changing the way the whole company operated, in addition to implementing a much more Agile process to rapidly develop new media content.

- They used an overall Agile transformation driven top-down by their CEO to transform the whole company to focus on delivering very-high-quality media content to the market quickly.[1]

   Valpak developed a more advanced implementation using a Lean approach, eliminating the iterations, and replacing them with more of a pull approach that allows work items to be addressed whenever there is capacity to work on them.

Valpak Franchisees

Order entry, office management, mobile/online sales tools, and CRM applications are developed to support Valpak's 170+ franchise locations located in the United States and Canada. These franchises are independently owned and operated locations that utilize Valpak's franchise system to sell Valpak print and digital products to local merchants.

Consumers

Savings/coupon applications and websites are developed that provide daily value to consumers looking to save money on their purchases. Consumers can interact with Valpak savings anywhere at any time, regardless of whether they are using Valpak's traditional "print" mailer or one of Valpak's several digital channels (web, mobile, SMS texting, e-mail). Valpak's savings content is also distributed to over 150 partner websites as well.

Merchants

Online websites and mobile applications are developed to allow merchants to manage their advertising campaigns with Valpak.

Corporate

Traditional back-office operations, including manufacturing, marketing, finance/accounting, order processing, and sales. The IT group develops and supports various ERP and custom application solutions to automate these back-office operations.

THE ROLE OF TECHNOLOGY AT VALPAK

Valpak's ability to utilize technology to transform their business is a very significant factor in their business success, and Valpak's IT group is an integral part of the business transformation and growth of the company.

- To compete with the quickly changing digital savings marketplace, Valpak transitioned the entire IT organization to Agile Scrum/Kanban processes with two-week sprint delivery cycles. They embraced this change and quickly adapted.

- This effort was so well done in the IT organization that Valpak is now driving the Agile culture throughout the company, heading toward "The Agile Enterprise."

As Director of Agile Leadership at Valpak, Stephanie Stewart (now Stephanie Davis)[2] was responsible for leading the Agile transformation. In this role, she was responsible for process facilitation, portfolio governance, program management, project management, and of course, oversight of related people, processes and tools.

Stephanie led the team of Agile project leaders, who handle everything from project management to Scrum mastering to leading Kanban teams.

A self-admitted Agile enthusiast, Stephanie has worked passionately to encourage and support the IT organization at Valpak in fully embracing Agile software development, to move Valpak toward a greater vision of "The Agile Enterprise."

Chris Cate, CIO, was the Agile executive champion working with Stewart and the executive leadership team in transforming the company over the past year. Cate also evangelizes "The Agile Enterprise" vision by encouraging the adoption of Agile values and the use of Agile methods for non-IT departments.

Bob Damato, Director of Software Engineering, led the adoption of Agile technical practices such as test-driven development, continuous integration, and evolutionary architecture. Strong leadership for technical practices across teams has been critical to maintaining and improving quality as part of the Agile transformation.

OVERVIEW

Valpak's overall enterprise-level approach is based on the Scaled Agile Framework.[3]

Scaled Agile Framework Implementation

The Scaled Agile Framework (SAFe) consists of three primary layers:

1. Portfolio Layer
2. Program Layer
3. Team Layer.

Portfolio Layer

The portfolio layer is the highest and most strategic layer in the Scaled Agile Framework, where programs are aligned to the company's business strategy and investment intent.

Program Layer

The Scaled Agile Framework recognizes the need to align and integrate the efforts of multiple teams that are engaged in large, complex enterprise-level development efforts to create larger value to serve the needs of the enterprise and its stakeholders.

Team Layer

The team layer forms the foundation of the Scaled Agile Framework and is where the fundamental design, build, and test activities are performed to fulfill the development requirements for each major area of business. At Valpak, there are actually two different development processes that are used in the team layer, as shown in Figure 27.1. In most cases, Scrum is used for more exploratory development, while a Kanban process is used to run the business kinds of development.

Valpak implemented the SAFe® from the bottom up:

- In October 2011, Valpak started with six Scrum teams and three Kanban teams at the Agile teams (bottom) layer of the SAFe® (Valpak currently has 10 Scrum teams and 3 Kanban teams).

- Shortly after the Agile teams were established, Valpak implemented road-mapping and release management with the middle layer of the Scaled Agile Framework in mind.

- Most recently, Valpak implemented the portfolio Kanban at the top layer of the Scaled Agile Framework with their leadership team of executive sponsors.

- Last but not least, Valpak added the architectural Kanban.

    There are two key things that are most significant about this case study:

    1. Valpak recognized the need to adapt the Agile development process at the team level into an overall enterprise model that is well integrated with their business. Valpak is one of the initial pioneers in the use of the Scaled Agile Framework to provide that integration.
    2. Valpak also recognized the need to use a Kanban process for *run the business* efforts, instead of Scrum, which is used for exploratory development, because the needs are very different.

    Both of these efforts show real thought leadership to fit a methodology (or combination of methodologies) to the business and projects instead of force-fitting the business and projects to a predefined, textbook approach.

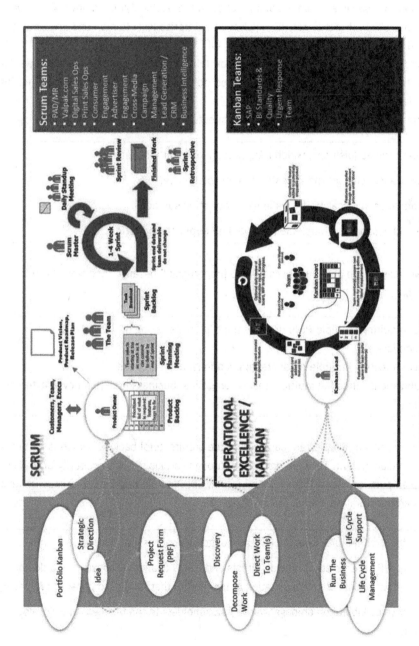

FIGURE 27.1 Valpak Agile process framework

Architectural Kanban

The *architectural Kanban* is a recommended practice in the Scaled Agile Framework that was implemented by Valpak. It recognizes the need at an enterprise level to plan and implement consistent and well-integrated architectures across all teams and projects.

This approach defined by the SAFe® consists of defining architectural epics, which are "large technology initiatives that are necessary to evolve portfolio solutions to support current and future business needs."[4]

According to Dean Leffingwell, sources of architectural epics can include the following:

- Mergers and acquisitions, which require technological integration.

- Technological change and infrastructure obsolescence.

- Performance and scalability challenges of existing solutions.

- Cost and economic drivers, such as avoiding duplication of effort.

Leffingwell says these are "initiatives of epic proportions," because they typically cut across three dimensions:

1. Time: requiring multiple PSIs to implement, perhaps taking up to a year or two to complete.
2. Scope: affecting multiple products, applications, and solutions.
3. Organizations: affecting multiple teams, programs, business units, and even external entities.[5]

At Valpak, architectural epics are captured in the architectural backlog, which is part of the architectural Kanban system. They are processed through various states of maturity until they are moved to implementation. According to Dean Leffingwell, the following are the primary motivators for using this architectural Kanban approach at Valpak:

- Make the Architectural Epic backlog and ongoing analysis visible to all.

- Provide WIP limits to ensure the architects and teams analyze responsibly, and do not create expectations for implementation or timeframes that far exceed capacity and reality.

- Help drive collaboration amongst the key stakeholders in the business and development teams.

- Provide a quantitative, transparent basis for economic decision-making for these most important technology decisions.[6]

TABLE 27.1   Architectural Kanban board

Swim Lane	Value Stream Stage					
	Queue	Research	Design	Prototype	Development	Done
Scalability						
Performance						
Reliability						
Technology Upgrades						
Frameworks & Infrastructure						
Innovation						

Table 27.1 shows how the architectural Kanban board is organized at Valpak. The "Value Stream Stage" indicates the stage of progress of each architectural Kanban epic ("Development" often involves the Scrum and Kanban teams helping to implement the architectural epic for their area of concern). The "Swim Lane" rows in the Kanban chart indicate how the architectural epics are grouped by area of focus.

The architectural Kanban board is made highly visible to all teams and provides work-in-progress (WIP) limits on each of the stages in the Kanban process. Weekly standups are held with the architects to review progress and discuss any issues.

Portfolio Kanban

The *portfolio Kanban* is a recommended practice in the Scaled Agile Framework that was implemented by Valpak. The purpose of the portfolio Kanban is to provide a way to plan, prioritize, and manage a portfolio of business epics. "It brings visibility to upcoming work as well as work in process, helps facilitate product development flow and can be a key factor in achieving enterprise—as opposed to team or program—agility, and thereby more fully optimized business outcomes."

According to Dean Leffingwell, the Kanban system is used in this context to accomplish several purposes:

- Make the strategic business initiative backlog (upcoming business epics) fully visible.

- Bring structure to the analysis and decision making that moves these initiatives into implementation, and make that process visible to all.

- Provide WIP limits to ensure the teams responsible for analysis do so responsibly, and do not create expectations for implementation or timeframes that far exceed capacity and reality.

> - Help drive collaboration amongst the key stakeholders in the business, Architecture and Development Teams.
> - Provide a quantitative, transparent basis for economic decision-making for these, the most important business decisions.[7]

The implementation of the portfolio Kanban at Valpak included:

- Highly visible (physical) Kanban board; however, no WIP limits were applied at the portfolio level because it wasn't found to be meaningful.

- Portfolio Kanban standups held with executive sponsors weekly. Each executive sponsor addresses any board movement of epics and any major decisions made.

- Definition of an epic:

  - ≥ 3 sprints of effort
  - ≥ 3 Agile teams to coordinate
  - Considerable corporate, franchise, or market value/impact.

- The portfolio Kanban is reset at the beginning of each new quarter based on the output from an executive quarterly planning meeting called the Portfolio Review Board (PRB) and based on retrospective improvements identified. PRB focuses on planning only; status and day-to-day collaboration is left to the portfolio Kanban process.

- Value stream (includes entrance/exit criteria) on portfolio Kanban includes *funnel, vet, design, build, deliver,* and *done. Build* is usually when the epic is being developed by the Scrum or Kanban teams.

- The *funnel* is divided into "current quarter," "next quarter," and "unplanned," so that everyone knows what epics have been planned for and what epics have come up that were not in the last quarterly plan. Epics not in the quarterly plan are subject to dot voting.

  Challenges with the portfolio Kanban have been:

- **Standup derailment risk:** There is too much talk about strategy or decision making or solutioning during the standup. The director of Agile leadership facilitates them; without that facilitation, they tend to fall apart.

- **Executive schedules:** Executive schedules are always busy, so not every executive sponsor shows up every week. Valpak tries not to cancel portfolio Kanbans if two or more executive sponsors are available.

- **Definition of an epic:** There was initially some confusion over what an epic was and if it deserved to be on the board (versus a strategy or a feature or a story). Valpak recently settled on a definition of an epic for this purpose.

- **Tying strategies to epics:** A list of all active business strategies has been established to tie to the epics on the portfolio Kanban. The goal is that strategies spawn one or more epics, and epics are managed via the portfolio Kanban process.

PROJECT MANAGEMENT APPROACH

Valpak has implemented the Scaled Agile Framework (SAFe®) very closely. The SAFe® does not recognize the concept of a "project"; however, Valpak does have a handful of projects wrapped around large cross-team efforts.

- For those projects, Valpak assigns an Agile project leader to work with the relevant Scrum Masters, Product Owners, and stakeholders to pull it all together, but there are only about a handful of these projects running at any given time.

- Most work (even work that crosses multiple teams) can be managed by collaboration among the product owners and Scrum Masters for the Agile teams.

At Valpak, projects are represented by epics, and each epic has an executive sponsor. Epics are managed via the portfolio Kanban process (described earlier). The portfolio is planned and prioritized quarterly using these epics. In most cases, epics require the work of three or more Scrum/Kanban teams to accomplish and therefore have more than one Product Owner involved.

Some epics also have important business tasks to be managed as well (not just a technology effort). For this kind of large epic, Valpak assigns an Agile project leader as a Project Manager to coordinate the work of the business and the Agile teams. A Waterfall type of schedule may be applied to a project (which is typically the case when you have to fold in the business tasks); however, it is still an Agile project management approach overall.

Where project management is needed for large cross-team efforts, the Agile project leader will create a project plan (schedule) or a roadmap to represent all the teams, sprints, and business tasks involved with the effort as well as significant milestones.

- Even though project management may be applied, the Agile values are still top of mind for the Agile project leaders.

- This might mean a light roadmap versus a lengthy project plan; in other words, the Agile project leaders plan and document to the needs of the project at hand.

- They apply the degree of project management that produces the most value and expect that the plan is certain to change.

Valpak uses four levels of planning:

1. *Daily*: Scrum team standups in front of task boards.
2. *Biweekly*: Scrum sprint planning (all teams run on a common sprint schedule with sprint planning on a Monday and the sprint reviews on a Friday, two weeks later).

3. *Monthly*: Release planning across Scrum teams (product owner-level planning with a six-month look-ahead).

4. *Quarterly*: Quarterly plan of epics (executive sponsor-level planning that fuels the portfolio Kanban for the upcoming quarter).

Valpak has learned to fit the project management methodology to the type of project:

- Scrum is used for most exploratory development and currently consists of 10 teams aligned with each of Valpak's major areas of business focus.

- Kanban is used for run-the-business type of work and/or areas that are not conducive to or don't require the highly prescriptive nature of Scrum.

- Infrastructure projects and large-scale corporate projects are managed as Waterfall with some of the Scrums and Kanbans producing work that supports those projects.

According to Stephanie Stewart:

> There is no one-size-fits-all methodology; at least not that I know of. . . Project Managers need to be more versatile and flexible than ever before. PMs need to be able to effectively pivot between command-and-control and servant leadership focused facilitators. Even though I love working with Agile teams right now, I also know your run-of-the-mill traditional project when I see it. I've learned to pivot between the two by applying different soft skills. Traditional is all about directing and managing whereas Agile is all about serving and facilitating.[8]

Tools, Communication, and Reporting

The following is a summary of Valpak's approach to tools, communication, and reporting:

- All Valpak Agile teams maintain physical boards in a common area. Scrum teams use a task board format to display their stories and tasks and show their progress each day. Kanban teams have boards with their custom value stream, and sometimes swim lanes for further classification of work. Additionally, Valpak uses Pivotal Tracker for Scrum teams to manage its backlogs and sprints.

- To assist with managing cross-team dependencies, each product owner tags stories (in Pivotal Tracker) with noted dependencies on other teams. A report is distributed each week that shows these dependencies so that product owners and teams can coordinate and collaborate accordingly.

- On the day after sprint planning Monday, two reports are distributed:
  - Sprint accomplishments (from previous sprint)
  - Sprint goals (for current sprint).

- On sprint review day, an email of sound bites of accomplishments for all Scrum and Kanban teams is sent to a broad group of stakeholders.

- After the product owners perform their monthly release planning process at the feature level, a one-page view is distributed with six-month look-ahead.

- After the Valpak executives meet for their quarterly planning of epics, a one-page view is distributed with a four-quarter look-ahead.

- After the last grooming (refinement) session of the sprint, each Scrum Master sends out a quick list of proposed stories, anticipated to be planned into the next sprint. These are highly subject to change but help product owners and stakeholders to see what's coming to better coordinate and collaborate across teams.

CHALLENGES

Cultural and Organizational Challenges

Tables 27.2–27.8 are a summary of the key cultural and organizational challenges faced by Valpak and how they were handled.

TABLE 27.2   Solutions to cultural and organizational challenges: managing cross-team dependencies

Challenge	Solution
Valpak has very highly integrated systems. In establishing the Agile teams, it was almost impossible to create teams that could operate independently of any other teams In most cases, teams are dependent on one another for a given feature or epic. The most common dependency at Valpak is between Valpak's BI Scrum teams and all other teams, since just about everything involves the ability to track and report	Besides continuous cross-team collaboration and communication, Valpak has implemented: - Cross-team dependencies report distributed each week based on dependencies tagged in Pivotal Tracker - Proposed stories distributed week prior to new sprint - A meeting referred to as the "Scrum Powwow" held each week with product owners and Scrum Masters to discuss current sprint, next sprint, and roadmap dependencies - Visibly flagging dependencies between teams on task boards - Shared acceptance criteria prior to or early in sprint

(continued)

TABLE 27.2 (*Continued*)

Challenge	Solution
	■ Better planning and coordination of handoffs between teams
	■ Allocation of time prior to the sprint for proper discovery and architecture across impacted teams for large/complex cross-team epics
	■ Cross-team post-planning standup with the impacted teams the day after sprint planning to sync up stories/tasks and collaborate early

TABLE 27.3 Solutions to cultural and organizational challenges: accountability at the top

Challenge	Solution
Since SAFe® was implemented bottom up, the Agile teams were in place well before the portfolio Kanban Prior to the portfolio Kanban process, there were major disconnects between the executives and the product owners Mixed directives and conflicting direction was coming from the top, leaving Product Owners and, therefore, teams with roadmap whiplash	■ Executive sponsors named for each Scrum team to establish accountability and ownership at the top ■ Product owners meet regularly with executive sponsors on their upcoming sprint plans and roadmaps ■ Portfolio Kanban established ■ Quarterly planning of epics ■ Weekly executive-level standups in front of portfolio Kanban board ■ Executive sponsors frequent the sprint reviews ■ Retrospectives held with executive sponsors to look for improvements

TABLE 27.4 Solutions to cultural and organizational challenges: Product Owner collaboration

Challenge	Solution
Collaboration can be just as difficult for Product Owners as it is for the teams At first, Product Owners were very comfortable working within their own teams	At first, a Scrum-of-Scrums meeting was held for 15 minutes each week in the task board common area with product owners and Scrum Masters to review dependencies and their progress

TABLE 27.4   (*Continued*)

Challenge	Solution
However, where there were dependencies with other teams, there were often collaboration issues between Product Owners	The Scrum-of-Scrums format becomes difficult beyond five teams, so Valpak evolved to the "Scrum Powwow" format. The Scrum Powwow is a one-hour meeting held each week with product owners, Scrum Masters, architects, and other IT leaders to collaborate on dependencies for current sprint, next sprint, and roadmap across teams
	Product Owners look for proposed stories distributed each week for dependent teams; Product Owners have access to one another's backlogs in Pivotal Tracker
	A cross-team dependencies report is distributed each week that helps Product Owners to manage dependencies with other teams

TABLE 27.5   Solutions to cultural and organizational challenges: franchise-based company

Challenge	Solution
Valpak is a franchise-based organization. This means that while products and features may be released more quickly under an Agile framework, they aren't necessarily adopted, utilized, or sold any quicker by the franchises	Valpak's sales and marketing organizations are becoming more and more Agile to quickly enable franchises with new products and features
Rolling out new products and features to Valpak's franchise organization takes considerable planning and support	Training and communication are happening more quickly and frequently than ever before to keep up with the Agile releases
	There will always be those franchises that are slow to adopt new products, new tools, and new features; however, Valpak is pushing as hard and fast as possible to roll out new products and features. If the product or feature proves to be valuable, most franchises will get on board

TABLE 27.6   Solutions to cultural and organizational challenges: managing stakeholders

Challenge	Solution
At Valpak, stakeholders are anyone who isn't the team, the Product Owner, the Scrum Master, or the executive sponsor This means that stakeholders are at all levels of the organization and all departments across the business Stakeholders also include Valpak's franchises and consumers; however, those particular stakeholders are represented indirectly by Valpak's sales and marketing organizations, respectively Product Owners had difficulty in managing the various, diverse stakeholders. Each product owner had dozens of stakeholders to deal with, each with their own needs	All stakeholders were given one-hour training on Agile, Scrum, and Kanban Over time, Product Owners refined their approach to stakeholder management. Each Product Owner has a slightly unique approach based on the needs of their stakeholders Some Product Owners began regular stakeholder meetings with core groups to elicit their needs and keep them apprised of progress Some Product Owners require that their stakeholders submit a project request form (PRF) to articulate their stories in proper format

TABLE 27.7   Solutions to cultural and organizational challenges: Agile culture shift

Challenge	Solution
An organization's culture can't be changed overnight with a simple announcement like "going Agile!" In fact, it can't be changed with just a kick-off and some training, either Starting out with Agile back in October 2011, Valpak had about 30% naysayers, 40% indifferent, and 30% enthusiasts	The culture that is *Agile* must be constantly developed and nurtured over time with every meeting, decision, action, and event that takes place across the organization To help implement this culture shift at Valpak, an Agile coach was involved (three to four days a week on site) for about eight months To continue the culture shift, Valpak has Agile excellence meetings with product owners, Scrum Masters, and IT leadership every other month to perform a retrospective of Agile (a retrospective of retrospectives, if you will) In addition, what was the old PMO was restructured to the Agile leadership office, led by the Agile leadership director supporting a team of Agile project leaders

TABLE 27.7   (*Continued*)

Challenge	Solution
	To continue the culture shift momentum in fun and memorable ways, Valpak holds events with the Agile teams and stakeholders like Xbox Kinect Fruit Ninja Tournaments. Most recently, Valpak held an Agile Roast for the first anniversary of Agile at Valpak

TABLE 27.8   Solutions to cultural and organizational challenges: developing high-performance teams

Challenge	Solution
With any Agile transformation, teams start off by "doing Agile," following the basic mechanics without necessarily understanding or embracing the values Such was the case at Valpak. All teams were trained and knew just enough to be dangerous	Moving the teams from the basic mechanics of Agile to truly high-performing teams; no longer "doing Agile" but "being Agile" Valpak has recognized that developing self-organizing, high-performance teams doesn't happen automatically and requires leadership and continued support "Agile is not about getting out of the way of your teams but rather staying involved as servant-leaders"[9]

Technical Challenges

Tables 27.9–27.13 are a summary of the technical challenges faced by Valpak and how they were handled:

TABLE 27.9   Technical challenges and solutions at Valpak: lack of Continuous Integration (CI)

Challenge	Solution
No organization or structure around automated testing. No automated execution of tests	Valpak implemented continuous integration where code is committed throughout the day with tests automatically run and errors sent to all developers to resolve

TABLE 27.10   Technical challenges and solutions at Valpak: build process

Challenge	Solution
Lack of structure and organization around build process. Build process was completely brute force and highly manual	Valpak's build process is now semi-automated and progressing toward fully automated with twice-daily builds to development and test environments

TABLE 27.11   Technical challenges and solutions at Valpak: legacy code base

Challenge	Solution
At the beginning of Agile, Valpak had a legacy code base consisting of over a half million lines of untested, unclean code	Valpak implemented Test-Driven Development (TDD) methodology, along with clean code practices Percentage of test coverage continues to increase with continuous support of TDD and clean code practices

TABLE 27.12   Technical challenges and solutions at Valpak: architecture role/involvement

Challenge	Solution
Early on, the architecture role was not well defined. In addition, as a leftover from the Waterfall days, teams did not feel empowered to make architecture decisions	■ Architectural Kanban ■ Architect and lead developer roles better defined ■ Improved collaboration between architecture and teams

TABLE 27.13   Technical challenges and solutions at Valpak: manual tests/quality/risk

Challenge	Solution
No real automated tests existed, which increased the risk of quality issues as Valpak moved faster in delivering working software to the business Releases required what Valpak calls "production checks" to ensure that everything was deployed correctly	Valpak is currently in the infancy stages of QA automation using Selenium Teams have begun to automate tests by including technical stories in each sprint

TABLE 27.14 Additional challenges and solutions at Valpak: planning sprints for sustainable pace

Challenge	Solution
Out the gate, no team had proper velocity measures. At first, teams planned too much into their sprints and ended up exhausted in trying to accomplish their sprint goals	Over time, the Agile leadership office worked with teams to establish some best practices for keeping within their sustainable pace Most teams have a pencils-down day built into their ground rules in which no new development can be completed toward the current sprint. Stories may be dropped as a result, but sustainable pace is preserved All teams use a custom capacity calculator (spreadsheet) that was created to take into account non-sprint work like meetings, administrivia, and support in order to better estimate sprint capacity for each team member So as not to over-commit to a sprint, most teams use the concept of extra-credit stories. Extra-credit stories are established during sprint planning by the product owner as a stretch goal of sorts; if time and capacity allow, the team will complete them

Other Challenges

Tables 27.14–27.20 are a summary of some other challenges faced by Valpak and how they were handled.

TABLE 27.15 Additional challenges and solutions at Valpak: decomposing stories

Challenge	Solution
Teams are often challenged to decompose stories to fit within a single sprint and without breaking them down by task (develop, test, etc.)	Proper story decomposition can be more art than science Through training and continuous coaching from Scrum Masters, Product Owners and teams learned to decompose stories better with each sprint

TABLE 27.16   Additional challenges and solutions at Valpak: story points

Challenge	Solution
Early into Valpak's Agile transformation some teams still struggled with the proper use of story points	Through training and continuous coaching from Scrum Masters, teams learned to think of story points as effort
Many teams were applying story points as a measure of complexity rather than effort	Story points are a relative measure of effort to be done, not complexity, not priority, not sequence
In addition, some teams were applying hours to tasks and skipping the story point step altogether, which basically negates the benefit of this relative estimating technique	Also, the point of improving Valpak's story point estimating is to some day stop having to think about hours

TABLE 27.17   Additional challenges and solutions at Valpak: team collaboration spaces

Challenge	Solution
At Valpak, the software development organization has the advantage of being co-located in the same building	Just a few months after Valpak's first sprint, a massive office cube move was organized across 70+ employees to bring together team members with their teams into cube quad areas
However, as a leftover from Valpak's Waterfall days, team members were scattered about, which meant there were no team collaboration spaces	Within the cube quads, the inner walls of the cubes were taken down to create a "bullpen" for each team
	To top it off, creative signs were created for each team and hung over their cube areas
	With open spaces for each team, collaboration and osmotic communication increased dramatically

TABLE 27.18   Additional challenges and solutions at Valpak: non-Agile tools

Challenge	Solution
Prior to Agile, Valpak used a traditional project management tool called @task	After an evaluation of several Agile tools, Pivotal Tracker was selected. Pivotal Tracker was considered just enough tool for Valpak's purpose, remembering the Agile Manifesto value of "Individuals and interactions over processes and tools"
Thus, Valpak's Agile process inherited a non-Agile tool. This meant that sprint planning and backlog management were tedious and time-consuming exercises for each team	Pivotal Tracker did not replace the need for physical task boards, which provide high visibility to the process that a tool can't match

TABLE 27.19 Additional challenges and solutions at Valpak: predicting release dates

Challenge	Solution
In some segments of the Agile community, it is not considered very Agile to predict a release date and be held responsible for meeting that schedule Because of the nature of the Valpak business, it was essential to be able to predict release dates with some level of accuracy	Valpak adopted a strategy to predict release dates ". . . by nature of time-boxing, the schedule is a fixed constraint. With the schedule being a fixed constraint (as is cost based on team size), it is scope that remains flexible. So, the scope that is included in any given release is flexible based on what the team was able to accomplish, what stories get dropped, what natural disaster impeded the team, who was out sick, and so on and so forth. Whether that scope accomplished is acceptable for release to production is always a product owner decision."[10]

TABLE 27.20 Additional challenges and solutions at Valpak: developing Agile project leaders

Challenge	Solution
Valpak recognized the need to define a new leadership role and job description for its project leaders so that someone in this position can assume the role of Scrum Master, Kanban leader, and/or Project Manager, depending on the work at hand	Valpak created a new job description for an *Agile project leader* oriented around leadership of technology-focused projects and teams relying on Agile values and principles. "The focus of this position is on delivering value over meeting constraints, leading the team over managing tasks, and adapting to change over conforming to plans."[11]

OVERALL SUMMARY

Key Success Factors

Four factors helped make Agile a success at Valpak:

1. Top-down support coupled with bottom-up drive
2. Hiring an independent coach
3. Continued support each and every day
4. Senior management engagement/business ownership.

Top-Down Support Coupled with Bottom-up Drive

Taking the Agile transformation seriously and supporting it at all levels of the business was essential to make the culture shift to the Agile mind-set. Stewart explains:

> Rather than continue to dabble with Agile, we went full-fledged, full-bore Agile (all software development teams all at once). Once our CIO declared that we were going Agile, there was no looking back.
>
> ■ He gained the support of our executive leadership team and they helped us to sell it to our stakeholders, including our franchise network.
>
> ■ At first, there were some hold outs (the resistance) but with each and every sprint we have managed to convert a few more into believers.
>
> ■ Whenever someone says, "Agile is not working for us," I like to respond with "No, we are not working for Agile."[12]

Hiring an Independent Coach

Attempting to DIY (do-it-yourself) Agile can sometimes backfire. You need to spend the money on bringing on an Agile coach for about six months to train the teams and support the process. Companies often tend to appreciate the advice of an independent expert more than a trusted member of the company's own staff.

Continued Support Each and Every Day

Where Agile has failed with other companies, it seems that it was done with a big one-time kick-off without continued support. Stewart continues:

> I think I read once that a culture change takes 10 years, so. . . We continue to support Agile each and every day. That's my primary role as director of Agile leadership, but I am well supported by it directors, product owners, Agile project leaders, and even team members that have become great evangelists in their own right. The more you talk about it, the more they talk about it. Ways in which we continue to support Agile here are things like:
>
> ■ Agile Excellence meetings every other month (the retro of all retros, so to speak).
>
> ■ Scrum Powwow each week (cross-team current sprint, future sprint, and roadmap discussions).
>
> ■ Team Building (with Xbox Kinect Tournaments for Agile teams to compete against one another).

- Celebrations (Valpak did an Agile Roast for their first anniversary this month, and we have cake a lot for important sprint/release milestones).

- Agile tours given to special groups of stakeholders or VIPs on request.

- Facility sponsor for Tampa Bay Agile meet-ups, along with helping to secure speakers and topics.

- Agile Manifesto and other Agile-related posters highly visible throughout the IT work space.[13]

  Stephanie Stewart said, "It's fun to hear people dropping the *A bomb* (that is, mentioning Agile) at large-scale events like our annual Coupon U with our franchises or corporate all-hands meetings. Agile is really a rock star around here!"[14]

Senior Management Engagement/Business Ownership

Getting Valpak's senior management engaged and committed to the effort was a significant factor in the success of the effort. Chris Cate describes the process:

Valpak has fundamentally changed the way we get work done. This is not just a new process that only involves technology teams but more of a mind-set across all groups that puts an emphasis on driving real business value to our customers in short iterative cycles.

Business is deeply engaged with their products and services that they are asking IT to build.

- Business (Exec Sponsor and Product Owner) control the backlog and work output—they own it!

- They now have complete ownership and accountability—no place to hide.

- They are actively engaged in managing priority discussions with their customers to identify the highest valuable feature that the team should work on.

- The business is getting more work done across all teams.

- None of our business owners would ever go back to the older Waterfall process.[15]

RESULTS AND CONCLUSIONS

More Strategic Management Focus

The Agile development process, using empowered, self-organizing teams, has enabled a major shift in the Valpak management approach. Cate explains it this way:

Conversations are fundamentally changing at the Sr. executive team level, focusing less on tactical implementation issues and more on strategic growth initiatives:

Do these things less:

■ No longer have to justify and/or explain extremely large project plans (the Executive Sponsor and Product Owner decide the sprint schedules and backlog priorities and own it).

■ No longer have to talk about why IT missed a date or did not set expectations correctly between project team and Executive Sponsor (ES and PO own the priorities and expectations).

■ Rarely have to play the 'peacemaker' role between various groups now that the business firmly owns the work product.

Started doing more of:

■ Strategic planning and execution to grow the business.

■ More involved in business development, partnership, and acquisition executions.

■ Partnering with other Sr. team members to help them think through growing their business lines versus fixing IT.[16]

Management of IT Resources

The shift to more empowered, self-organizing teams has also made the management of IT resources much easier. Cate says:

■ Less time needed to manage personnel issues since the process helps to correct this naturally.

■ Teams are empowered to solve their own problems through collaboration, thus reducing functional manager activities.

■ Much more open and transparent environment where the employees, or contractors, not fully contributing are identified and coached up or out quicker.[17]

Time-to-Market

Valpak releases software to production at the end of every sprint. Time-to-market was significantly improved as a result of shorter iterations. With two-week sprint cycles, new/enhanced software was available more quickly than ever before to stakeholders, including internal, franchises, merchants, and consumers.

Alignment and Collaboration

Alignment and collaboration between business and IT were increased as a result of a highly visible process. Product Owners became truly accountable for their products and teams, and the "us versus them" mentality quickly disappeared.

Employee Productivity and Morale

Productivity and morale among the teams improved as a result of the empowered and self-organizing team's principle of Agile. In fact, team morale and pride are greater than they've ever been at Valpak.

Delivering More Frequent Value to Customers

A value-driven process is now in place. Decisions are made on what to work on based on value—that value is derived from stakeholder input. With two-week sprint cycles, teams deliver the highest value stories each sprint. Value is being delivered to stakeholders more frequently than ever before.

Openness and Transparency

Teams, at the portfolio and architecture levels use highly visible task boards to track progress—anyone on the team or in the company can easily see a visual status of what stage of work each story/task is in at any time. In addition, teams know what they're going to be working on for the next two weeks, and all the teams have visibility into what the other teams are doing.

Responsiveness and Adaptivity

Changes can be implemented more quickly than ever before. It is easier to change strategy and pivot. This makes experimentation and risk taking much more feasible. Being able to try out new ideas and turn on a dime is critical, not only for startups but also for larger companies that do not want to become obsolete.

Software Quality

There is increased emphasis on Test-Driven Development (TDD), unit testing, and clean code. Continuous integration allows recognizing the impact of a change quickly.

LESSONS LEARNED

Forming Projects Around Teams

Prior to Valpak's Agile transformation, when a project was initiated, a team would be formed around the project. In most cases, the team members were already assigned to other projects. Stewart says:

> So, with project plans changing as they do, a team member would typically end up being pulled in multiple directions, reporting to multiple masters, across many projects all at the same time. From a people perspective, this is not a fun place to be (remember, "multitasking" is the new four-letter word). The result was low morale from all the project whiplash going on.[18]

With the new Agile approach, Valpak shifted to "forming projects around teams." There are 10 Scrums and three Kanbans. Stewart says:

> We have naturally experienced a shift in our project initiation process. Now, when a project is requested, we evaluate the request based on the existing Scrum and Kanban teams, and *we form the project around the teams.* Sometimes this means breaking up the project to fit the vision and/or skill of the teams. So, as opposed to forming a team around a project, we are now forming a project around teams.[19]

Planning Team Capacity and Developing a Sustainable Pace

Valpak's Scrum Masters needed a better approach to refine team capacity calculations during sprint planning and to allow for non–sprint activities (such as product support) that might require some of the team's time during a sprint. Valpak developed and implemented a team capacity calculator (spreadsheet) for this purpose, which factored in the amount of time each team member was actually available to work on sprint activities. Stewart describes the process:

> During the planning session, we mitigate this risk by having team members predict their personal sprint capacity (50%, 80%, etc.), taking into account non-sprint meetings and a predictive measure of support. Also, we can further manage this by only working on support that our Product Owner deems immediate work. Any other support can be backlogged and planned into a future sprint. Now, should all hell break loose in the support arena, some stories may be dropped or some stories might not meet the team's Definition of Done.[20]

Using Sprint Reviews and "Science Fairs"

There was a lot of difficulty associated with scaling sprint reviews to an enterprise level with multiple teams needing to hold sprint reviews with some of the same stakeholders concurrently.

> Our first take at Sprint Reviews was pure madness! Each Scrum Master would schedule a separate 1- to 2-hour meeting on the same Friday for each Scrum team. That's nine 1- to 2-hour meetings within an 8-hour business day [—] and on a Friday no less![21]

Valpak developed a common sprint review process accompanied by a "science fair" approach.

> Reviews [were] shortened to 90 minutes, with each team presenting for just 10 minutes each. Within those 10 minutes, the Scrum Master provides a condensed version of the Sprint accomplishments and metrics, and the team demonstrates only the *sexy* stuff . . . Following that 90-minute period, we hold a one-hour Science Fair in the cube areas where the teams sit. Just like parents viewing student experiments at a school Science Fair, this hour is for stakeholders . . . to make the rounds at their own pace to each team they have an interest in for any given sprint.[22]

DISCUSSION TOPICS

1. Why did the Scaled Agile Framework provide value to Valpak over a standard Scrum approach?
2. What are the strengths and weaknesses of the Valpak approach?

NOTES

1. Jochen (Joe) Krebs, Presentation to the Agile Boston Group, February 2010.
2. http://www.linkedin.com/pub/stephanie-stewart/31/317/a01.
3. http://scaledagileframework.com/.
4. Dean Leffingwell, "Architectural Epic Abstract," http://scaledagileframework.com/architectural-epic/.
5. Ibid.
6. Ibid.
7. Ibid.
8. Stephanie Stewart, comments from "Podcast This Week."
9. Ibid.
10. Blog, "Release Dates in Agile Not Taboo," posted by Stephanie Stewart, June 25, 2012, http://iamagile.com/2012/06/25/release-dates-in-agile-not-taboo-4/.

11. Jim Highsmith, *Agile Project Management* (Reading, MA: Addison-Wesley, 2010).
12. Stephanie Stewart, personal email correspondence.
13. Ibid.
14. Ibid.
15. Chris Cate, Cox Target Media CIO, personal email correspondence, November 12, 2012.
16. Ibid.
17. Ibid.
18. Blog, "An Agile Light Bulb Moment: Forming Projects Around Teams," posted by Stephanie Stewart, April 20, 2012, http://iamagile.com/2012/04/20/an-agile-light-bulb-moment-forming-projects-around-teams/.
19. Ibid.
20. Blog, "Accountability to Sustainable Pace and Work/Life Balance," posted by Stephanie Stewart, April 10, 2012, http://iamagile.com/2012/04/10/accountability-to-sustainable-pace-and-worklife-balance/.
21. Blog, "Our Evolution of Sprint Reviews," posted by Stephanie Stewart, June 13, 2012, http://iamagile.com/2012/06/13/our-evolution-of-sprint-reviews/.
22. Ibid.

28 Case Study: Harvard Pilgrim Health Care

BACKGROUND

HARVARD PILGRIM HEALTH CARE (HPHC) IS A FULL-SERVICE HEALTH BENEFITS COMPANY serving members throughout Massachusetts, New Hampshire, Maine, and beyond. HPHC's mission is "To improve the quality and value of health care for the people and communities we serve."[1] For over 40 years, HPHC has built a reputation for exceptional clinical quality, preventive care, disease management, and member satisfaction and has consistently rated among the top plans in the country.

HPHC has a strong reputation for being customer-focused and having a commitment to excellence. HPHC was ranked the number-one private health plan (HMO/POS) in America according to an annual ranking of the nation's best health plans by the National Committee for Quality Assurance (NCQA) for 2012/2013.[2] HPHC is the only private health plan in the nation to be named number one for member satisfaction and quality of care for nine consecutive years.[3] Harvard Pilgrim Health Care of New England, HPHC's New Hampshire affiliate, is the top-ranked plan in New Hampshire.

Summary of HPHC's business

- Not-for-profit health plan, based in Wellesley, Massachusetts
- 1,100 employees across eight locations
- Over one million members in Massachusetts, New Hampshire, Maine, and beyond
- Full range of health insurance choices, funding arrangements, and cost-sharing options.

Key business challenges faced

- Emerging requirements for complex benefits (flexibility)
- Increasing populations and complex processing push the extremes of capacity (scalability)

- Intensifying demand for information and results of processing (transparency)

- Escalating need for faster deployments of new products (speed to market)

- Aging application suite not up to the challenge.

Michael Hurst (now retired from HPHC)[4] was responsible for leading the Agile Project Management adoption effort for HPHC. Hurst is the director of the HPHC PMO, which is in the fifth year of a hardware, software, and "project-ware" migration of its core systems for health care and claims management. He started in IT as a programmer for the aeronautical engineering department at MIT 45 years ago. Subsequently, he became a multistate licensed psychologist and an associate professor at Boston University School of Medicine for 25 years. He founded five health care and health IT companies, with one going public.

Vijay Bhatt,[5] deputy CTO of HPHC, has been the Agile executive sponsor working with Hurst and the joint project management office (JPMO) in implementing this overall program over the entire five-year period.

Deborah Norton, CIO, has been the supporting executive sponsor for the entire IT strategy program, including the Scrum program track. She also has encouraged the use of Scrum for business projects and for technical but non-IT strategy projects. A key directive in the adoption process was her declaration that all IT strategy projects were to be Agile Scrum by default with exceptions made only as requested and approved by her office.

OVERVIEW

HPHC adopted the following IT strategy principles in 2008 to deal with replacing its entire operational software and hardware systems:

1. Create strong business operations and IT partnerships.
2. Develop flexible business architecture to support change (manage risk with component solutions, implemented during a five-year window).
3. Partner with vendors interested in our outcome.
4. Implement component solutions for health care, such as

   - Oracle Health Insurance (OHI) for claims processing
   - Access and Identity Management (AIM) for security
   - Oracle E-Business Suite (EBS) for finance, sales, billing, service
   - Edifecs for (Electronic Document Interchange (EDI) platform capabilities
   - Other application vendors as appropriate.

5. Introduce Agile approach and philosophy without disrupting the four years of NCQA #1 standing as the best health care plan in America.[6]

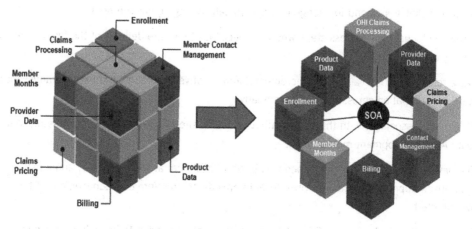

FIGURE 28.1   HPHC architectural transformation.

Source: Courtesy of Harvard Pilgrim Health Care.

Implementation of this plan required HPHC to totally revamp its internal systems infrastructure to move from a legacy big-box hardware/software environment that was no longer supportable to a totally new service-oriented architecture approach, as shown in Figure 28.1.

- This effort involved the replacement of the majority of the company's most business-critical software applications.

- The company set a goal to complete this entire effort within a five-year period by the end of 2012.

- However, new federal mandates and business priorities resulted in an extension into 2013 for the full scope of capabilities.

- The flexibility of the Agile Scrum process allowed them to do this while ensuring that the highest business priorities always were at the top of the development and funding queue.

It was recognized early on that the classical Waterfall approach couldn't deliver results in that timeframe and would have an unacceptable level of risk. The key drivers for change were:

- Business users didn't want to wait an extensive period of time between requirements and UAT to have something delivered.

- The technical team wanted much greater business input all along the way—they didn't want to discover at the end of the project that they hadn't got it right.

- It was too risky without frequent and continuing business feedback given 10 strategic programs spanning 14 functional areas across five years and a very significant financial investment.

- It was too slow to respond to change in a dynamically changing environment.

- It was too focused on process steps when the need was to ensure delivery of the highest business priority capabilities.

As a result, HPHC developed an incremental and iterative approach—the Scrum framework for an Agile transformation across the whole company.

- All projects associated with this major initiative, called the *IT strategy*, were, by default, to move to an Agile development methodology.

- The new approach that was developed was a hybrid of Scrum at the team level and a more plan-driven approach at the enterprise level to provide program/project management of the overall effort.

- The total effort involved over 200 different projects that all needed to be well integrated to successfully achieve completing the overall business objectives.

There were a number of cultural obstacles to overcome in implementing the Agile transformation.

- Initially the cultural issues and resistance to an entirely new way of managing projects overwhelmed the intensive training that was received, and a couple of the initial six major projects failed in their first attempts at creating and executing using Agile Scrum processes.

- However, those challenges have been overcome with continuing training and emphasis, resulting in the overall IT strategy being on track for completion with planned business priority-driven extensions.

- The implementation, which itself has been iterative and incremental, has been very successful in spite of the far-reaching impacts of all the new development and applications.

Impact of Outsourcing and Vendor Partnering

The effort was further complicated by the fact that a large portion of the responsibility for HPHC's IT infrastructure maintenance and application development is outsourced to Dell Services (formerly Perot Systems). New SOA application development, as opposed to developing extensions to the legacy monolithic application or maintenance and updating work, was a new role calling for new skills along with increased resources. Some of these were provided through off-shore resources and some through in-shore resources at Dell. The additional skills, coordination, communication, monitoring, and overall collaboration were a challenge that took considerable time and effort to achieve. The two companies really applied the Agile principle of "Customer collaboration over contract negotiation."[7]

In addition, Oracle Healthcare Systems played a significant role in providing a large amount of application software developed in collaboration with HPHC's Scrum teams to fit HPHC's needs. A number of other application vendors also were involved, and a working relationship that meshed with the HPHC environment had to be devised for each one.

As a result, in addition to the normal challenges of an Agile transformation of building cross-functional collaboration across different functional departments, HPHC needed to build an equivalent level of cross-functional collaboration among these different companies.

Role of the PMO

Prior to the Agile transformation, HPHC had built a classical plan-driven PMO organization implementing Waterfall software development life cycles (for IT projects). Even before the company moved to a more Agile approach, many of the traditional PMO artifacts and processes fell into disuse. It was recognized that a major shift was needed in the role of the PMO. As a result:

- **The PMO in today's organization is relatively small:** It plays more of a consultative role on the methodology and provides training to the rest of the organization on the methodology. It truly has a "servant-leader" role as opposed to a directive, command-and-control role.

- **Most of the Project Managers are distributed into the operational business units with which they are associated:** These project managers are not part of the PMO organization. Only the director, five enterprise-level project/program managers, two senior consultants, a reporting analyst, and a training coordinator are part of the current PMO organization.

Responsibilities of the PMO include:

- Cultural education in the Agile Manifesto principles and philosophy

- Both team and individual training in Scrum practices

- Selecting tools and providing training in how to use the tools for team purposes

- Coaching Scrum behaviors, reinforcing best practices; supporting collaborative learning mechanisms

- Facilitating portfolio, program, and project schedule and resource planning and conflict resolution

- Training and implementation of Kanban release management process

- Training, implementation, and reporting of cross-team dependencies

- Reporting on portfolio programs and projects to all levels of the organization.

A joint PMO (JPMO) between HPHC and Dell Services was created to deal with the Agile rollout, training, coaching, and overall implementation of the IT strategy. The objectives of creating a joint PMO with Dell were:[8]

- to reduce redundancy, cross-messaging, and project management costs

- to promote by team example, increase communications, share work

- to ensure project success of five-year IT strategic portfolio

- to support new software development process.

Project Governance

In addition to the JPMO, several other project governance mechanisms were implemented to provide project coordination and consistent implementation of the methodology, and to provide a framework for ongoing, continuous improvement. The project governance framework included:

1. Product Owners' Forum (the POF), where all business and technical product owners meet with the CIO to coordinate their individual efforts. The POF was chartered with:

   - improving product owner communications about team progress, hindrances, and cross-team dependencies
   - identifying opportunities for inter-team collaboration, sharing of resources, development needs and efficiencies
   - identifying and resolving potential mismatches or conflicts in resources and needs
   - escalating to IT Strategy Committee for review or ratification if needed.

2. Scrum Masters' Forum, where all Scrum Masters meet to coordinate their individual efforts:

   - issues and answers about the Scrum Master role
   - tools for meetings and progress tracking
   - resources and training
   - shared best practices.

3. Business Analysts' Forum, where business analysts meet to discuss:

   - issues and answers about the business analyst role on Agile/Scrum teams
   - shared experiences and problem solving
   - resources and training
   - shared best practices.

Figure 28.2 shows the structure of the overall Project Governance Model that was implemented by HPHC. The three listed business unit columns are only examples of the 14 IT strategy initiative business units that were actually in play. There were 10 IT strategy initiatives, 14 business units, and 40+ Scrum and hybrid Agile teams; in addition, there were usually 40–50 other corporate-level projects being conducted at the same time as the IT strategy ones.

At the end of the second and third years of the Agile implementation, a survey was implemented to assess the effectiveness of the methodology and how completely and consistently it was being implemented. The survey was administered to all members of 46 teams in flight that were using Agile/Scrum. The survey content was created by Vijay Bhatt and the joint PMO, modeling some of the items on the Nokia survey but modifying as appropriate for the HPHC focus on compliance with the Scrum methodology.[9] Bob Sullivan, senior PMO project manager consultant, conducted the survey and analyzed the results:

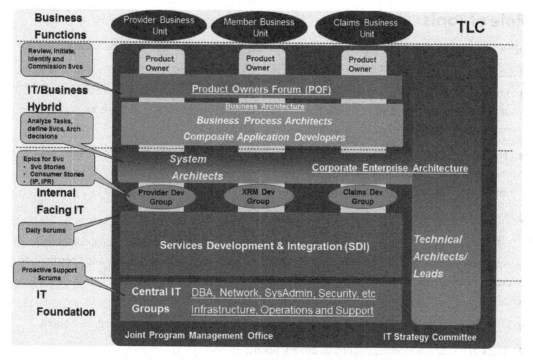

FIGURE 28.2   HPHC IT strategy governance structure.

Source: Courtesy of Harvard Pilgrim Health Care.

- The 2009 results across the many different facets of a Scrum framework showed that these teams achieved 80–90% compliance in 75% of the teams and to a much lesser degree in the other 25% (ones that were much more hybrid or even traditional).

- A repeat of the survey in 2010 (obtained in 2011) showed essentially the same results, with some improving and others dropping elements of the Scrum framework.

- Other, less-structured surveys of the product owners and the business units showed a high degree of satisfaction with the new applications and functionality meeting their needs. The IT department also received far fewer requests from the business units for modifications of the functionality they received. A defect tracking program with a daily Scrum was put in place and continually reviewed with the business and development teams to improve the process even further.

In 2012, HPHC had some perceived loss of Agility and Scrum compliance at the team level. Although it hasn't been deemed necessary to do a formal survey again, the HPHC estimates that one-third of the 100+ total project teams are Agile Scrum; one-third are hybrid; and one-third are classical plan-driven. This kind of variation in project methodologies stems from the widespread types of projects ranging from IS upgrades and maintenance through application configurations on up to full new development projects.

Role of Tools

HPHC recognized early on that the scope and complexity of this effort could not be successfully accomplished without the appropriate tools to manage the effort.

- Initially, as suggested by the Scrum trainers and Agile coaches, teams used spreadsheets and story wallboards. However, as the number of teams grew, the complexity of the work and coordination grew exponentially.

- The simple approach did not work from the portfolio management and reporting perspective, though it was fine for the individual teams until they had to deal with the tremendous number of inter-team dependencies.

- Nor did the team-to-team Scrum-of-Scrums work.

   There were over 1,200 inter-team dependencies identified. After a selection process, HPHC selected Rally software and now relies heavily on it to provide overall management and coordination of dependencies, reporting, and integration of the entire effort. The Rally software is used to do the following:

- Provide teams with a common tool for tracking their stories and tasks in a Scrum framework.

- Track and manage interdependencies among teams.

- Provide overall status reporting at both the enterprise level and the project/team level. Teams can use paper-based story cards within their teams if they want to, but all teams are required to use the Rally software for managing user stories and milestones that have inter-team dependencies either as predecessors or successors. Some larger programs have created federated projects in Rally that contain all the stories and milestones from multiple teams that have external dependencies but have the individual teams managing everything else via wallboards, spreadsheets, Microsoft Project or other tools.

- Provide a Kanban board management approach for monthly promotion to production from all the teams.

- Provide a Kanban team management approach for environment development and test data management.

Project Methodology Mix

The company has a range of different kinds of projects—most are Agile, either Scrum and/or Kanban, but there are also some plan-driven projects.

- Projects in the Corporate Information Management (CIM) group (Extract, Transform and Load (ETL) projects, report development projects, and the like) are managed with their own highly structured and predictable process and tools. They maintain a single Rally *project*, which is the repository of all CIM stories with milestones, deliverables, and dependencies with other team projects.

- The product group (insurance products, not software products) also has a highly structured and predictable path to creating their product deliverables. It has largely traditionally structured projects, though it, too, is now finding that Agile might be the better process. As a result, these projects have become hybrid projects.

- At the other end of the spectrum are projects doing new code development for new or existing but incomplete applications. These are all Agile Scrum projects.

Also Agile (but using Kanban rather than Scrum as the primary implementation framework) are those building environments for testing groups of applications (called *development integrated test environments*) and those creating sets of test data for the different applications, projects, and environments. Both of these teams are highly limited by capacity in staff, hardware, and software that have to be brought together in usable stacks and then rearranged as needed. The Kanban approach fits them very well.

Project Portfolio Management

Every year, HPHC's senior executives review a list of proposed project initiatives and prioritize and rank those initiatives to maximize the overall ROI of the portfolio.

- Each project is required to have a one-and-a-half-page Project Opportunity Statement (POS) that summarizes the project objectives, costs, schedule, and expected ROI. Behind this minimum requirement, each project may have a variable level of backup supporting details, depending on the nature of the project.

- For some projects, a two- or three-page POS may be sufficient; for other projects, much more detailed information is needed to support the request.

The PMO Senior Consultants provide assistance to the Executive Sponsors, Project Leads, and Product Owners in completing the POS. These are published for community consumption so that teams and the general business community can understand the goals, rationale, and expected outcomes and return to be expected.

The following is a summary of the management approach used to monitor progress against these plans:

- Project Managers begin weekly reporting once projects have been approved with a designated budget and staff (HPHC, Dell Services, and other third parties when it is a technical project). Progress against the projected cost and schedule is tracked in Dell's Project Reporter (PR) database reporting tool.

  - It is used for all projects and provides the source of truth for all budgets, resource plans, and progress reports against the approved project milestones and resources.

- Project and Program Managers update PR every Friday with basic summaries and detail updates as needed.
  - The PMO then extracts the information and creates an executive dashboard that is published every Wednesday for the leadership team.

- For teams with both a Scrum Master (SM) and a Project Manager (PM), the PM normally is responsible for the external team reporting. There are some teams where the SM also has the PM responsibilities, normally assisted by a project coordinator or assistant for the administrative duties. There are relatively few PMs who can adapt to both roles at the same time simply due to the external demands, but there are some who have no problem.

- The PMO creates all needed reports from the team entries in Rally and PR.

- Once a month the CIO presents a comprehensive report on the entire portfolio to the executive leadership committee (ELC) and to the board of directors. This comprehensive report is provided as a dashboard along with detailed backup.

- Once a quarter, corporate leadership reviews the status of all the corporate initiative programs and projects in the portfolio. Then the reports are assembled and communicated to the corporate community. Any adjustments in priority or revisions in timelines and milestones are also set at this time as part of the process.

PROJECT MANAGEMENT APPROACH

Project Methodology

The HPHC project methodology for the IT strategy was designed with the following principles. It was to be the following approaches:

- **Business-based approach:** The projects were not to be just IT projects; they were to be operational improvement projects with business value that met a defined market needs.

- **Development approach:** The Agile methodology meant business collaboration, increased flexibility, responsiveness, speed to market, and productivity.

- **Component (SOA) approach:** This focused on standards and reuse of services.

- **Approach with regular reviews:** Reviews of capability requirements for course corrections and realignment to the Corporate and IT Strategies over the five-year plan.

Figure 28.3 shows a high-level overview of the HPHC life cycle model. The following is a description of the phases in the life cycle model.

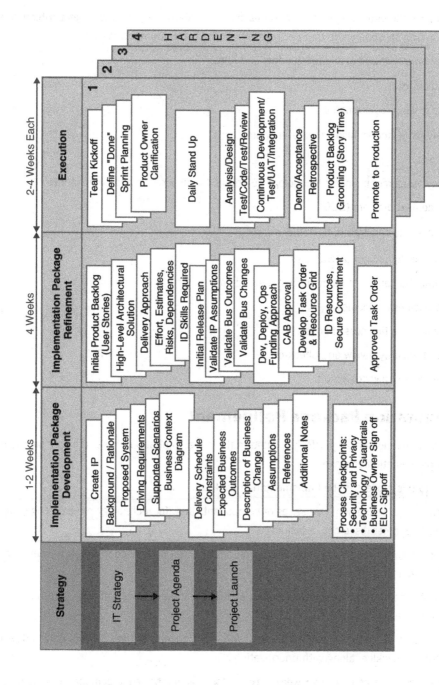

FIGURE 28.3 HPHC project life cycle model.

Source: Courtesy of Harvard Pilgrim Health Care.

Implementation Package Development

Each project goes through a project initiation phase prior to startup to plan the project, which consists of completing an *implementation package* containing the following items:

- Identification of owner and all stakeholders and roles

- Vision, scope, goal

- High-level driving requirements

- Epics (very large stories, also called scenarios)

- Assumptions

- Technology assignments and/or technology guardrails for all services and components

- Business context diagram (BCD)

- Delivery schedule constraints-flexibility matrix

- Description of the project deliverables

- Recommendation of the implementation methodology

- Expected business outcomes

- Description of the business change

- Privacy and Security review and check-off

- Executive sponsor owner sign-off.

Implementation Package Refinement

After the implementation package (IP) is created in the project initiation phase, the next step is to refine the implementation package:

- The input to this process is the implementation package.

- The output of that refinement is a series of components and, for technology projects engaging Dell Services or other resources, a task order.

Table 28.1 shows a summary of the inputs and outputs that take place in the implementation package refinement (IPR) process:

- The IP is developed by the business, normally a HPHC functional business area, with the assistance of the IT department.

- The IPR process is conducted by Dell Services with assistance from the business unit in developing the product backlog, along with prioritization.

- The entire process takes four to six weeks, except for projects that are a component or continuation of a very large product backlog.

TABLE 28.1 Implementation package inputs and outputs

Inputs	Implementation package refinement outputs
Content	■ Product backlog (PBL)
	■ PBL Prioritization
	■ Approximate PBL sizing
	■ High-level architectural solution
	■ High-level release plan timeline
	■ Expected sprint dates, cumulating in interim releases ("Schedule," "Milestones")
	■ Interim releases ≫ Final release
	■ Multi-team coordination within program and PBL
	■ Resource grid
	■ Release plan
Participants	■ Product owner
	■ Technical leads; more of team as needed
	■ Scrum Master and/or project manager
Outputs	Task order

The most successful process HPHC had for Agile project initiation was the IP, then the IPR; however, for new teams that did not have an established velocity, both the IP and IPR were limited to only proposing three initial sprints. The team would establish its velocity and could then forecast its release plan (at a high level) for its product backlog with a much better degree of accuracy. After that point, an experienced Agile team would use its past velocity in creating subsequent and follow-on projects.

Project Reporting

The following is a summary of the reporting tools used by HPHC to manage the project:

■ The Rally software was used heavily for project reporting at a detailed ("micro") level, primarily for team reporting but also for dependency coordination and management.

■ Weekly project status reports were used for executive high-level reporting (Table 28.2).

TABLE 28.2   Project status report

Project name	Executive sponsor/ project lead	Project Manager	Baseline start	Baseline end date	Revised end date	Product effective date	Overall status
							Green
							Yellow
							Red

- A risk, assumptions, issues, and decisions (RAID) report was also used by almost all teams to track these items and manage their resolution.

- Dependencies between teams' user stories were tracked in Rally, and a dependency report was used to track and report the status of all dependencies.

- Project *dashboard* reports were also updated weekly to provide a high-level status of all projects as shown below:

- A special "Yellow–Red Status Report" was used to report on any project that was in red or yellow status during the daily stand-ups for all 100+ projects in flight at any given time.

  - There simply were too many participants to allow oral reporting on the standard "What I've done, What I'm doing, and What is standing in my way" Scrum process.
  - Those details were made available through Project Reporter for everyone's review.
  - The Yellow–Red report was used to focus on what was standing in the way for teams.
  - This report did provide the details for just those teams so that the Scrum could focus on resolving blocking issues.
  - For any given Scrum, 1–15 teams might be providing their status and their blocking issues. For a team needing assistance or wanting to escalate to everyone's attention, these reports were one of the best ways to do that.

Contractual Relationship with Dell Services

One of the most significant complicating factors that HPHC had to deal with was the contractual relationship with Dell Services.

HPHC outsourced a substantial amount of responsibility for maintenance of its IT infrastructure and application development to Dell.

In fact, the number of Dell people involved in the effort was significantly larger than the number of HPHC IT people involved. HPHC had about 80 IT people engaged in the effort, along with several hundred from the business units (as SMEs and stakeholders), and another several hundred Dell technical staff, SMEs, and stakeholders from the operating units it maintains for HPHC. In total, about 800 people were participating between Dell and HPHC.

TABLE 28.3   Results of Joint PMO

Before Joint PMO	After Joint PMO
Dual project reporting systems and data entry	A single reporting system
Two project administration systems (project creation, approval, administration, and change process)	A single project administration, process tracking, and compliance tool
Dual training programs	HPHC management of both internally and externally sourced training for new programs with joint PMO approval
Dual project repositories	Single project repository linking to each PMO's website and the project reporting and project administration systems
Infrequent and informal coordination meetings between PMOs	■ Joint PMO email distribution group for communications ■ Initial biweekly meetings are now weekly to address JPMO backlog ■ Thrice-weekly joint pre-review meetings for new projects and change requests ■ Weekly project reviews ■ HPHC CIO and Dell Client Executive issue escalation meeting

It was essential to develop a close collaborative relationship with Dell to make this work, and both companies decided to form a joint PMO to facilitate sharing of responsibility for the effort. The formation of a joint PMO was initiated to enhance communications and collaboration between the two companies while reducing duplication of effort in a number of areas. Significant results are summarized in Table 28.3.[10]

CHALLENGES

Cultural and Organizational Challenges

Tables 28.4–28.7 are a summary of the key cultural and organizational challenges faced by HPHC, and how they were handled.

TABLE 28.4 Solutions to cultural and organizational challenges: cultural change

Challenge	Solution
Many people felt comfortable with a plan-driven approach and had difficulty adopting the shift in thinking that was required to move to a more Agile approach. Other cultural issues included:	It was recognized that some people were not going to be able to make the transition to an Agile approach, and some of those people moved on to other assignments that were more plan-driven
■ Resistance to change and opposition from employees on both the business and technical side to the change	About 800 total business, IT, and Dell folks were put through training in the first two years; and subsequently all IT and Dell staff routinely take the Agile training courses
■ Cultural differences between HPHC and Dell in terms of time and activity tracking needs and requirements	Special training and coaching are made available to HPHC business units to assist their transition (where desired and possible)
■ Focusing on completion of deliverables rather than process and task steps	
■ The "just enough" approach of Agile completely butted against the HPHC culture for 120% planning, effort, and thoroughness	
■ Territorial (stovepipe) attitudes toward responsibility and collaboration and differences of opinion on resolutions	
■ Instead of an HPHC business unit dictating requirements, they had to assign SMEs to work on teams with Dell associates to define and refine stories, acceptance criteria, test cases, and test data. The collaboration was daily rather than quarterly	

TABLE 28.5 Solutions to cultural and organizational challenges: level of business involvement

Challenge	Solution
Although business leaders wanted much closer involvement in the development of their needed applications, the need for continual work with the team, the reviews, the demonstrations, and the re-prioritizations every two to four weeks with Agile teams had a far greater impact on them than expected	More continual involvement of the business with the team (even if only through product backlog prioritization meetings, sprint reviews, and demonstrations) is a far more productive and useful process for both the development team and the business. The business really gets what it wants most in a timely fashion, and that reality has led to acceptance of the additional demands on its time
The business leaders were very pleased with the results and the delivery time, but in the beginning it was a very foreign process that did not fit in with their traditional day-to-day operations	The business has had to really change its allocation of staff time and duties from just their functional orientation to the collaborative team. As many of the coaches and trainers predicted, this transition has taken years, but it is clearly becoming the norm instead of the exception
After all, they were not product managers or developers. They were business people running operations with their staff who had to be assigned to these teams as subject matter experts to assist in story writing, requirements, test scenarios, test data creation, and demonstrations	

TABLE 28.6 Solutions to cultural and organizational challenges: project portfolio management

Challenge	Solution
Agile does not provide any guidance on how to do project portfolio management of a large number of projects like this at an enterprise level	HPHC had to develop an overall approach to determine how projects were prioritized and selected for implementation to maximize the overall ROI of those projects and provide a way of measuring progress once those projects were selected
	HPHC also had to devise different levels of reporting that minimally affected teams yet provided the kind of information required at higher levels in the organization

TABLE 28.7 Solutions to cultural and organizational challenges: Scrum Master and Project Manager roles

Challenge	Solution
HPHC had to reconcile the Scrum Master role with the Project Manager role: Those who had training either with HPHC or outside of the company realized that the Scrum Master was not the director of everyone's time and activities but was much more facilitative Scrum Masters have some activities similar to a project manager role, such as removing obstacles to progress, setting up meetings outside of the team workspace and time, recording progress, and reporting to outside parties The traditional Project Managers were more comfortable with their responsibilities for schedule and budget but much less so with the increased facilitative role and decreased command and control	Some Scrum Masters did take on more of the Project Manager tasks without becoming a "dictator" to the team Similarly, those who were more project manager-oriented had to be more facilitative, let team members choose and prioritize tasks rather than be assigned them, and empower team members to ask lots of questions to clarify requirements and tests The recruitment and selection process of both HPHC and of Dell Services had to change to emphasize individuals with experience and comfort in both the Scrum Master and project manager roles. If individuals only had experience in one role or the other and not both, they needed to provide evidence that they could work with a cohort (either Scrum Master or project manager) in managing a project team Some Agile projects would have a Scrum Master and a 10% Project Manager for Dell timekeeping and reporting responsibilities Eventually, it became more common to have a project coordinator or administrator assigned to handle administrative duties, relieving a Scrum Master to focus on team facilitation

Contractual Challenges

Tables 28.8–28.11 are a summary of the key contractual challenges faced by HPHC and how they were handled.

TABLE 28.8 Solutions to contractual challenges: cross-functional collaboration across companies for individual performance reviews

Challenge	Solution
Creating cross-functional collaboration across multiple companies (HPHC and Dell and various vendors) was essential because most teams were composed of representatives of both companies	The companies had to agree to change some of their normal performance measurement systems to allow feedback and inputs from outside the company to support a collaborative cross-functional team approach
For both HPHC and Dell employees assigned to teams, a large part of their performance review had to come from the teams, but neither company's HR systems originally allowed such input	Over time, both companies made changes to allow this cross-company input and feedback

TABLE 28.9 Solutions to contractual challenges: team assignments and resource sharing

Challenge	Solution
Initially, there were too many shared resources, and with that level of resource sharing, it becomes very inefficient for the people to do "context switching" from one team to another that they participate in	Over time, both Dell and HPHC were able to concentrate team assignments, but it is still a struggle since all of the Dell staff time is either billed to projects or core maintenance work, while HPHC staff is not accounted for in the same way
	HPHC's technology partners, especially Dell Systems, had to change from assigning staff to five or six teams in slices of 16–20% time to just one or two teams. However, this is one area that has significantly gone backward, as the pressure on budgets and time accountability has risen

TABLE 28.10 Solutions to contractual challenges: contracting approach

Challenge	Solution
Because the Agile development approach relied heavily on outside contractors (Dell and Oracle), a contracting approach needed to be developed that was consistent with the Agile development approach. It wasn't possible to force another company involved in a contract to completely change its own project methodology	The contracts with Dell and Oracle were treated as high-level agreements. It was explicitly agreed and understood that the detailed requirements would be further elaborated as the projects progressed. Broad schedule milestones were set along with very high-level budgets that would be refined as the details became known in the iterative Agile process
Standard IT contracting normally specifies a very detailed list of requirements, if not a full work breakdown structure, along with detailed release planning, often months or even years in advance of the work	Some compromises needed to be made to come up with a consistent way of managing the effort at a project level while meeting the overall corporate goals and contract obligations
In this five-year program, that simply was not possible since nearly all the applications were brand-new and were being delivered into a brand-new architecture (SOA)	Both Dell and Oracle, and other vendors, supplied staff to participate locally in teams in addition to their offsite resources. They also supplied coordinators to provide the overall schedule and resource coordination
In addition, there was urgency to begin, since both the hardware and software platforms were no longer supported; many new types of features and functions were needed that could not be developed in the old environment, and change was the name of the game in health plan capabilities	With Dell, HPHC created the joint PMO
Nearly everything was changing with urgency, so the long lead time for detailed prospective planning simply did not exist	With Oracle Health Insurance (OHI), the effort was large enough (10+ projects at any given time) that the HPHC OHI Product Owners created a program office. Then the HPHC joint PMO provided the same kind of coordination and collaboration with that program office

TABLE 28.11   Solutions to contractual challenges: compensation, billing, and multidisciplinary roles

Challenge	Solution
The HPHC reimbursement system paid for job activities according to designated position codes, and Dell staff had to record the activity code along with the project and other information in their timekeeping system These were contractual terms that were embedded in the financial systems of the two companies It was a real cultural change for folks to take on multidisciplinary roles and activities	Since Agile Scrum philosophy includes the concept of the team being responsible for completing work and team members assisting others who need additional help, HPHC and Dell Services both had to change their respective time accounting, billing, and payment systems to reflect the fact that developers could write requirements and stories (previously that activity had been restricted to business analyst and subject matter expert job codes) Similarly, Business Analysts might write test cases, which previously would have been restricted to test architects and QA specialists

Technical Challenges

Tables 28.12–28.15 are a summary of the technical challenges faced by HPHC and how they were handled.

TABLE 28.12   Solutions to technical challenges: service-oriented architecture interdependencies

Challenge	Solution
The implementation of a service-oriented architecture introduced some significant dependencies between the teams designing the services and the teams developing the applications that consumed the services The service-oriented architecture required the creation of 14 major services with over 100 interfaces	The Rally software was used to track and manage interdependencies among teams. In addition there were several forums created for product owners and Scrum Masters to coordinate the efforts of their teams

TABLE 28.13   Solutions to technical challenges: automated regression testing

Challenge	Solution
Because of the fast pace of the effort and the complexity and interdependencies in the overall solution, it was essential to implement some form of automated regression testing as the five-year program progressed	HPHC used an offshore team as a centralized resource to design and implement an extensive automated regression test capability. Teams now have to provide their regression test scenarios and test data before putting their work into the promotion to production process

TABLE 28.14   Solutions to technical challenges: enterprise-level architecture planning and design

Challenge	Solution
HPHC has a very complex environment, and it is essential to create an integrated approach for all enterprise systems to work together in a well-designed service-oriented architecture. To achieve that goal, it is essential to have an architecture that is consistently implemented across all project teams	All projects go through an architectural review early in the project
The risk associated with a team developing its portion of the solution in a way that is not consistent with the rest of the architecture and then having to refactor the design and the affected applications to become more consistent is unacceptable	An architect is assigned to each team to provide additional guidance and coordination. A SOA Center of Excellence (SOA COE) and Corporate Architecture Board review all projects technical designs and are responsible for approving them and any changes made over time
No single project team has sufficient visibility into the overall architecture at an enterprise level to make decisions unilaterally regarding designing its particular portion of the application to be consistent with that overall architecture	A formal change management process is conducted through the SOA COE for changes involving dependencies and Services. The process involves all affected teams and in the last year has provided turnaround answers or solutions in less than a week for all cases
Some level of upfront architectural guidance has to be provided to the teams, and some level of coordination and review is needed to ensure that the architecture is consistently implemented	

TABLE 28.15   Solutions to technical challenges: software release process

Challenge	Solution
The process for releasing software had to be carefully managed to provide some level of configuration management and release control to ensure that any applications being released were consistent with other applications that were already in production or being released at the same time	A staged release process was developed to test compatibility of a given application in combination with other applications in different environments prior to release. For example, groups of related applications (ones that "touched" one another) would have an environment created for that level of testing
Initially, the process for releasing applications was too Agile (basically "when ready" according to team standards) and needed to be better controlled	Over time, a larger systems integration environment was created to allow testing of end-to-end scenarios. Work continues on this very complicated issue
	A Kanban process was devised to control the flow of applications through this staging process prior to release. A monthly promotion to production schedule was devised and then used across all teams. In the future, all iterations and releases will be coordinated across all of the teams
	As the applications and SOA capabilities became more mature, an incremental process of selectively releasing a limited number of components at a time was adopted that allowed migrating business lines dependent on those capabilities
	This approach was adopted, rather than a "big bang" approach of releasing a large number of components simultaneously and trying to migrate the data and records of millions of customers simultaneously. This more iterative approach to migrating to the new environment has proven highly successful

Other Challenges

Tables 28.16 and 28.17 are a summary of some other challenges faced by HPHC and how they were handled.

TABLE 28.16   Solutions to other challenges: office space

Challenge	Solution
To provide co-location for the teams, a number of conference rooms were converted into "bullpens" to provide a working area for the teams This presented a challenge because in most cases the team members had their own individual cubicle plus the shared bullpen This temporary doubling of space needs presented a challenge because it consumed a lot of floor space unnecessarily until adaptation could be made	To more efficiently utilize the office space, many of the people who work on Agile teams had to give up their individual cubicles and use the Agile workrooms as their primary work area For most on these teams, this approach worked well since HPHC provided "drop-in" cubes for those needing temporary individual space

TABLE 28.17   Solutions to other challenges: government regulatory requirements

Challenge	Solution
HPHC is required to comply with a number of government regulatory requirements for the healthcare industry One of those requirements is to use industry-standard procedure codes for all medical procedures The government has recently required converting from the ICD-9 system of about 1,400 procedure codes to the ICD-10 system of over 140,000 procedure codes The Department of Health and Human Services (HHS) published a regulation requiring the replacement of the ICD-9 code set with ICD-10 as of October 1, 2014. This was a very ambitious goal with an end-date that could not slip, and it had a significant impact on most of HPHC's applications	In order to meet this very difficult requirement in the time required, over 40 different Agile teams were created, and the effort required to meet this requirement was partitioned among those teams so that they could work in parallel to get this done within the mandated timeframe. This very large mandated program also created its own program PMO to work with the larger JPMO Many government mandates had to be given priority over other business needs, and the corporate initiative team helps set the priorities and resolve the disputes

KEY SUCCESS FACTORS

1. Immediacy and persistency

- Addressed issues as they arose (very frequent interactions/follow through to resolution)
- Individual commitment to success and trust in the other's commitment to success
- No fear of losing one's job
- Persisting through trial balloons to useful practices and tools.

2. Leadership support

- Positive feedback on successes
- Joint communications announcing implementation of new methods.

3. Well-trained people

Having the right people on the teams who are properly trained in the Agile methodology is essential. It is important to recognize that some people will not be able to make the transition, and having alternative teams and work is important for morale in the workforce.

4. Cultural change

Changing the culture of the organization to enable a more collaborative, cross-functional approach was essential. Some projects will fail if cultural change is not addressed early on and continually reinforced thereafter.

5. Integrated hybrid methodology

Successfully accomplishing this effort required a combination of an Agile approach at the team level and more of a plan-driven approach at the enterprise level.

6. Tools

The use of tools such as Rally has been essential for managing an effort of this scope and complexity.

7. Collaborative approach with other companies

The overall project could not have been successfully accomplished without developing a collaborative approach with the other companies involved (Dell, Oracle. and other vendors).

8. Early planning

- Get executive buy-in upfront
- Five-year high-level roadmap (plan for delivery of capability along the way)
- Fit within an achievable and supportable funding profile
- Ensure project planning happens to at least some degree in terms of architecture and data management.

9. Execution

- Accountability (clear lines of responsibility)—Senior business and IT operations staff as Product Owners
- Flexibility (don't be afraid to change what's not working)

- Dependency management (SOA begets intricate dependencies that need significant oversight)
- Daily but short meetings to monitor and manage progress, and find/resolve impediments.

10. Bringing it home

- It took real courage of conviction to begin the migration.
- The IT strategy is still a work-in-progress, but well over 80% of the way there, even when another major corporate initiative had to take priority, extending the five-year program by six months.

CONCLUSIONS

1. Hybrid project and portfolio management methodology[11]

- Found a hybrid portfolio approach that was conducive to finding and fixing issues early
- Delivered value early and all along the way
- Delivered usable capabilities in year 1 that gave the strategy momentum and credibility
- Enabled a strong business and IT partnership
- Adapted more readily to changing priorities (inevitable over five years).

2. Risk management

- Started to cross the bridge before it's finished—very challenging step that pushes everyone in a very interdependent environment
- Began migration with the simplest accounts to support; then added capabilities just ahead of the migrating accounts (online billing, EDI enrollment, administrative services only billing)
- Segmented and managed risk along the way.

3. Architectural approach

- Provided reusable enterprise-wide integration components (services)
- Extended the capabilities of key applications (e.g. eBusiness suite)
- Had a corporate architecture board and center of excellence to provide overall perspective and planning for the highly integrated, service-oriented environment.

4. Agile contracting approach

- The Agile contracting approach was very successful and created strong collaborative relationships with key partners such as Oracle and Dell.

LESSONS LEARNED

1. Enormous culture shift

- There was an enormous cultural shift required for this effort to be successful. Help is needed to make this transition; don't transition to Agile without help.

2. Adapting the methodology to fit the business

- Don't be afraid to adapt the methodology but use caution with eyes wide open. Agile principles can be applied in many ways.
- Consider the scope and complexity of the effort. As the scope and complexity of an enterprise-level project increases, more planning and structure are needed, and a purely team-based, rather than program, approach may break down.

3. Release management

- Allowing teams the freedom to release to production without synchronizing across dozens of projects with interdependencies among them proved to be very challenging. It eventually required moving to a common release schedule to synchronize capability delivery across many projects.

4. Assigning projects to teams

- The pre-existing traditional process created new teams for every project. As Agile became more embedded, it was clear that projects should be assigned to teams, especially teams that had been working any element of the product backlog. This revised process presented more challenges for the outsourced IT vendor, Dell Services, since there was and is less flexibility in assigning resources according to their needs, but it led to much more effective teams.

- Preserving the learning associated with teams is important and, for that reason, it is best to keep existing teams intact as much as possible. This may mean fitting the project (e.g. partitioning the product backlog) to existing teams as much as possible rather than reorganizing and reforming teams to fit a project.

5. Architectural design planning

- Agile does not typically involve a significant amount of planning and design planning of the architecture of the system up front. The architecture is expected to evolve and code be refactored as needed as the project progresses.
- HPHC found that the costs and complexity of this method were untenable. It was more efficient and gave teams better coordinated direction to have more upfront planning and sign-off of the architectural design approach, with reviews by an architectural board for changes. The level of effort required for reworking and refactoring the design as the dozens of projects progressed for a program of this cost and complexity is both financially prohibitive and disruptive to development.

6. Estimating project schedules

- HPHC developed a life cycle model that included an implementation package (IP) stage followed by an implementation package refinement (IPR) stage. In both of these stages of the process, an attempt is made to estimate the schedule for completing the effort. HPHC learned that new teams that did not have an established velocity had a very difficult time estimating a schedule.

■ For that reason, the IP and IPR stages of the process were limited to proposing only the first three sprints of a project for new teams. After the team gained proficiency and had an established velocity, the team would use its past velocity in creating subsequent and follow-on projects for the same product backlog.

7. QA testing

■ The complexity of the QA testing effort at HPHC, with as many as 30–40 applications or functional components going into the production environment at the same time, was significant and required a great deal of thought and planning. HPHC needed a new way of thinking about QA and the path to production (e.g., synchronizing promotions of dependent parts) of many different but interdependent applications (quite different than the software companies who develop a single product or application).

■ Creating and maintaining the environments for the extensive integrated testing in a SOA environment required a new approach. The costs to create a complete replica of the entire production environment for testing purposes were prohibitive. As a result, HPHC and Dell created the following:

■ a team responsible for managing test environments.

■ development of integrated test environments that consist of the application to be tested and its immediate neighbors—this is a continually ongoing process.

■ a system-integrated test environment, which was a subset of the full production environment that would allow end-to-end testing of a variety of scenarios. This environment included all of the applications upstream and downstream (not just the immediate neighbor applications) that were part of the end-to-end scenarios being tested.

■ a team responsible for creating test data needed in the environments. It was simply too much to expect project teams to be responsible for the collection of all the needed data, assembling it in a way consumable by the applications in a test environment, and then providing a "reset" capability for continued use. Teams provide the scenarios and descriptions of the data needed to this team, who then get it, assemble it, and proof it with the teams prior to the actual testing work.

■ Building automated regression testing scenarios and continuing functional capability (updated datasets) to teams is critical to having teams stay productive both in their own environment and in the integrated environments.

8. CIO retrospective

■ Cultural change is difficult—far more difficult than we imagined. It is extremely challenging to create an effective case for change when what people have been doing historically has been working and working well. They simply do not see anything it in for them. Identifying the believers at all levels is essential, as is continuous communication and reinforcement of the imperative for change.

- The new world order takes a lot of getting used to. Reordering of priorities, reshuffling of resources, and refactoring were not viewed as essential and positive elements historically; rather they were viewed as near failures. Changing the mind-set of "That's not a bug, that's a feature" required senior-level involvement and, on occasion, still does.

- HPHC's IT strategy would not have succeeded without the transition to Agile. That said, it was important to make the change not with the mind-set of a religious zealot but, rather, allowing folks to adopt and adapt. That fosters a sense of ownership among the business participants and technologists alike.

- A program as large as HPHC's most assuredly requires infrastructure, support, and coordination (and at times direction) from the project management office (PMO). While the product owners and Scrum Masters have significant autonomy, orchestration of this massive effort is critical. The PMO plays a central role in this orchestration. An Agile transformation brings tremendous value; portfolio management of an effort this large provides the complementary skills and tools to deliver capabilities effectively.

DISCUSSION TOPICS

1. What do you think is the biggest risk in this kind of project?

2. How would you overcome the risk?

NOTES

1. 2011 HPHC Annual Report.
2. NCQA's Private Health Insurance Plan Rankings, 2012–13.
3. NCQA's Private Health Insurance Plan Rankings, 2011–13, HMO/POS. NCQA's Health Insurance Plan Rankings 2010–11—Private. *U.S.News*/NCQA America's Best Health Insurance Plans 2005–2009 (annual). America's Best Health Insurance Plans is a trademark of *U.S.News & World Report*. NCQA The State of Health Care Quality 2004.
4. http://www.linkedin.com/pub/michael-hurst/0/541/909.
5. http://www.linkedin.com/in/vbhatt.
6. https://www.harvardpilgrim.org/portal/page?_pageid=213,56268&_dad=portal&_schema=PORTAL.
7. www.agilemanifesto.com.
8. Michael Hurst and Hope Krakoff, "Measuring Effectiveness of Project Management: HPHC and Perot Systems," presentation to the Babson College, Center for Information Management Studies, Waltham, MA, October 21, 2009.

9. Maarit Laanti, Outi Salo, and Pekka Abrahamsson, "Agile Methods Rapidly Replacing Traditional Methods at Nokia: A Survey of Opinions on Agile Transformation," *Information and Software Technology* 53 (3) (March 2011), pp. 276–290, http://dl.acm.org/citation.cfm?id=1937326.

10. Ibid.

11. Able Workshop, Bentley University Conference Center, Waltham, Massachusetts, October 12, 2011.

29 Case Study: General Dynamics, UK

BACKGROUND

GENERAL DYNAMICS, HEADQUARTERED IN FALLS CHURCH, VIRGINIA, employs approximately 95,100 people worldwide.[1] The company is a market leader in business aviation; land and expeditionary combat systems; armaments and munitions; shipbuilding and marine systems; and information systems and technologies.

General Dynamics, UK Limited, is a leading prime contractor and complex systems integrator working in partnership with government, military and civil forces, and private companies around the world. General Dynamics, UK has 50 years' experience in technical leadership, manufacturing expertise, and prime contract management skills to deliver C4I communications solutions, armored fighting vehicle (AFV) technology, and security systems for critical national infrastructure and deployable infrastructure. General Dynamics, UK has 11 world-class facilities across the United Kingdom and internationally, with over 1,650 highly skilled members of the team.

Due to the nature of the government contracting environment that General Dynamics, UK has to operate in, projects have to work within two key constraints—time and cost—while other Agile methods don't always do so. This case study illustrates how a company can successfully develop a hybrid Agile approach that blends together elements of a plan-driven approach for managing time and cost constraints and still provide flexibility to their customers. Nigel Edwards at General Dynamics, explains it this way:

> From a customer perspective, the approach provided flexibility since it allowed the detail of the technical requirement to evolve without the risk of cost and schedule overruns. The approach did, however, require a greater investment on the part of the customer as they had to be an embedded part of the day-to-day working of the delivery team to help ensure the requirements evolved in a way that ensured the end product met the end user's needs. Traditional project delivery approaches allow a more arm's-length relationship.[2]

The DSDM (Dynamic Systems Development Method) framework is a hybrid, project-level Agile framework that was successfully used by General Dynamics, UK Limited to deliver the Combat Identification Server (CIdS) Technology Demonstrator Project (TDP) for the Ministry of Defence (MOD) in the United Kingdom. The objective of the project was to help clear "the fog of war" by providing a picture of the position of nearby friendly forces on the ground in the cockpit of an aircraft.[3] The following is a brief summary of some key aspects of the project and is reproduced with the permission of the DSDM Consortium and General Dynamics, UK.

Nigel Edwards of General Dynamics, UK was responsible for the management of this program. Nigel has 25 years' experience in engineering development projects and 10 years' experience as a program/project manager. Core to this experience has been leading and managing multidiscipline engineering teams and partner company relationships.

OVERVIEW

The following is an overview of some of the key attributes of the approach used by General Dynamics, UK on this project.

Requirements Prioritization and Management Approach

Traditionally, project management in the defense sector has focused on meeting technical requirements, sometimes at the expense of project duration and cost. In a DSDM project, the performance requirements are prioritized and expressed as *must, should, could,* and *won't* (this time) or *MoSCoW*, which was discussed in Chapter 4.

In the DSDM model, trading out requirements provides the flexibility to ensure on-time and on-cost delivery of an acceptable and fit-for-purpose solution rather than a perfect one. The CIdS project requirements were categorized using the MoSCoW approach. Through a trading process during project execution, this technique enabled a few *should* and *could* requirements to be removed from the project solution to prevent potential cost and schedule overruns, without any customer penalties being incurred.

In practice, the development of the CIdS project capability was divided into increments, which provided checkpoints where capability could be demonstrated.

- Each of these increments was divided into fixed-length time-boxes. If, in the delivery of any time-box, it was apparent that there was risk to the achievement of the must requirements, trade-offs were made to defer requirements from the should and could requirements to assure delivery of must requirements.

- If it became apparent that the must requirements were in jeopardy, then the time-box and potentially the whole project could have been stopped or replanned; extension of a time-box is not permissible. This discipline is important in the "fail early" principle of DSDM.

Contract Negotiation and Payment Terms

Implementation of this type of approach requires a certain amount of trust between the government contracting agency (or any other customer) and the contractor.

■ The government contracting agency had an initial concern that the contractor would abandon *should and could* requirements at the first opportunity, and an incentivization mechanism was sought.

■ However, this approach contradicted DSDM principles where cost—and hence, price—is fixed, with requirement trading offering the project contingency.

■ During contract negotiation, this incentivization position was relaxed as the government contracting agency recognized that the contractor would aim to maximize the delivered capability.

   One aspect that required ongoing negotiation through the project was the payment plan.

■ Once the project started, it became apparent that the detailed project schedule that was developed during the initial stages of the project no longer fit the payment plan that had been agreed on prior to contract award.

■ By taking a pragmatic and flexible approach and recognizing that this is the nature of DSDM, the government contracting agency and the contractor were able to align the program plan and the payment schedule.

   A major challenge was that the initial stages of the project required a fair degree of tolerance and patience with lack of clarity and uncertainty.

Planning Approach

Instead of including a detailed schedule of the project delivery with the bid submission, expert advice was that this process should be performed jointly with the government contracting agency after contract award as part of the project-level planning phase (DSDM Foundation phase) of the project. Before the team launched into detail, it needed to put the project into context by focusing on the fundamental project objectives. This was achieved through the collaborative development of a *Single Statement of User Need*, which stated:

> Report blue force track information to authorised requesting entities on demand in the Close Air Support (CAS) mission.

   As the planning progressed, so did the collective understanding of the technique. A key objective of the DSDM approach is to progressively reduce risk through incrementally delivering a solution. As such, establishing how the technical solution would be elaborated within the DSDM framework was also a key part of planning the overall strategy for delivery of the CIdS project. Using DSDM, the CIdS project comprised three distinct phases:

1. *The foundation phase* developed the requirements, technical design, and also planned in detail the subsequent exploration/engineering phases.
2. *The exploration/engineering phase* was where the CIdS solution was incrementally developed through a series of time-boxes where MoSCoWed requirements were logically grouped. These time-boxes constrained the implementation to time, cost, and quality.
3. *The final deployment phase* demonstrated the CIdS solution to end users.

The high-level plan for the project, laid out during the foundation phase, became the outline that defined the rest of the project.

■ Throughout the CIdS project, focus was maintained on control and metrics regarding achievements within each time-box.

■ While all project time-boxes were planned ahead at an outline level, as individual time-boxes ended, the performance within that time-box enabled the detailed planning of the next time-box.

■ This incremental and iterative development approach, where achievement informs ongoing plans, was at the heart of the project.

Personnel Management

Selection of the right personnel to lead and work on the project was critical. DSDM requires people with the ability to work effectively within collaborative teams and with a tolerance of ambiguity. This was not a project for task-focused, left-brained project managers! Equally the assignment of the technical coordinator role was pivotal; this role requires a combination of technical domain knowledge, foresight, and leadership to ensure the technical solution achieved its requirements.

Communication

Traditionally there is reticence to share "warts and all" information between partners and/or customer. As the team relationship developed, aided by significant periods of co-located working, an open and no-surprises culture of communication developed. A clear no-blame mandate was established to promote open and honest communication. Where frustrations between groups did develop, they were effectively addressed to prevent any damage to the day-to-day working relationships. Other communication aids were also developed by the team.

Through time-box management worksheets and clear communication of requirements to be implemented, associated estimates and the allocation of resources were provided. Daily progress of time-boxes was reviewed by team leads such that exceptions were managed as they arose instead of later. A key facilitator of effective communication was an electronic shared working environment, supported by e-mail and telephone/video conferencing.

Management and Leadership Approach

CIdS project success was underpinned by the foundation phase outputs and the associated facilitated workshops.

- Through the foundation phase, the project developed the plans collaboratively, ensuring consensus.

- Strong leadership, however, ensured that there was no management by committee that could have resulted in indecision and deviation from the project objectives.

- Establishing the detailed technical requirements and partitioning these into a logical development strategy was challenging, but the detailing of the time-boxes was eventually achieved.

Emphasis through the exploration and engineering phase was to ensure that the project objectives were delivered. Managing schedule adherence was relatively straightforward since it was bound by the time-boxing. However, metrics were devised to monitor performance against technical outcomes to help inform the planning of subsequent time-boxes. These also addressed concerns that incomplete time-boxed requirements could be deferred into later time-boxes, potentially resulting in a bow wave—a technical solution only meeting the *must* requirements.

PROJECT MANAGEMENT APPROACH

The overall project management approach was based on the DSDM (Dynamic Systems Development Method) framework. The following is a high-level overview of the DSDM framework. Please refer to the DSDM website for more details.[5]

- **Project startup:** At the start of the project, it was important that everybody involved understood the fundamental principles of DSDM and the mechanics of how to use it as a project management technique. This was achieved through facilitated workshops led by a DSDM subject matter expert. Apart from having the DSDM expertise, the added benefit was that the facilitator was neutral, which was helpful when setting up management team structures and the project delivery approach.

- **Project metrics:** Ensuring the project had a measure of delivery performance was essential to ensure the project outcomes could be achieved. Metrics were put in place to measure time-box achievement in terms tasks/requirements achieved, schedule, and cost. While time-boxes could not be allowed to overrun, they could finish early. Initially it was important to gauge how well time-boxes were estimated to ensure they weren't being overloaded, and to inform subsequent time box planning. The metrics also provided a mechanism to demonstrate to the customer the project was achieving its objectives.

- **Team structure:** The fact that the industry team had contractual responsibilities could not be ignored and informed the allocation of key roles and responsibilities in the DSDM team structure. The DSDM independent facilitator provided a crucial role in the exercise of allocating individuals to roles during one of the team workshops. The fact that the team was involved in the process helped reinforce understanding of the role and responsibilities as well as the team relationships. Developing the team structure and relationships was helped by the fact that the same people allocated to the projects delivery had previously worked together in preparing the proposal for the opportunity in the first place.

- **Real-time decision-making:** Due to the fixing of time in a DSDM project, real-time decision-making was crucial. Key team roles were therefore empowered to make decisions on various tasks and subjects.

- **Project negotiations:** Negotiations were necessary during the project from various perspectives, for example, at a technical level with requirement/scope trading. Empowerment was crucial in this regard.

DSDM Overview

DSDM Atern is a hybrid Agile approach that is used primarily outside the United States. Figure 29.1 shows the key differences between DSDM Atern and a classical Waterfall approach:

In the traditional approach to project management (left-hand diagram), the feature content of the solution is fixed while time and cost are subject to variation.

If the project goes off track, more resources are often added or the delivery date is extended. However, adding resources to a late project just makes it later. A missed deadline can be disastrous from a business perspective and could easily damage credibility. Quality is often a casualty and also becomes a variable, accompanied by late delivery and increased cost.

Atern's approach to project management (right-hand diagram) fixes time, cost, and quality at the Foundations phase while contingency is managed by varying the features to be delivered. As and when contingency is required, lower priority features are dropped or deferred with the agreement of all stakeholders in accordance with MoSCoW rules.[6]

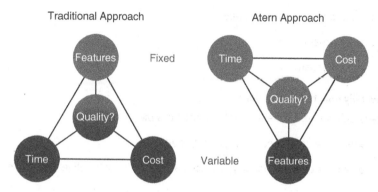

FIGURE 29.1   Project variables of DSDM and DSDM Atern

DSDM Principles

The following are the key principles behind DSDM:[7]

1. Focus on the business need

- Understand the true business priorities.

- Establish a sound business case.

- Seek continuous business sponsorship and commitment.

- Guarantee the minimum usable subset (of features).

2. Deliver on time

- Time-box the work.

- Focus on business priorities.

- Always hit deadlines.

3. Collaborate

- Involve the right stakeholders, at the right time, throughout the project.

- Ensure that members of the team are empowered to make decisions on behalf of those they represent.

- Actively involve the business representatives.

- Build a one-team culture (with the business).

4. Never compromise quality

- Set the level of quality at the outset.

- Ensure that quality does not become a variable.

- Design, document, and test appropriately.
- Build in quality by constant review.
- Test early and continuously.

5. Build incrementally from firm foundations

- Strive for early delivery of business benefit where possible.
- Continually confirm the correct solution is being built.
- Formally reassess priorities and ongoing project viability with each delivered increment.

6. Develop iteratively

- Do enough design upfront to create strong foundations.
- Take an iterative approach to building products.
- Build customer feedback into each iteration.
- Accept that most detail emerges later rather than sooner.
- Embrace change—the right solution will not emerge without it.
- Be creative, experiment, learn, evolve.

7. Communicate continuously and clearly

- Run daily team standup sessions.
- Facilitate workshops.
- Use rich communications techniques, such as modeling and prototyping.
- Present instances of the evolving solution early and often.
- Keep documentation lean and timely.
- Manage stakeholder expectations throughout the project.
- Encourage informal, face-to-face communications at all levels.

8. Demonstrate control

- Use an appropriate level of formality for tracking and reporting.
- Make plans and progress visible to all.
- Measure progress through focus on delivery of products rather than completed activities.
- Manage proactively.
- Evaluate continuing project viability based on the business objectives.

CHALLENGES

Cultural and Organizational Challenges

Table 29.1 is a summary of the key cultural and organizational challenges faced by General Dynamics, UK and how they were handled.

TABLE 29.1 Solutions to cultural and organizational challenges: creating an environment of trust

Challenge	Solution
The customer/industry and industry team needed to be transparent while respecting each of the industry partners was a separate commercial organization	The team, in spite of residing in different parts of the South of England, worked very closely together, either through face-to-face meetings or via telephone conferencing
The Ministry of Defence (MoD) needed to be comfortable that the government was not being short-changed and that the industry team was making the best possible use of the funding available	A lot of the interaction was at the technical level, as that underpinned the project

However, higher-level project management interaction was also required so that the customer was assured the project was achieving its objectives |

Contractual Challenges

Tables 29.2–29.4 are a summary of the key contractual challenges faced by General Dynamics, UK and how they were handled.

TABLE 29.2 Solutions to contractual challenges: using the contract as a framework

Challenge	Solution
While having a contract between parties was necessary, it needed to allow for a flexible working relationship and ensure all organizations remained committed to common project objections	The contracting arrangement was such that all parties worked to the same contract milestones. These milestones were linked to payment

The net effect was that all parties were motivated by the project's achievements and success rather than the delivery of individual products and services into the project

The approach aimed to foster a "one for all, all for one" team ethic |

TABLE 29.3 Solutions to contractual challenges: business expectations—risk but no contingency

Challenge	Solution
The DSDM concept of trading scope as a contingency for cost/schedule impacts is alien, but once it was witnessed, General Dynamics bought into the process	The metrics produced to demonstrate project performance to the customer similarly helped to reassure General Dynamics, UK's management the project would deliver its objectives within budget
The management of cost and schedule through clearly defined time-boxes was crucial to minimize risks coming to fruition and hence the need to trade scope	In addition, the monthly financial reviews conducted on all projects within General Dynamics, UK provide further confidence in the project delivery

TABLE 29.4 Solutions to contractual challenges: keeping to the fundamental rules

Challenge	Solution
Preserving fixed cost and schedule while delivering a product at the commensurate quality was a challenge	The foundation phase of the project was all about defining an achievable scope of work within the time and cost constraints to an acceptable quality
It was important to recognize that using DSDM and having the opportunity to trade out scope could not be used as an excuse for delivery minimal capability and/or poor quality	The quality was underpinned by the various engineering development standards produced during the foundation phase

Technical Challenges

Tables 29.5 and 29.6 are a summary of the technical challenges faced by General Dynamics, UK and how they were handled.

TABLE 29.5 Solutions to technical challenges: delivering the required scope/capability

Challenge	Solution
Ensuring that the ability to trade out requirements was not used as a cop-out for not delivering capability was driven by a robust project management approach	The delivered project scope/capability and product quality are related in our domain, as the quality is driven by the robustness of the engineering development process
	A more robust development process will reduce the scope/capability of the delivered project

TABLE 29.6 Solutions to technical challenges: deriving the technical requirements

Challenge	Solution
The capability delivered built on existing technology from three independent organizations	The formation of an integrated project team was key
The project required further development of the existing technology, as well as new technology across the organizations	Although remotely located for significant amounts of time, the use of technology and ensuring proactive behaviors across the team for effective and timely communication mitigated a significant technical and project risk
	Organizational boundaries were removed by the individuals involved to ensure, as far as was possible, a seamless, integrated project team

OVERALL SUMMARY

Key Success Factors

1. Relationships, Professionalism, and Transparency

The Ministry of Defence (MoD) is a key customer for all the industry partners in their own right. To not recognize the key factors could be potentially damaging to the reputation of the industry partners and have a direct impact on future business. It was important for General Dynamics, UK, as the lead partner, to create the project environment to develop the relationships and ensure professionalism and transparency.

2. Coaching and Mentoring

The DSDM project delivery technique was new to everyone involved in the project and meant that everyone needed to learn the process together. Throughout the project, it was important to ensure that all involved maintained the same common view of the DSDM process that was used for managing the CIdS project. Through team mentoring, a common understanding was established.

3. Teamwork

Successful delivery of the CIdS project was achieved through the behaviors and motivations of the people involved—the teamwork.

- A new project delivery method required key personnel with a pragmatic, flexible, and open-minded approach, coupled with a tolerance for ambiguity and excellent team-building skills.

- The teamwork has come about through the desire to collaborate, learn together, communicate, and build the necessary relationships to successfully deliver the CIdS project.

- Any stresses in the team dynamic were recognized and dealt with early, before they become issues.

4. Conflict Management

Through the highly collaborative and open-team approach, with a no-surprises culture, conflict has been kept to a minimum.

- As in most projects there have been stress points during the project, but because of the team relationships, these have been anticipated and quickly resolved.

- In classical plan-driven projects, with their fixation on the achievement of all technical requirements, a defensive posture can often arise if problems in achieving those requirements emerge.

- The CIdS project, through the highly collaborative approach and safety net of requirements contingency, resulted in a far more open environment where good and bad issues were openly discussed.

5. Risk Management

The CIdS project benefited from a risk management strategy that used an incremental development approach in which high-risk requirements were tackled early, with requirements' flexibility protecting on cost, on schedule, to quality delivery.

- A joint risk register was established to inform time box management and ongoing engineering development, but not to identify any required financial contingency.

- Instead of holding financial contingency, any additional funds that were required to ensure delivery of *must have* requirements were drawn from the budget associated with delivery of lesser priority tradable requirements.

CONCLUSIONS

1. Collaborative Approach to Contract Management

Through using the DSDM framework on the project, the traditional customer/supplier and prime contractor/subcontractor divides have been bridged to form what truly has been an integrated project team. The one-for-all, all-for-one ethos has prevailed throughout, resulting in a team that focused on delivering on the project objectives rather than what is best for individual organizations.

- Although the DSDM framework has been central to the delivery of the project, it is not a panacea.

- The CIdS project has ultimately been successful because of the team—their professionalism, their technical capabilities, and their commitment to the principles of DSDM.

    The MoD sponsor stated:

> The use of the Dynamic Systems Development Method has been key to the success of the Combat Identification Server TDP: Agile project management techniques, the close involvement of stakeholders throughout, including interim demonstrations, and a constant focus on the deliverables have ensured that the final product will truly hit the mark. This methodology has clearly worked extremely well and I would hope to see lessons from this project applied to future projects.[8]

2. More Effective Project Management

Key benefits that resulted from the CIdS project management approach include the following:

- More believable plans, schedules, and budgets, and likelihood of adherence were a fundamental outcome as a result of the DSDM approach with its fixed cost and time approach.

- Using DSDM led to a contract that eliminated financial contingency, delivering maximum possible capability for the funds available.

- The time-boxing approach resulted in multiple decision points at the end of each time-box and each increment, at which alternative ways forward were compared.

- Through the *Single Statement of User Need*, a common vision of the project objectives was collaboratively developed.

- With short time box spans, risk assessment has been aided by the increased focus that resulted.

- Greater risk taking was enabled as, by principle, the project could not go over budget, or slip schedule. This allowed desirable but higher-risk requirements to be included as *should* or *could* requirements without fear of penalty to the project. As a result, greater capability has been delivered to the end user for the same money than would otherwise have been possible.

    With government funding under continual pressure, successfully demonstrating incremental capability was also a way of reducing the risk of project cancelation by keeping stakeholders engaged and supportive of the CIdS project.

LESSONS LEARNED

Tailor the Agile Delivery Technique as Part of Early Project Planning

From earlier experience of using DSDM, the process and principles initially seemed quite daunting.

- In order to successfully apply the technique, it was important to fully understand the objectives of the project through the foundation phase.

- Having understood the objectives, it was then important to understand DSDM and how DSDM could be used to iteratively deliver the objectives.

- The lesson was not to rush into the project delivery, and that the early detailed planning is a fundamental part of enabling project success.

Agile Techniques Can Be Applied to New Project Environments

DSDM's traditional application is internal change projects, particularly IT-related projects. The team led by General Dynamics, UK demonstrated the technique can be applied in a multi-organizational environment, on a project that involves bespoke engineering development. The lesson was, with an open and pragmatic team approach, Agile techniques can be adapted for use outside their traditional applications.

DISCUSSION TOPICS

1. What would be the difficulty of getting this approach to work with a government contract?

2. How would you do it? What would be the most critical success factor?

NOTES

1. Information provided courtesy of General Dynamics, UK Limited.
2. Nigel Edwards, email comments to author, September 11, 2012, and November 2, 2012.
3. "DSDM Case Study—Improving Outcomes Through Agile Project Management," http://www .dsdm.org/case-studies.
4. MoSCoW Method, http://www.dsdm.org/dsdm-atern/atern.
5. http://www.dsdm.org.
6. *DSDM Atern Handbook*, http://www.dsdm.org/atern-handbook/flash.html#/15/.
7. Ibid.
8. Major Fiona Galbraith, MoD Sponsor.

30 Agile Hardware Development

MUCH HAS BEEN SAID ABOUT APPLYING AGILE PRINCIPLES AND PRACTICES TO SOFTWARE DEVELOPMENT,
but comparatively little has been said about Agile hardware development. Joe Justice has been
a pioneer in the area of applying Agile to hardware projects and has graciously given me
permission to summarize some of his materials in this chapter. This entire chapter is based on
Joe Justice's work. Joe is a real thought leader in this area. He's been working with Agile and Scrum
for a long time and has been teaching Scrum Master and Product Owner training for about 10 years.

In his training, he recognized something I completely agree with that most Scrum Master training and Product Owner training simply focuses on the "mechanics" of how to do Scrum. The standard training doesn't typically go into examples of how to apply it in different situations at all which is
what is really important.

Instead of the typical approach of simply teaching the mechanics of how to do Scrum, Joe
spends a very limited amount of time on that and goes much deeper into how to apply it in
real-world situations.[1]

In his Product Owner class, he goes even further and breaks up the training into three sections:

- The first third of the training is on Scrum.

- The second third is on how to run a more advanced enterprise-level Agile company.

- The final third is on how to run an Elon Musk company like Tesla which is very advanced.[2]

In this chapter, we will go over the general approach that Joe has developed for Agile Hardware
Development and discuss a very advanced enterprise-level hardware development approach that is in
use by Tesla.

AGILE HARDWARE DEVELOPMENT OVERVIEW

Applying Agile to hardware development is not necessarily a matter of using Scrum as you might find in software development; in many cases, it is more a matter of taking a very innovative approach to using Agile principles and practices as well as Lean principles and practices to develop a very efficient and effective approach for doing hardware development.

Hardware Development Challenges

Joe has summarized some of the challenges that you might encounter in a hardware development environment:

Version and Change Control

"In software, we can delete the previous version or deploy right on top of it. In hardware, how do you embrace agility?"[3] When you build a car and deploy it to people, you have a responsibility to support it including compliance with any applicable automotive product safety requirements. If there is a fault found in the design of any vehicle, you need to know exactly which vehicles that fault might apply to and take corrective action to resolve it. That is not as simple as issuing a bug fix to patch a software defect.

Regulatory and Safety Requirements

Another challenge is complying with the various automotive safety regulations that automotive vehicles must comply with to be considered "road-legal." In 2006, Joe launched his own company and started building cars using Agile principles and practices. His company built a total of 14 "road-legal" cars, and proved that you could "design, analyze, build, test, and deploy 'road-legal' cars in less than a week."[4] That led to four world records and opened the opportunity to work with a number of major automotive companies including Toyota, BMW, Volkswagen Group, and others to apply those same principles and practices on a much larger scale to automotive development and manufacturing. Eventually, that also led to Joe working with Tesla.

Innovative "Out-of-the-Box" Thinking

Another major challenge is associated with developing a culture of innovative, "out-of-the-box" thinking. Joe used the example associated with Bitcoin. At one time, Tesla bought an investment in bitcoin worth over $1.5 billion which was very successful. It is highly unlikely that other major automotive companies would take such a risk. It illustrates the fact that Tesla has an innovative culture where Agile change agents in a leadership capacity can play a significant influence role and where the Product Roadmap is willing to be disruptive.[5]

The Speed of Change Is What Is Important

The Speed of Change at Tesla

In 2020, Joe joined Tesla where he developed and managed Agile@Tesla as a program in collaboration with Lean@Tesla. The primary role of that effort was to "reduce the cost to make change and to make it even faster and cheaper to do new product introduction and new product development."[6]

The challenge that Joe took on was: "How do you build cars like you build apps? When you want to make a change to an app, you tie into your automated tests and your automated deployment and your continuous integration and continuous deployment. How do you do that in hardware?"

When Joe joined Tesla, they were already having a lot of success in the area or accelerating speed of change and they were making on average about twenty-seven updates in hardware per model per week where it is typical in the automotive industry for most manufacturers that a significant hardware change could be a four-year project. That process is typically based on Waterfall phases to verify that:

- You have the correct suppliers in queue.

- The suppliers can deliver at the price you want.

- You're ready to receive and inspect the goods.

- You can process and store them appropriately.

- You can add value to them in a sequence of steps.

- The factory is configured, and the staff is trained.

- The support staff is also trained.

The Importance of Pace of Innovation

Joe pointed out that there is a direct correlation between an automotive company's ability to make change and the company's total market capitalization. And the executives of major automotive companies realize the significance of how rapidly that they can make change:[7]

Volkmar Denner, Chairman of the Board at Robert Bosch GmbH:

- For Bosch, agility is crucial, it allows us to adjust to the increasing speed of change around us. Agility allows us to remain in a position as an innovation leader.

Akio Toyoda, CEO, Toyota Motor Company:

- When you look at a product, going to the actual sites, going to the genba, what is important is to fuel the change point.

> Herbert Diess, Chairman of the Board, Volkswagen Group:
>
> ■ The big question is: Are we fast enough? If we continue at our current speed, it is going to be very tough.
>
> Elon Musk, Chief Executive Officer, Tesla:
>
> ■ Pace of innovation is all that matters in the long run.

How Much Can We Do in Parallel?

Joe pointed out that an important factor in achieving the speed of innovation is to do many tasks concurrently that might normally be done sequentially. Doing those tasks concurrently rather than sequentially means that there needs to be some method of coordinating the work of teams that are concurrently working on changes that may be related. That requires teams to have a known stable interface so that teams can iterate their work completely independently. Beyond that, there are Agile practices that need to be assimilated into the company's culture and environment that are important to accelerate the speed of innovation even further, including:

■ **shared ownership practices** to allow round-the-clock development.

■ **continuous integration and continuous deployment** to provide direct feedback to the work in progress. Continuous integration is often roboticized to automate the effort.

■ **integration with suppliers** to allow changes that impact suppliers to allow real-time price adjustments to the total bill-of-materials.

Joe says: "It's all about speed of feedback loops which reduces cost of change . . . Deployment of parallel executable chunks really does seem to be the entire name of the game and trying to do that with as little hierarchy as possible."[8]

How to Put This Into Practice

Flattening the Company Hierarchy

Joe indicates that a very important factor in enabling this kind of environment is to flatten the company hierarchy and streamline the decision process. As an example,

■ Elon Musk has been known to sleep in a sleeping bag near his office where he is readily available to make decisions quickly. Tesla doesn't have other layers of management to do that.

■ He is also known to work on the assembly line, building cars occasionally, which keeps him in touch with the day-to-day operations of the company and reduces the cultural gap between senior executives and line employees.

- In addition, employees are empowered to make decisions. "If anyone needs to make a large capital expense, you turn to the people around you and say 'Is this a good idea?' and people say 'Yes or No' and you're handed a credit card and you go."[9] That works because the people required to make decisions are right there and there is limited cultural gap between the people who make decisions and the people who carry out those decisions.

Organizing Around Independent Teams

Work is broken up into functional groups that are organized around a particular module of functionality that group is responsible for. There are shared services of a general nature to support their work; however, each of those groups works independently and has sufficient autonomy and empowered decision making to do whatever is needed to successfully accomplish their work.

For example, these groups can do their own procurement, their own software and hardware testing, and they do their own test fixtures to enable them to work independently without waiting for resources to be available from outside groups. The biggest risk in this is integration risk. Will all these modules really work together when their assembled into a finished automobile? Continuous integration is important to manage and mitigate that risk.

Small Team Sizes Are Important

Joe quoted Jeff Bezos from Amazon as saying, "Never have a meeting where two pizzas couldn't feed the entire group."[10] That means about six to nine people and that's essential for effective decision making. Any more people than that runs the risk of the group getting bogged down and unable to make decisions.

The Importance of Configuration Management

Many automotive companies like to track changes by model year. That is, all cars built within a given model year will have essentially the same configuration. Some kind of configuration management is important, of course. If there is any kind of issue such as a safety issue, you have to know which particular cars are impacted by that issue. However, if configuration changes are limited to model years of cars, that can have a significant impact on slowing down the pace of innovation.

Modern configuration management is much more sophisticated than that. At Tesla, every single car potentially has a unique and well-defined configuration that is managed and tracked in a relational database. For example, if there was a problem with a particular kind of adhesive used in building cars, Tesla would know exactly which cars might be impacted by that problem.

The Importance of Employee Empowerment

Having a company culture where employees and teams are empowered to make decisions is very important. At Tesla, "From the minute you're hired, you are empowered. You can buy a robot, you can buy land, you can buy a truck from the minute you're hired. And, it's the funding team's job to make sure that there's a pool of money that you can pull from."[11]

Product Owner Mindset

Having the right mindset is important. There is no significant marketing department or competitive analysis department. Instead of some independent group doing some kind of analysis to determine what should be done to keep up with other competitors everyone is empowered to do whatever is physically possible to make the product better and no one needs to wait for approval to do something that makes sense.

Group Ownership

In order to work in an environment with continuous, round-the-clock, round-the-world development, the idea of "group ownership" is important. It changes how production works. No one can exclusively own anything. When someone leaves at the end of a 12-hour shift, nothing stops and their work needs to be handed off and seamlessly overlapped with the work of the next shift.

Automation Is the Answer

Integration must be automated so that you know very quickly if new parts work together in a given configuration. That means that the parts have known stable interfaces and robots need to be adaptable to be able to handle different parts easily. "Humans are for creative problem solving; automation is for everything else."[12]

At a process level, automation is also important to automate the process flow and "routers" are used to automatically control the flow of raw materials in the assembly line. New processes might be tested on separate pilot assembly lines before being phased into normal production operations. That is exactly the way that many enterprise level software capabilities are tested in a staging environment before they are phased into a production environment.

HOW IT'S DONE AT TESLA

The approach for doing Agile Hardware Development at Tesla is very advanced and goes well beyond a typical Agile/Scrum approach. In fact, Tesla doesn't really use Scrum at all, but the company is extremely dynamic and Agile. In this section, we'll discuss how Tesla goes well beyond a normal

Agile/Scrum approach. There are definitely pros and cons associated with the Tesla approach and it is definitely not the right approach for all companies but it gives you an idea of how to take an Agile approach to an extreme.

The Tesla Approach

1,000-Year Goal

Tesla has a 1,000-year goal to "Spread the light of consciousness out among the stars." This seems to be an extremely broad and nebulous goal but it is sufficient to unify the company around a long-term vision.

Around-the-Clock Development

All employees at Tesla work in two 12-hour shifts that run continuously for 24 hours a day, every day of the week. Employees typically work three 12-hour shifts one week followed by four 12-hour shifts the following week.

Real-Time Budgeting

Many companies do budgeting on an annual basis which somewhat establishes the cadence for the company's ability to make major changes and restricts their ability to using inspection and adaptation to make significant changes on a more frequent basis. Most major automotive companies, introduce a minor model change every 2.5 years and a major model change every 5–7 years on average. In that environment, a 1-year budget cycle is very appropriate. It might be difficult for Tesla, which introduces sixty new parts into production every day.

Cross-Functional Employee Participation

All employees at Tesla are considered to have the same role and are intended to be cross-functional. "Whether you're a designer, a tester, a builder, a coder, or a welder, you're a developer and that increases the opportunities for employees to be open to cross-functionality."[13] In Scrum, that has always been an ideal goal, but it is rarely done in practice. For example, in a Scrum software development team; in theory, Developers might be capable of testing and testers might be capable of doing development.

Even some roles which are required by government regulations such as a Health and Safety Officer and a Plant Manager are rotated among employees. Any employee who has been through the training required to be a Plant Manager could conceivably go into that role and the training is provided by apps on everyone's phones. This keeps the company feeling flat "where anyone feels that they can do anything."[14] Some of these roles may change more than once in a week or even in a day.

Employee Accountability

Employee results are managed automatically by software called "Digital Self-Management" instead of a manager.

- There are no performance reviews or promotions, and employees make more money from options. Elon Musk has tried to set up an environment with almost zero management.

- He also has strived to set up an environment with no waiting, where no one needs wait an extensive amount of time for a management decision. To accomplish that goal, artificial intelligence software is widely used for automating some management decisions. For example, Tesla has a software application called "Auto-bidder" that helps to determine if the company is getting the best price for a procurement item.

Scrum Roles

Tesla does not use the standard Scrum roles of Scrum Master and Product Owner, instead the teams are completely self-organizing and completely share responsibility for results.[15] The only Scrum role that is recognized in Tesla is the role of a Developer. As previously discussed, every employee in Tesla is considered to be a "Developer."

Some of the functions that normally might be performed by a Scrum Master or Product Owner are automated in software:

- The equivalent of a "Digital Scrum Master" creates and maintains the Scrum Board.

- Some of the functions that might be performed by a Product Owner are also performed by software.

Scrum Events

Scrum normally includes five events:

- Sprint
- Daily Scrum
- Sprint Planning
- Sprint Review
- Sprint Retrospective.

Joe points out that the five events of Scrum are like a "30-day Waterfall." For many companies, that would be a significant improvement over a standard Waterfall development process, but it would be way too slow for Tesla. For that reason, advanced Agile companies like Tesla do not use the five

Scrum events anymore. The employees and artificial intelligence software propose new goals every day.

Scrum Product Backlog and Sprint Backlog

Tesla also does not use the Scrum Product Backlog and Sprint Backlog. The work to be done is changing much too frequently and dynamically for that to be useful. In Tesla, work is done continuously without being broken up into sprints. Tesla also focuses on automating safety, quality, and certification testing which allows the teams to use a continuous integration and continuous deployment process.

Development Teams

Another significant difference from Scrum is that assignment of employees to teams is much more fluid. In a normal Scrum software development environment, the assignment of people to Scrum teams is fairly stable and doesn't normally change. In Tesla, teams are much more dynamic.

Tesla has a principle called "The Law of Two Feet" which says that "You should use your own two feet to walk to any project that you can add the most value to. You might stay on the same project for days or even years, but you could walk to another project at any time without being told to do it."[16]

Mob Work

All work in Tesla is done by "Mobs." A "mob" is a small ad hoc team that is focused on a particular task. A mob is typically two to five people. The Digital Self-Management software provides real-time feedback on the productivity of a Mob and the decisions being made by the Mob. Budgets are assigned to a Mob on the basis of the work to be done and reallocated to Mobs dynamically as needed.

The Human Element

With all this software and automation, you might wonder, "Where is the human element? What happens when something goes wrong and someone raises an alarm?"

"Employee alerts, complaints, and requests are received anytime by chat, email or face-to-face. If a request appears valid and important to the mission, the request is added to the 'Justice Board.'" (The Justice Board is essentially a high-level Product Backlog of work to be done. It is the large agenda of all projects which are funded.) "Teams self-organize around the Justice Board to reduce risk and accelerate the mission, and this includes addressing valid employee requests."[17]

OVERALL SUMMARY

The Trade-Off Associated with Creativity and Innovation

For Elon Musk, pace of innovation is the only thing that matters, and the implementation of Agile at Tesla clearly reflects that. However, in any company, there is a trade-off between:

- an emphasis on planning and control, and

- an emphasis on creativity and innovation.

Tesla has clearly swung that pendulum heavily toward the creativity and innovation extreme and it is a wonder that the company can be profitable without at least some level of planning and control. In fact, in the early days of Tesla, the company showed very weak operational management results.[18] And, at one point, Elon Musk was criticized for putting too much emphasis on starting new companies and not enough emphasis on operational management to make them profitable once the companies were established.

- In the most recent five years of this data, Tesla's overall Return on Equity went from a dismal −89% to +13% in the most recent period over a five-year period.

- Tesla's Return on Equity is still at the bottom end of other major automotive manufacturers.

So, there is definitely some trade-off between an extreme emphasis on creativity and innovation and some emphasis on planning and control to achieve some level of operational performance in terms of profitability and Return on Equity. That extreme emphasis on creativity and innovation has worked for Tesla up to this point and provided them with a significant technology leadership position in the market for electric vehicles. However, as that market matures, that balance may shift somewhat.

Does the Tesla Agile Hardware Development Model Work for All Companies?

The Agile Hardware Development model used at Tesla is clearly an extreme version of Agile that goes beyond the normal Agile/Scrum development process. It has worked for Tesla, but there is some concern that too much of the company's operations are dependent on the personal direct management of Elon Musk and it is evident that he has many interests (some would call them distractions) beyond electric vehicles. Some analysts and fund managers are hoping that Elon Musk will finally hire a Chief Operating Officer.[19]

It would be difficult to get this extreme of a model to work in other companies that don't have a single, strong person like Elon Musk as a leader to keep the company moving in the right direction. Such an extreme model may also not be completely appropriate for a company that doesn't need to have such extreme pressure on product innovation. Nonetheless, many of the ideas in this model would likely be worthwhile to other companies who are interested in implementing an Agile Hardware Development process.

DISCUSSION TOPICS

1. Is the Agile approach described here a good fit for Tesla? Why or why not?

2. What are the strengths and weaknesses of this approach?

3. Would the same approach work at another automotive company like Ford?

NOTES

1. Joe Justice, Intellias Agile Days, https://m.youtube.com/watch?v=tx7v164fHmc.
2. Ibid.
3. Joe Justice, Business Agility Institute, Agile Hardware, https://businessagility.institute/learn/agile-hardware/514.
4. Ibid.
5. Ibid.
6. Ibid.
7. Ibid.
8. Ibid.
9. Ibid.
10. Ibid.
11. Ibid.
12. Ibid.
13. Joe Justice, Intellias Agile Days, op. cit.
14. Ibid.
15. Ibid.
16. Ibid.
17. Ibid.
18. Seeking Alpha, "Comparing Ford's Operational Efficiency to Tesla and Others," https://seekingalpha.com/article/4451079-comparing-ford-operational-efficiency-to-tesla-and-others.
19. "Tesla Desperately Needs a No. 2 for Elon Musk," *CNN Business*, June 9, 2021, https://www.cnn.com/2021/06/09/investing/elon-musk-tesla-coo/index.html.

③ 31 Non-Software Case Studies

AGILE AND SCRUM ARE NOT LIMITED TO SOFTWARE DEVELOPMENT; although that is where they are most commonly used. In general, Agile and Scrum can be used in any project that has some level of uncertainty; and in many areas, the principles behind Agile can be used without a formal Scrum approach. I will go through two examples of that in this chapter.

AGILE HOME REMODELING

The first project that I want to talk about is how I went about applying Agile principles to a home remodeling project of my own. I have been a Project Manager for a good portion of my career and I have managed projects of every shape and size you can imagine, including some large and complex multi-million dollar projects but even though this project was not very large, it is one of the most challenging projects I've ever had to manage and it really required adapting Agile principles to fit this situation.

Background

Here's some background on this project.

- We've owned this house for about 15–20 years but for most of that time it was a rental property that we had rented out to other people. It is only in the last year or two that we've been able to stop renting it and have it for our own use.

- Because it was a rental property, it only had the basic essentials and has never been significantly remodeled.

- My wife was not happy with it especially since immediately prior to this project some of the major appliances in the kitchen stopped working.

- We considered selling it and moving to a new house but that was just too expensive, and we really like the community that this house is in.

Why Was This Project So Difficult?

You might naturally ask: Why was this project so difficult? Here are some of the factors that made it difficult:

- My wife was the major stakeholder in the project, she is a perfectionist, and she has a habit of changing her mind frequently about what she wants. (Her response to that is "She doesn't change her mind, she just decides as she goes along.") In fact, that is exactly the way an Agile approach works—you try to defer making detailed decisions until the last possible moment but that presents a big challenge in managing the project especially when outside contractors are involved.

- All of the work in this project was done by multiple outside contractors which made the selection of a contractor extremely important.

- In spite of these difficulties, a major challenge was to try to manage the costs and schedule of this project within reasonable levels.

Project Planning and Inception

The planning and inception of this effort took about 2–3 months to decide on what we wanted to do and that included:

- Looking at the results of similar home improvement projects in our community.

- Interviewing potential contractors.

- Doing a preliminary selection of appliances and construction materials, including cabinets, counter-tops, and lighting.

- The result of that effort was what I would call a conceptual plan for doing the project. We had a pretty good idea of what we wanted to do but there was still a lot of uncertainty in the plan that could only be resolved after we had selected a contractor and the project was in progress.

Project Scope

The project scope included:

- Knocking down a wall that separated the kitchen from the rest of the house to create a more open environment.

- Ripping up the concrete floor to re-route electrical and plumbing connections.

- Replacing all the existing kitchen cabinets and appliances.

- Installation of new lighting fixtures.

- Moving the entranceway to the master bedroom to be more consistent with the new floor plan.

- Removing a pantry and replacing it with a new pantry cabinet which required knocking down a wall and moving an intercom system.

- Repainting the entire area and many other cosmetic enhancements.

Contractor Selection

The selection of a contractor was very important. The contractor needed to have some level of flexibility and adaptivity but be well qualified to do the work at a reasonable cost. The first task was to select a contractor (or contractors) to do the work. Table 30.1 shows the choices we were faced with.

Contractor "A"

- Was the most widely known contractor in this area—they advertise widely on television.

- They have a good reputation for delivering a high-quality result.

- They would also take full responsibility for the overall solution.

- However, their approach is fairly rigid and controlled—once you sign a contract with them, it is very difficult to make any changes.

Contractor "B"

- Was much less widely known.

- Offered much more flexibility and willingness to work with us to come up with a design that was customized to meet our needs.

- They would also take overall responsibility for managing the overall solution.

TABLE 30.1 Contractor selection matrix

Criteria	Contractor "A"	Contractor "B"	Contractor "C"
Reputation	Very well-known	Less well-known	Least well-known
Risk	Low risk	Medium risk	Highest risk
Overall solution	Takes overall responsibility	Takes overall responsibility	Responsibility split between two contractors
Flexibility/Adaptivity	Very limited	Medium	Best
Cost	Highest cost	Medium cost	Medium cost

Contractor "C"

- Offered the most flexibility to meet our needs but was actually two different contractors so it was not really possible for either of them to take overall responsibility for the overall solution.

Selecting a contractor was difficult:

- Contractor "A" was probably the lowest risk choice from a traditional project management perspective. It would require less management on my part but offered little flexibility to adapt the solution to meet our needs.

- Contractor "C" was the highest risk and involved coordinating the work of two different contractors but offered the most flexibility to meet our needs.

- Contractor "B" was a compromise between those two extremes. The advantage that they had over Contractor "C" was that they were a single contractor who would take overall responsibility for the solution, but their costs were considerably higher than Contractor "C".

We ultimately chose Contractor "C" primarily because it provided the most flexibility and adaptivity at a reasonable cost. There was a big difference between Contractor "A" and Contractor "C".

- In a relationship with Contractor "A", I would have relied on a very clear and well-defined contract to deliver the solution, but I would have had little or no flexibility to make changes. (That's what many people might call "Waterfall.")

- In a relationship with Contractor "C", there was a statement of work, but it was understood to be flexible and subject to change and much of the relationship relied on a spirit of trust, partnership, and collaboration. (This relationship was much more similar to Agile.)

A major downside of Contractor "C" was that the work was actually split between two contractors which made it more difficult to manage and probably had the highest risk. However, these two contractors had a history of working together successfully on other similar projects and I had a good feeling about these individuals that I could trust and partner with them to manage the overall solution. Here's how the work was split between the two contractors:

- Contractor "C1" did the demolition and prep work, including electrical and plumbing to prepare the new kitchen.

- Contractor "C2" provided the kitchen cabinets and counter-tops and installed them after the initial demolition and prep work had been completed.

- Following the installation of the cabinets and counter-tops, Contractor "C1" returned to do the finish work which included final installation of new lighting fixtures and repainting of the entire area.

How Did the Project Work Out?

There were a lot of difficulties associated with managing this effort:

- **The scope of the project changed numerous times:** my wife decided that we couldn't remodel the kitchen without replacing all the living room furniture and carpets; and, of course, there had to be changes to the rest of the house as well which included repainting the master bedroom, replacing pictures, and reupholstering other furniture. and enhancements to other areas of the house.

- **Nailing down the design requirements was very difficult:** as I mentioned, my wife changes her mind frequently, and insists on perfection in the end-result. I can't tell you how many different kinds of granite counter-tops and how many different floor tiles we looked at before making a final selection; and how many times what I thought was a "final selection" changed before it really became a "final selection."

- **This was a project management nightmare:** this was not a large project, but it was one of the most difficult ones that I have ever had to manage. For a traditional plan-driven Project Manager, this would have been a nightmare attempting to control all these changes and being caught in the middle between a very demanding stakeholder (my wife), on the one hand, and contractors, on the other hand, who have to deliver the work within a given cost.

    However, this is a perfect example on a small scale of what an Agile Project Manager must do—you have to learn how to balance flexibility and adaptivity to maximize the business value of the solution with some level of planning and control.

What Were the Results?

The project turned out to be enormously successful:

- It was completed in a little over three weeks from the time the work started.

- It went over the budget that we expected to spend but the costs were still at a reasonable level.

- Most importantly, my wife was delighted with the way it came out and she is the most important stakeholder I needed to satisfy.

Overall Conclusions and Lessons Learned

I know this is an unusual situation, but I like to use unusual situations to encourage "out-of-the-box" thinking rather than viewing standard, stereotypical Agile case studies. Here's how I think these lessons learned can be applied to a business situation:

- **Contractual relationships:** Most businesses could not survive without some kind of contractual relationships with outside contractors and many businesses have significant supply chains that are critical to the success of their business.

- The traditional way of managing those contracts is to develop a firm, fixed-price contract and use a competitive process involving multiple bidders to get the lowest possible price.

- That is a relatively low-risk approach from a cost-management perspective but doesn't necessarily result in the best overall solution.

- When there is a lot of uncertainty in the requirements and it is important to maximize the business value of the solution, a different approach is needed that is based on more of a collaborative partnership with a contractor to work together to maximize the value of the solution.

- **Trust:** Developing that kind of relationship with contractors requires trust. Naturally, it will not be possible to develop that kind of relationship with just any contractor. That's why it is important in business to have strong relationships with a selected number of contractors who can be regarded as close partners rather than a broader number of contractors based on price alone.

- **Risk management:** I took a significant amount of risk in this project going with contractors that I thought were the highest risk from a project management perspective, but that risk paid off in terms of the overall quality of the solution. I think a similar thing is true in a business environment—many times you have to take a risk to maximize the value of the solution.

- **Productivity:** This project was completed amazingly fast once the work was started and that was largely due to the fact that I empowered the contractors to get the job done the best way they knew how, and I didn't attempt to micro-manage what they were doing.

AGILE BOOK PUBLISHING

I've published five books in my career. The first two books were on Business Excellence and the last three (including the first edition of this book) were on Agile Project Management. For the last three books, I actually used Agile principles to publish the books.

How Was the Agile Approach Different?

Here are some of the major aspects of what I did differently:

- **Started quickly without getting bogged down in planning:** I had a general outline of how the book was going to come out, but the details evolved a lot as the development of the book was in progress.

- **Used an incremental and iterative approach to writing:** I broke up the work to be done into individual sections and chapters within each section and used an incremental and iterative approach for finishing each one.

- **Relied heavily on feedback and inputs to guide the writing as it was in progress:** I had a team of people who had volunteered to review the book and provide feedback and inputs as it was in progress.

The key message I want to make is:

- Agile does not necessarily require implementing Scrum.

- Sometimes, it is just applying good common-sense principles to your work.

Lessons Learned

Just Get Started

One of the most important principles I've learned is "just get started." When you're faced with writing a book or developing a major online training course, it can be a daunting experience and just getting started is sometimes the hardest part:

- We may have just a basic idea of what we want to do.

- We're not sure how the final result is going to come out.

- We're not certain how the final result will be structured—what should come first, etc.

- We don't want to produce something that is going to be a failure.

You have to stop worrying about all of that and have the courage to just get started. I think of this kind of effort like developing a fine art sculpture. You start with a lump of clay, and you just keep molding and shaping it until it becomes a work of art.

I did some software development for a long time and learned a lot from that work. The way I developed software was many times to just start writing some code and reorganizing it and refactoring it as I went along. Initially, it might have been pretty messy but as it progressed, it got better and better and became very well organized, but it certainly didn't start out that way.

If you never get started, it will remain just a lump of clay.

- It takes some courage and confidence in yourself to do this. You've got to have courage and confidence that if you just get started, that somehow the final result is going to come out OK if you keep working at it.

- It also takes patience and commitment because you may have to go through a large number of iterations to get something useful out of it. You may even have to throw something away completely and start over again.

Fail Early, Fail Often

One of my favorite Agile slogans is "Fail early, fail often." You must be willing to try things without fear of failure. I've known some people who have been crippled by fear of failure. Failure should be seen as an opportunity for learning.

Another favorite quote of mine is from Thomas Edison who said "*I have not failed. I've just found 10,000 ways that won't work.*" That quote was in reference to his repeated attempts to invent the electric light bulb. Think of where we would be today if Thomas Edison hadn't persisted despite repeated failures.

Use an Incremental and Iterative Approach

I used an incremental and iterative process for writing the books I wrote. Many people don't understand the difference between the words "incremental" and "iterative":

- **"Incremental"** means that you break up a solution into pieces and develop one piece (increment) at a time.
  - Using an incremental approach is very important. In any large effort like writing a book or developing an online course, it's best to break it up into "bite-sized pieces."
  - If you try to take on too much at once, you'll never finish it. The effort to write a book can easily take well over a year and it's easy to get discouraged in that period that you will never finish if you don't see progress in the work. There's another saying that I like that says: "*How do you eat an elephant?—One bite at a time.*"

- **"Iterative"** means that if you're not sure what a given piece should be, you develop something and then continue to refine that piece until you meet the customer's expectations.
  - Taking an iterative approach is also important. A close corollary to "Just Get Started" is "Don't Expect Perfection."
  - A major reason for not getting started sometimes is that we're afraid to produce something that is less than perfect.

  We must accept that whatever we produce on the first iteration is certainly not going to be perfect and the final result may not be perfect either. Get something done quickly and then continue to refine it as needed to meet customer expectations.

Work at a Sustainable Pace and Do a Little at a Time

When you're doing a long project like writing a book that can take well over a year and requires a lot of creativity and innovation, working at a sustainable pace is very important.

- You can easily get burned out by trying to do too much too quickly and when that happens, your creativity can go downhill quickly.

- Sometimes you need to put it down, walk away from it for a while, and come back when you're refreshed to start work again.

Why Do People Have Trouble with This?

I think all of this is just good, common-sense things to do—why do people have trouble doing this?

- I think many people think of Agile as Scrum and also think about doing it mechanically and aren't sure how they would go about applying a Scrum process to this kind of effort.

- Agile is not just Scrum—it is a way of thinking, and we need to understand the principles behind it rather than attempting to follow a well-defined process and doing it mechanically and "by-the-book."

DISCUSSION TOPICS

1. What are some examples of situations in your own work or personal life that you could apply Agile principles to?

2. What would be the benefits?

3. What would be the difficulties to overcome?

(32) Overall Summary

HERE ARE SOME OF THE KEY OVERALL MESSAGES that I believe are most important.

EVOLUTION OF THE PROJECT MANAGEMENT PROFESSION

The Future of Project Management

The project management profession is at a major transition point:

- The project management profession is beginning to go through rapid and profound changes due to the widespread adoption of Agile methodologies.

- Those changes are likely to dramatically change the role of Project Managers in many environments, as we have known them, and raise the bar for the entire project management profession.

1. Don't Ignore the Impact of Agile and Lean

The project management profession has been somewhat slow to recognize the impact of Agile and Lean over the years:

- Prior to 2013, Agile wasn't really recognized as a legitimate form of project management and there was a big chasm between the Agile and project management communities. It was almost considered heresy for a Project Manager to even think about an Agile approach. With the introduction of the PMI-ACP certification in 2013 by PMI, PMI at least recognized Agile as an important form of project management.

- Agile and classical plan-driven project management were essentially treated as separate and independent domains of knowledge with little or no integration between the two. It's been up to individual Project Managers to figure out how (and if) the two approaches could be integrated.

PMBOK® has always been considered to be the "bible" of project management and up until PMBOK® version 7 which was released in 2021:

- PMBOK® has been heavily associated with a classical plan-driven approach with little or no mention of Agile.

- PMBOK® tried to define a fairly prescriptive approach for doing project management. Earlier versions of PMBOK® have been over 500 pages long and have attempted to define a checklist of things to consider in almost every conceivable project management situation that you could be in.

PMBOK® version 7, released in 2021, is the first version of PMBOK® that has really attempted to develop a more integrated approach:

- It has moved away from an approach based on telling you how to do project management towards an emphasis on understanding the principles behind the project management approach at a deeper level.

- It recognizes that this approach requires a lot more judgement and skill and doesn't lend itself well to following checklists.

It should be no surprise that many Project Managers have been deeply schooled in a classical plan-driven approach to project management and may not know any other way to do project management. If a classical plan-driven approach is the only project management approach that you know, you will likely have difficulty in a number of important areas such as software development that require a different approach.

2. Important Shifts in Thinking

Becoming an Agile Project Manager may require important shifts in thinking. Some of these shifts in thinking include:

- emphasis on maximizing value versus control

- emphasis on empowerment and self-organization

- limited emphasis on documentation

- managing flow instead of structure

- cross-functional leadership approach.

3. Becoming an Agile Project Manager Is a Journey

This book is only the beginning of that journey. You should commit yourself to ongoing continuous learning. Think of the Shu-Ha-Ri analogy of how people learn the martial arts:

- **Shu:** In the "Shu" stage, the student learns to do things more-or-less mechanically, "by the book," without significantly deviating from the accepted rules and practices and without improvising any new techniques.

- **Ha:** In the "Ha" stage, the student begins to understand the principles at a deeper level and learns how to improvise and break free from rigidly accepted practices, but it's important to go through the "Shu" stage and gain mastery of the foundational principles before you start improvising. You have to learn the foundational principles before you can improvise. Improvisation without knowledge is just amateurish experimentation.

- **Ri:** Finally, in the "Ri" stage, the student gets to the highest level of mastery and is able to develop his/her own principles and practices as necessary.

What Does It Take to Become a Good Agile Project Manager in This New Environment?

Here are some thoughts and recommendations on this.

1. Don't Position Yourself as a "General Purpose Project Manager"

First, it is important to understand that there are two areas of knowledge required to be a good Project Manager:

1. Knowledge of project management principles and practices (both Agile and classical plan-driven).
2. Knowledge of a domain to apply it to (for example, construction, software development, etc.)

People often don't understand or overlook the need for going beyond knowledge of project management principles and practices and also developing some domain knowledge to apply it to. In this new environment that is particularly important.

- A Project Manager who only understands project management principles and practices is essentially a high-level administrator.

- Project Managers need to go beyond that if they are to focus on delivering business value and that requires some understanding of the business that you're operating in.

- Focusing on a particular business domain will also help you to tune your project management approach around that particular business area. How you would do project management in a construction company would likely be very different from how you might do project management in a software development environment.

2. Don't Be a "Cookbook" Project Manager

Some Project Managers are used to following well-defined project management processes even to the extent of using prescribed checklists of what to do in a given situation and perhaps also using fill-in-the-blanks document templates to complete.

- "Cookbook" approaches no longer work very well.

- A good Project Manager needs to understand the principles at a deeper level behind the project management approach in order to adapt the project management approach to fit the situation.

  In my books, I've often talked about the difference between a "cook" and a "chef":

- A good cook might have the ability to create some very good meals, but those dishes might be limited to a repertoire of standard dishes, and his/her knowledge of how to prepare those meals might be primarily based on following some predefined recipes out of a cookbook.

- A chef, on the other hand, typically has a far greater ability to prepare a much broader range of more sophisticated dishes using much more exotic ingredients in some cases.

  - The chef's knowledge of how to prepare those meals is not limited to predefined recipes, and in many cases, a chef will create entirely new and innovative recipes for a given situation.
  - The best chefs are not limited to a single cuisine and are capable of combining dishes from entirely different kinds of cuisine.

  In this new project management environment, we need more Project Managers who are "chefs", not "cooks."

WHAT TO DO DIFFERENTLY

1. Focus on Results and Business Value

Focus on results and value and adapt the approach as needed to optimize the overall business value of the project. In the past, many projects have been regarded as successful if they delivered well-defined requirements within an approved budget and schedule. That's a very narrow definition of success and there have been many projects that have met their cost and schedule goals but failed to deliver an acceptable level of business value.

- The primary emphasis in a project should be on creating business value and simply meeting cost and schedule goals may or may not be an effective measure of that.

- Also, an excessive emphasis on planning and control to achieve cost and schedule goals can stifle the creativity and innovation that are needed to maximize the value of solutions.

  This does not mean that costs and schedules are no longer important, but they are only one component of business value. In a particular project, creativity and innovation may be more important than predictability, planning, and control.

2. Fit the Approach to the Project

There is no longer just one standard way of doing project management. Don't make the mistake of force-fitting a project to some predefined, textbook methodology. The right approach is to go in the other direction and fit the methodology to the nature of the project.

Rather than force-fitting all projects to a classical plan-driven project management approach, Project Managers need to learn how to fit the project management approach to the nature of the project based on:

- the level of uncertainty in the project
- the relationship with the customer
- the training and sophistication of the project team
- the organizational environment and culture that the Project Manager is part of.

That requires an understanding of the principles at a deeper level in addition to the mechanics of how to get things done.

3. Use a Flexible and Adaptive Approach

A typical plan-driven project management approach doesn't provide sufficient flexibility and adaptivity to manage projects in an uncertain environment and it can waste a lot of time in trying to develop detailed plans which only become outdated and no longer relevant as the project is in progress.

- The planning approach should be adapted to fit the level of uncertainty in the project.
- It can be foolish to attempt to do highly-detailed planning for a project with a high level of uncertainty and a more flexible and adaptive approach to evolve the plan as the project is in progress is typically needed for projects with a higher level of uncertainty.

4. Deliver Incrementally and Efficiently

A classical plan-driven project management approach typically attempts to deliver the entire solution all at once. That can be a very inefficient approach because:

- it delays the release of value
- it increases the risk in the project
- it isolates the customer from the solution until it is complete when it is often too late to make significant changes.

5. Develop a Close Partnership with the Customer

Don't rely on a typical, "arm's-length," contractual relationship with the customer. Actively involve the customer in the project in a spirit of collaboration, trust, and partnership and share information openly and transparently.

6. Build Quality into the Solution

Don't rely on a typical Quality Assurance approach to find and fix defects by an independent organization after the solution is complete; make an emphasis on quality an integral part of the development effort. "Quality" should be everyone's responsibility; it should not be the responsibility of only a separate Quality Assurance organization.

7. Focus on Empowered Teams and Leadership

Many Project Managers in the past have been noted for what is called a "command-and-control" approach to management that provides strong overall direction to teams. That approach is appropriate in a classical plan-driven environment where predictability, planning, and control are important. In an environment with a high level of uncertainty where creativity and innovation may be more important that an emphasis on planning and control, a softer approach is needed.

GENERAL RECOMMENDATIONS

1. Delivering Faster Is Important

Produce value as quickly as possible:

- Don't get too bogged down in planning, particularly on projects with a high level of uncertainty.
- Deliver results incrementally if necessary to show progress quickly.
- Projects generally should not be longer than six months—anything longer than that should probably be considered as a major initiative and broken down into smaller projects as necessary.

2. Do the Right Thing

Focus on delivering business value in addition to managing costs and schedules.

- It can take courage to do the right thing when there is extensive pressure to meet schedule and budget goals.
- The actual purpose of all projects is to deliver value—budget and schedule are just constraints.

3. Don't Get Lost in the "Mechanics of Doing Agile"

Too many people implement Agile and Scrum mechanically and "by the book."

- It's easy to do that, particularly for someone who has a limited level of experience.

- It takes some level of skill to go beyond that and focus on understanding the principles at a deeper level rather than simply following the mechanics of doing Agile/Scrum.

4. Take a Systemic, Enterprise-Level View

Focus on Governance, People, Process, and Tools

When projects are in trouble and aren't going well, many people try to push harder to make the process work. A better approach is to focus on the systemic factors that are at the core of making an Agile project successful:

- **Governance:** An effective system of project governance to provide oversight over projects to ensure that they effectively fulfill customer needs and manage the company's business interests.

- **People:** Having the right people on the project where the resources are well trained in the process, technology, and tools, coupled with an approach based on high-performance teams that are cohesive, cross-functional, empowered, and self-organizing.

- **Process:** Having the right methodology in place that is well designed and appropriate to the nature of the project and is also well integrated with the customer for the project so that the customer is fully engaged collaboratively in the process in a spirit of partnership and trust.

- **Technology and tools:** Having the right tools in place to support the process and people is extremely important to maximize the efficiency of the people and process, especially in cases that involve distributed teams that are not co-located, and the projects are fast-paced and demanding.

Organizational Culture Is Very Important

An Agile approach that is not well integrated with the culture of the company it is part of is not likely to be fully successful. However, changing the company's culture so that it is more conducive to Agile is desirable if that's possible. It's often not that simple.

- A company's culture should be shaped around whatever makes sense to drive its primary business, and that might or might not be in alignment with an Agile development process.

- Changing the culture of a company is not easy to do and is typically not something that can be done quickly.

Appendices

THE APPENDICES TO THE BOOK contain additional information that is intended to supplement the primary information in the rest of the book.

Appendix A

Additional Reading and Resources:

This appendix contains a list of recommended additional reading to provide a broader and deeper understanding of some of the material in this book.

Appendix B

Glossary of Terms:

This appendix contains a list of definitions and terms used throughout the book.

Appendix C

Example Project/Program Charter Template:

This appendix contains an example of a project charter template that can be used in the macro-level of the Managed Agile Development framework. This charter template is a general model that is intended to be customized as necessary to fit a given project and business environment.

Appendix D

Suggested Course Outline:

This appendix contains a suggested course outline for a graduate-level Agile Project Management course based on this textbook.

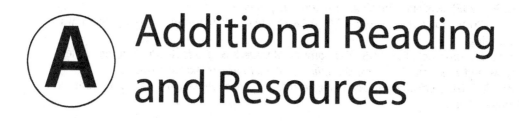

Additional Reading and Resources

ADDITIONAL READING

Appelo, J. (2011) *Management 3.0: Leading Agile Developers, Developing Agile Leaders*, Addison-Wesley.

Boehm, B. and Turner, R. (2003) *Balancing Agility and Discipline: A Guide for the Perplexed*, Addison-Wesley.

Cobb, C.G. (2011) *Making Sense of Agile Project Management: Balancing Control and Agility*, Wiley.

Cohn, M. (2006) *Agile Estimating and Planning*, Prentice Hall.

Cohn, M. (2010) *Succeeding with Agile Software Development Using Scrum*, Addison-Wesley.

Goldratt, E. (2012) *The Goal: A Process of Ongoing Improvement*, North River Press.

Griffiths, M. (2018) *PMI ACP Exam Prep : A Course in a Book for Passing the PMI Agile Certified Practitioner (PMI ACP) Exam* (Updated Second Edition).

Highsmith, J. (2010) *Agile Project Management*, Addison-Wesley.

Larman, C. (2003) *Agile & Iterative Development: A Manager's Guide*, Addison-Wesley.

Laufer, A. (2012) *Mastering the Leadership Role in Project Management: Practices that Deliver Remarkable Results*, Pearson Education.

Managed Agile Development (Blog Site) https://managedagile.com/ Chuck Cobb.

Pichler, R. (2010) *Agile Product Management with Scrum*. Addison-Wesley.

PMI, (2021) *A Guide to the Project Management Body of Knowledge (PMBOK® Guide)*, Seventh Edition, PMI.

Poppendiek, M. and Poppendiek, T. (2003) *Lean Software Development: An Agile Toolkit*, Addison-Wesley.

Poppendiek, M. and Poppendiek, T. (2007) *Implementing Lean Software Development: From Concept to Cash*, Addison-Wesley.

Poppendiek, M. and Poppendiek, T. (2010) *Leading Lean Software Development*, Addison-Wesley.

Rhoades, A. and Bass, J. (2011) *Built on Values: Creating an Enviable Culture that Outperforms the Competition*. Jossey-Bass.

Ries, E. (2011) *The Lean Startup*, Crown Publishing.

Rubin, K. (2013) *Essential Scrum: A Practical Guide to the Most Popular Agile Process*, Addison-Wesley.

Schwaber, K. (2003) *Agile Project Management with Scrum*, Microsoft Press.

Schwaber, K. (2007) *The Enterprise and Scrum* (Developer Best Practices), Microsoft Press.

Senge, P. (1996) "Systems Thinking," *Executive Excellence*.

Senge, P. (2006) *The Fifth Discipline: The Art & Practice of the Learning Organization*, Crown Publishing.

Sliger, M. and Broderick, S. (2008) *The Software Manager's Bridge to Agility*, Addison-Wesley.

Treacy, M. and Wiersema, F. (1995) *Discipline of Market Leaders*, Addison-Wesley.

Wysocki, R. (2019) *Effective Project Management: Traditional, Agile, Extreme, Hybrid* (8th Edition), Wiley.

ADDITIONAL SOURCES FOR CASE STUDIES AND PAPERS

The following is a list of additional resources containing case studies, articles, and research papers provided by Dr. Winston Gonzalez, D.M., SPC. These illustrate the application of Agile practices to areas outside of software development:

Agile Education, Agile Based Learning Environment (ABLE). Agile Alliance. https://www.agilealliance.org/resources/sessions/agile-education/.

Agile Education, Learning Education Agile Framework (LEAF). Learning Education Agile Framework. https://www.l-eaf.org/.

Agile Hardware, The Scrum in Hardware Guide. Scrum Inc. https://www.scruminc.com/scrum-in-hardware-guide/.

Agile Manufacturing, Difference Between Lean and Agile Manufacturing. Redwood Logistics. https://www.redwoodlogistics.com/difference-between-lean-and-agile-manufacturing/.

Agile Manufacturing, Lean Production. Vorne. https://www.leanproduction.com/agile-manufacturing/.

Agile Marketing, Agile Marketing Manifesto. Agile Marketing Manifesto.org. https://agilemarketingmanifesto.org/.

Agile Marketing, Agile Marketing: A Step-by-Step Guide. McKinsey. https://www.mckinsey.com/business-functions/growth-marketing-and-sales/our-insights/agile-marketing-a-step-by-step-guide.

Agile Marketing, What Is Agile Marketing? Agile Marketing. https://agilemarketing.net/what-is-agile-marketing/.

Boston Consulting Group, General Beyond Software. Taking Agile Way Beyond Software. https://www.bcg.com/publications/2017/technology-digital-organization-taking-agile-way-beyond-software.

Human Resources, Agile HR Manifesto. AgileManifesto.Org. https://www.agilehrmanifesto.org/.

Johnson, A., Agile Architecture. Agile Architect Website. https://www.agilearchitect.org/agile/index.php.

Justice, J., Agile Hardware. Agile Hardware Institute. https://businessagility.institute/learn/agile-hardware/514.

Sutherland, J. et al., Agile Defense Procurement. Owning the Sky with Agile: Building a Jet Fighter Faster, Cheaper, Better with Scrum. https://www.scruminc.com/wp-content/uploads/2015/09/Release-version_Owning-the-Sky-with-Agile.pdf.

Sutherland, J. et al., Agile Sales. Scrum in Sales. https://www.scruminc.com/wp-content/uploads/2014/05/Scrum_in_Sales.pdf.

Sutherland, J. et al., Agile Venture Capital. Take No Prisoners: How a Venture Capital Group Does Scrum. https://www.scruminc.com/wp-content/uploads/2014/05/Take-No-Prisoners.pdf.

Sutherland, J. et al., Agile in Religion. Scrum in Church: Saving the World One Team at a Time. https://www.scruminc.com/wp-content/uploads/2014/05/Scrum-in-Church-Agile.pdf.

Swingler, K. and Human Resources, Agile Human Resources: Creating a Sustainable Future for the HR Profession. https://www.amazon.com/Human-Resources-Kelly-Swingler/dp/1947441337.

(B) Glossary of Terms

Agile The word "agile" has two different connotations.

1. As a general adjective, and

2. As a more specific reference to a methodology or framework that is based on principles and practices based on the Agile Manifesto.

In the context of a general adjective, the Merriam Webster Dictionary defines the word *agile* as:[1]

> Marked by ready ability to move with quick easy grace <an agile dancer>.
> Having a quick resourceful and adaptable character <an agile mind>.

Dictionary.com defines the word *agile* as follows:[2]

> Quick and well-coordinated in movement; lithe: *an agile leap.*
> Active; lively: an agile person.
> Marked by an ability to think quickly; mentally acute or aware: *She's 95 and still very agile.*

Both of those definitions are probably very appropriate in the context of an Agile Project Manager.

- A cross-functional Agile team moves quickly and autonomously with fluidity and grace without an excessive amount of external direction and orchestration.

- An Agile project approach is very adaptive to volatile and uncertain environments as necessary rather than being locked in to a well-defined plan that might be difficult to change.

- Agile people are very resourceful, high-energy people who are able to think quickly in a variety of different circumstances.

- You will often see Agile capitalized. In this context, the word "Agile" refers to a specific methodology or framework that is based on the principles and practices based on the Agile Manifesto.

- In actual practice, the word "agile" is very loosely used. In many cases, the word "agile" has become conflated with Scrum because Scrum is the most widely used Agile framework. That usage is not accurate and can be misleading.

Agile Modeling   Agile Modeling is a methodology developed by Scott Ambler that is a collection of values, principles, and practices for modeling software that can be applied in a software development project in an effective and lightweight manner.[3]

Agile Project Management   Agile Project Management is a style of project management that recognizes the need to fit the project management approach to the nature of the project and blend the principles and practices of:

(1) a classical plan-driven project management approach with an emphasis on planning and control to achieve predictability over project costs and schedules, with

(2) a more flexible and adaptive Agile (value-driven) project management approach with an emphasis on maximizing the business value a project produces in an uncertain environment in the right proportions to fit the nature of the project.

Agile Unified Process (AUP)   The Agile Unified Process is a simplified version of the Rational Unified Process (RUP), developed by Scott Ambler. It describes a simple, easy-to-understand approach to developing business application software using Agile techniques and concepts yet still remaining true to the RUP.[4]

Burn-Down Chart   A burn-down chart is a chart often used in Scrum Agile development to track work completed against time allowed. The x-axis is the time frame, and the y-axis is the amount of remaining work left that is labeled in story points and manhours, etc. The chart begins with the greatest amount of remaining work, which decreases during the project and slowly burns to nothing.[5]

Burn-Up Chart   A burn-up chart is a graphical representation that tracks progress over time by accumulating functionality as it is completed. The accumulated functionality can be compared to a goal such as a budget or release plan to provide the team and others with feedback. Graphically, the x-axis is time and the y-axis is accumulated functionality completed over that period of time.

- The burn-up chart, like its cousin the burn-down chart, provides a simple yet powerful tool to provide visibility to the sprint or program.

- The burn-up chart can be thought of as the mirror image of the burn-down chart but is generally extended over multiple sprints to show the strategy being followed as the project builds toward release and product delivery.[6]

Code Refactoring   Code refactoring involves removing redundancy, eliminating unused functionality, and rejuvenating obsolete designs and improving the design of existing software, in order to improve reliability and maintainability of the software. Refactoring throughout the entire project life cycle saves time and increases the quality of the software. Code refactoring is more commonly used with Extreme Programming—it is less commonly used with Feature-Driven Development.

Continuous Integration   Continuous integration is the practice of frequently integrating new or changed software with the code repository. It is a way of early detection of problems that may occur when individual software developers are working on code changes that may potentially conflict with each other. In many typical software development environments, integration may not be performed until the application is ready for final release.

Daily Standup   A daily standup meeting is a short organizational meeting that is held each day in a typical Agile/Scrum project. The meeting, generally limited to between 5 and 15 minutes long, is sometimes referred to as a stand-up, daily standup, or a daily scrum. The purpose of the meeting is for each team member to answer the following three questions:

1. What did you do yesterday?

2. What will you do today?

3. Are there any impediments in your way?

The participants typically stand rather than sit during the meeting to reinforce the idea that the meeting is intended to be short and to discourage wasted time. The standup is not meant to be a place to solve problems, but rather to make the team aware of current status. If discussion is needed, a longer meeting with appropriate parties can be arranged separately.[7]

Delphi Method   The Delphi method is a forecasting method based on the results of questionnaires sent to a panel of experts. Several rounds of questionnaires are sent out, and the anonymous responses are aggregated and shared with the group after each round. The experts are allowed to adjust their answers in subsequent rounds. Because multiple rounds of questions are asked and because each member of the panel is told what the group thinks as a whole, the Delphi Method seeks to reach the "correct" response through consensus.[8]

Dynamic Systems Development Model (DSDM)   Dynamic Systems Development Method (DSDM) is a framework based originally on Rapid Application Development (RAD), supported by continuous user involvement in an iterative development and incremental approach, which is responsive to changing requirements, in order to develop a system that meets the business needs on time and on budget. DSDM was developed in the United Kingdom in the 1990s by a consortium of vendors and experts in the field of Information System (IS) development, the DSDM Consortium, combining their best-practice experiences.[9] It is most frequently used outside of the United States.

Epic   An epic is a large user story that needs to be broken down into smaller user stories prior to the start of an Agile iteration.

Extreme Programming (XP)   "Extreme Programming is a discipline of software development based on the values of simplicity, communication, feedback, and courage. It works by bringing the whole team together in the presence of simple practices, with enough feedback to enable the team to see where they are and to tune the practices to their unique situation."

■ "In Extreme Programming, every contributor to the project is an integral part of the whole team. The team forms around a business representative called 'the Customer', who sits with the team and works with them daily."

- "Extreme Programming teams use a simple form of planning and tracking to decide what should be done next and to predict when the project will be done. Focused on business value, the team produces the software in a series of small fully integrated releases that pass all the tests the customer has defined".[10]

Feature-Driven Development   Feature-driven development (FDD) is a model-driven approach that puts more emphasis on defining an overall model of the system and a list of features to be included in the system prior to starting the design effort.

Gantt Chart   A Gantt chart is a type of bar-chart that shows both the scheduled and completed work over a period. A time-scale is given on the chart's horizontal axis and each activity is shown as a separate horizontal rectangle (bar) whose length is proportional to the time required (or taken) for the activity's completion. In project planning, these charts show start and finish dates, critical and non-critical activities, slack time, and predecessor-successor relationships.[11]

IT Strategy   An IT strategy is a comprehensive plan that information technology management professionals use to guide their organizations. An IT strategy should cover all facets of technology management, including cost management, human capital management, hardware and software management, vendor management, risk management and all other considerations in the enterprise IT environment. Executing an IT strategy requires strong IT leadership; the chief information officer (CIO) and chief technology officer (CTO) need to work closely with business, budget and legal departments as well as with other user groups within the organization.

Many organizations choose to formalize their information technology strategy in a written document or balanced scorecard strategy map. The plan and its documentation should be flexible enough to change in response to new organizational circumstances and business priorities, budgetary constraints, available skill sets and core competencies, new technologies and a growing understanding of user needs and business objectives.[12]

Kanban Board   A Kanban board is a columnar status board where each of the columns represents stages in a progression of stages in the flow of a process and items are shown in the column of the board corresponding to their current status in the process flow. A Kanban board could be implemented in the form of a physical board such as a whiteboard with 3 x 5 cards or "yellow stickies" representing the items in the process or it could be implemented in an online tool.

Lean Manufacturing   According to the Learning Center:

Lean manufacturing is a comprehensive term referring to manufacturing methodologies based on maximizing value and minimizing waste in the manufacturing process. Lean manufacturing has evolved in North America from its beginnings in the Toyota Production System (TPS) in

Japan. Many of the most recognizable phrases, including *kaizen* and Kanban, are Japanese terms that have become standard terms in lean manufacturing.

At the heart of lean is the determination of value. Value is defined as an item or feature for which a customer is willing to pay. All other aspects of the manufacturing process are deemed waste. Lean manufacturing is used as a tool to focus resources and energies on producing the value-added features while identifying and eliminating non value added activities.[13]

Lean Software Development   Lean Software Development is a translation of Lean manufacturing and Lean IT principles and practices into the software development domain. It is heavily based on the work of Tom Poppendiek and Mary Poppendiek.

Pair Programming   Pair programming is an Agile software development technique in which two programmers work together at one workstation—one developer plays the role of an observer while the other developer in the pair writes the code.

Pert Chart   A Pert chart is a project management tool that provides a graphical representation of a project's timeline. PERT, or Program Evaluation Review Technique, was developed by the United States Navy for the Polaris submarine missile program in the 1950s. PERT charts allow the tasks in a particular project to be analyzed, with particular attention to the time required to complete each task, and the minimum time required to finish the entire project. A PERT chart is a graph that represents all of the tasks necessary to a project's completion, and the order in which they must be completed along with the corresponding time requirements. Certain tasks are dependent on serial tasks, which must be completed in a certain sequence. Tasks that are not dependent on the completion of other tasks are called parallel or concurrent tasks and can generally be worked on simultaneously. [14]

Plan-Driven Project Management   Plan-driven project management is the predominant project management approach that has been in effect for many years. It has an emphasis on planning and control to achieve predictability over the costs and schedule for a project with relatively well-defined requirements.

Portfolio Kanban   A portfolio Kanban is a Kanban board that is used as a tool in a portfolio management process to represent the various projects and initiatives that are being managed through each stage of the portfolio management process.

Product Backlog   The Product Backlog is a high-level document list of all required features and wish-list items prioritized by business value in an Agile project. It is the "what" that will be built. It is owned by the Product Owner and continuously prioritized and reprioritized as the project progresses, work is completed, and detailed requirements are better understood.

Prototype Model   A prototype model is a type of software development life cycle model that is used to progressively define the requirements for a product or application. It involves developing the requirements as much as possible from user input, then a prototype model is built based on the known requirements. User feedback is then used to progressively refine the model as necessary until it satisfies the user. Prototyping can also be an effective method to demonstrate the feasibility of a certain approach.

- The basic reason for the limited use of prototyping is the cost involved in this build-it-twice approach. However, with modern software development tools, prototyping need not be very costly and can actually reduce the overall development cost.[15]

QA Testing   QA testing or software testing is a set of processes aimed at investigating, evaluating, and ascertaining the completeness and quality of computer software. Software testing ensures the compliance of a software product in relation with regulatory, business, technical, functional, and user requirements. The objectives of these processes can include:[16]

- verifying software completeness in regards to functional/business requirements

- identifying technical bugs/errors and ensuring the software is error-free

- assessing usability, performance, security, localization, compatibility and installation.

Rational Unified Process (RUP)   The Rational Unified Process (RUP) is a software development process from Rational, a division of IBM. It divides the development process into four distinct phases that each involve business modeling, analysis and design, implementation, testing, and deployment. The four phases are:

- **Inception:** The idea for the project is stated. The development team determines if the project is worth pursuing and what resources will be needed.

- **Elaboration:** The project's architecture and required resources are further evaluated. Developers consider possible applications of the software and costs associated with the development.

- **Construction:** The project is developed and completed. The software is designed, written, and tested.

- **Transition:** The software is released to the public. Final adjustments or updates are made based on feedback from end users.[17]

Scrum   Scrum is an Agile software development framework based on multiple small teams working in an intensive and interdependent manner. It is, by far, the most widely-used Agile framework. It is considered to be a framework rather than a methodology because it is not intended to be very prescriptive.

- The term is named for the *scrum* (or *scrummage*) formation in rugby, which is used to restart the game after an event that causes play to stop, such as an infringement.

■ Scrum employs real-time decision-making processes based on actual events and information. This requires well-trained and specialized teams capable of self-management, communication and decision making. The teams in the organization work together while constantly focusing on their common interests.[18]

Software Development Life Cycle (SDLC)   A software development life cycle (SDLC) is a process or framework that describes the activities performed at each stage of the software development process and provides an overall roadmap for the project. It provides a template for executing a project that can be tailored to fit a particular project. An example of an SDLC would be the Waterfall model; however, as used in this book, an SDLC would include any project framework, such as Scrum.

Spike   According to Erik Phillipus:

A spike is an experiment that allows developers to learn just enough about the unknown elements in a user story, e.g. a new technology, to be able to estimate that user story. Often, a spike is a quick and dirty implementation or a prototype, which will be thrown away. When a user story on the product backlog contains unknown elements that seriously hamper a usable estimation, the item should be split into a spike to investigate these elements plus a user story to develop the functionality.[19]

Sprint   In product development, a sprint is a set period of time during which specific work has to be completed and made ready for review.

Story Points   Story points are a method of estimating the level of effort associated with implementing a user story. The typical form of story points is a Fibonacci series such as 1, 2, 3, 5, 8, 13, with 1 being a minimal level of effort and 13 being a maximum level of effort for a user story in an iteration.

Test-Driven Development (TDD)   Test-driven development (TDD) is commonly used in Agile methodologies to integrate testing directly into the software development effort. According to agiledata.org:

Instead of writing functional code first and then your testing code as an afterthought, if you write it at all, you instead write your test code before your functional code. Furthermore, you do so in very small steps—one test and a small bit of corresponding functional code at a time. A programmer taking a TDD approach refuses to write a new function until there is first a test that fails because that function isn't present. In fact, they refuse to add even a single line of code until a test exists for it. Once the test is in place they then do the work required to ensure that the test suite now passes.[20]

Total Quality Management (TQM)   Total Quality Management (TQM) is a holistic approach to long-term success that views continuous improvement in all aspects of an organization as a process and not as a short-term goal. It aims to radically transform the organization through progressive changes in the attitudes, practices, structures, and systems.

TQM transcends the product quality approach, involves everyone in the organization, and encompasses its every function: administration, communications, distribution, manufacturing, marketing, planning, training, etc. Coined by the US Naval Air Systems Command in the early 1980s, this term has now taken on several meanings and includes:

- Commitment and direct involvement of highest-level executives in setting quality goals and policies, allocation of resources, and monitoring of results;

- Realization that transforming an organization means fundamental changes in basic beliefs and practices and that this transformation is everyone's job;

- Building quality into products and practices right from the beginning;

- Understanding of the changing needs of the internal and external customers, and stakeholders, and satisfying them in a cost effective manner;

- Instituting leadership in place of mere supervision so that every individual performs in the best possible manner to improve quality and productivity, thereby continually reducing total cost;

- Eliminating barriers between people and departments so that they work as teams to achieve common objectives; and

- Instituting flexible programs for training and education, and providing meaningful measures of performance that guide the self-improvement efforts of everyone involved.[21]

User Persona   A user persona is a description of a specific type of user that impacts the requirements. For example, a banking customer would be an example of a user persona. User personas are useful ways of characterizing the users of the system and keeping the development effort focused on satisfying their needs.

User Stories   User stories are one of the primary development artifacts for Scrum and Extreme Programming (XP) project teams. A user story is a high-level definition of a requirement, containing just enough information so that the developers can produce a reasonable estimate of the effort to implement it.

User stories are much smaller than other usage requirement artifacts such as use cases or usage scenarios. User stories provide a way of breaking up the project into individual work items that can provide a way of estimating and tracking work to be done. Each user story may be further refined as necessary as the design progresses.

Velocity   According to Agile Sherpa:

Velocity is the measure of the throughput of an agile team per iteration. Since a user story, or story, represents something of value to the customer, velocity is actually the rate at which a team delivers value. More specifically, it is the number of estimated units (typically in story points) a team delivers in an iteration of a given length and in accordance with a given definition of "done."[22]

Waterfall Model   The Waterfall model is a software development life cycle model that describes a linear and sequential development method. Waterfall development has distinct goals for each phase of development and once a phase of development is completed, the development typically proceeds to the next phase and there is no turning back.

The advantage of Waterfall development is that it allows for departmentalization and managerial control. A schedule can be set with deadlines for each stage of development and a product can proceed through the development process sequentially. Development moves through the phases of the model, such as concept, design, implementation, and testing, and ends up at deployment and operation and maintenance. Each phase of development normally proceeds in a prescribed order, without any overlapping or iterative steps; however, tailoring is often done to make the process more efficient.

The disadvantages of Waterfall development are that:

- It does not allow for much reflection or revision—once an application is in the testing stage, it is very difficult to go back and change something that was not well thought out in the concept stage.

- It assumes that all requirements can be defined upfront prior to any design work and that is not a realistic assumption in many situations and it doesn't provide an approach that is flexible and adaptive to customer needs.

- It may require an excessive amount of documentation artifacts.[23]

The word "Waterfall" is very loosely-used in actual practice. In many cases, the word "Waterfall" is used to refer to any methodology that is plan-driven and not completely Agile.

Work Breakdown Structure (WBS)   A Work Breakdown Structure (WBS) is a project network-modeling step in which the entire project is graphically subdivided into manageable work elements (tasks) that is typically shown in a hierarchical structure. WBS displays the relationship of each task to the other tasks, to the whole and the end product (goal or objective).[24]

XP   See Extreme Programming.

NOTES

1. Merriam-Webster dictionary, "agile," http://www.merriam-webster.com/dictionary/agile.
2. Dictionary.com, "agile," http://dictionary.reference.com/browse/agile.
3. Agile Modeling, http://www.agilemodeling.com/.
4. Agile Unified Process, http://www.ambysoft.com/unifiedprocess/agileUP.html.
5. Techopedia, "What Is a Burn-down Chart?," http://www.techopedia.com/definition/26294/burndown-chart.

6. Thomas F. Cagley, "Metrics Minute—Burn Up Charts," http://tcagley.wordpress.com/2011/05/09/metrics-minute-burn-up-charts/.

7. "What Is a Daily Standup Meeting?," http://searchsoftwarequality.techtarget.com/definition/daily-stand-up-meeting.

8. Delphi Method definition, http://www.investopedia.com/terms/d/delphi-method.asp.

9. Select Business Solutions, "What Is DSDM?," http://www.selectbs.com/adt/process-maturity/what-is-dsdm.

10. Ron, Jeffries, "What Is Extreme Programming?", https://ronjeffries.com/xprog/what-is-extreme-programming/.

11. Gantt Chart definition, http://www.businessdictionary.com/definition/Gantt-chart.html#ixzz3ArL59bly.

12. "What Is IT Strategy?," http://searchcio.techtarget.com/definition/IT-strategy-information-technology-strategy.

13. Learning Center, "Lean Manufacturing," http://www.vorne.com/learning-center/lean-manufacturing.htm.

14. Pert Chart definition, http://www.investopedia.com/terms/p/pert-chart.asp.

15. Prototyping Software Lifecycle Model, http://www.freetutes.com/systemanalysis/sa2-prototyping-model.html.

16. "What Is Software Testing?," http://www.techopedia.com/definition/17681/software-testing.

17. RUP (Rational Unified Process) definition, http://www.techterms.com/definition/rup.

18. "What Is Scrum?," http://searchsoftwarequality.techtarget.com/sDefinition/0,,sid92_gci1230820,00.html.

19. Erik Phillipus, "Architecture Spikes," http://www.agile-architecting.com/Architecture%20Spikes.pdf.

20. "Introduction to Test-Driven Design," http://www.agiledata.org/essays/tdd.html.

21. "What Is Total Quality Management?," http://www.businessdictionary.com/definition/total-quality-management-TQM.html.

22. Agile Sherpa, "Velocity," http://www.agilesherpa.org/agile_coach/metrics/velocity/.

23. "What Is a Waterfall Model?," http://searchsoftwarequality.techtarget.com/sDefinition/0,,sid92_gci519580,00.html.

24. "What Is a Work Breakdown Structure?," http://www.businessdictionary.com/definition/work-breakdown-structure-WBS.html#ixzz3ArUoGY3m.

C Example Project/ Program Charter Template

THIS APPENDIX CONTAINS AN EXAMPLE of an optional Project Charter template that can be used to define the macro layer in a hybrid, managed agile development approach. This template is provided as an example and is intended to be customized to fit the project and business environment that it is used in.

PROJECT OVERVIEW

Background

Provide a brief description of the background behind the problem that the project or program is intended to address to a sufficient level to allow the reader to understand the context of the problem.

Problem Statement

Provide a brief description of the problem that the project or program is intended to address from a business or operational management perspective.

Project Vision

Write a concise vision statement that summarizes the purpose and intent of the project and describes what the world will be like when the project is completed. The vision statement should reflect a balanced view that will satisfy the needs of diverse customers as well as those of the developing organization. It may be somewhat idealistic, but it should be grounded in the realities of existing or anticipated customer markets, enterprise architectures, organizational strategic directions, and cost and resource limitations. Consider using the following template:

- For (target customer)
- Why (statement of the need or opportunity)
- The (product name)
- Is a (product category)
- That (key benefit, compelling reason to buy or use)

Success Criteria

What are the success criteria for the project? How do you know if the project has been successful?

Project Approach/Development Process

Identify the development process and/or any deviations from the standard methodology that will be used for this project or program.

PROJECT PLAN

This section outlines the plan for managing the project.

Scope

The project scope defines the range of the proposed products and services the project will deliver. Scope can be represented using a context diagram, an event list, and/or a feature table. Scope might be subdivided into the scope of the initial product release and planned growth strategies for subsequent releases. It is also important to define what the project will not include, so describe limitations and exclusions, such as product features or characteristics that a stakeholder might anticipate, but which are not planned to be included in the project.

In Scope

The project scope provides an overview of the user stories that the project will deliver. Scope might be subdivided into the scope of the initial product release and planned growth strategies for subsequent releases.

Release	Priority	Story #	Story Name	Description

Out of Scope

It's also important to define what the project will not include, so describe limitations and exclusions, such as product features or characteristics that a stakeholder might anticipate, but which are not planned to be included in the project.

- Out of Scope Item #1

- Out of Scope Item #2

- Etc.

Related Projects and Systems

Identify any related projects or systems and describe the interrelationship.

Project or system	Interrelationship

Project Participants

Identify key project participants and the role that they will play in the project.

Role	Name	Organization
Business Sponsor		
Product Owner		

Role	Name	Organization
Business Process Owner		
Subject Matter Experts		
Stakeholders		
Project Manager		
Business Analyst		

CONSTRAINTS, ASSUMPTIONS, AND RISKS

Constraints

Constraints are impediments and other factors that must be considered when executing the project or program. The constraints that are applicable to this program include: _____

    Identify known project constraints, such as products to be reused, components to be acquired, interfaces to other projects or products, or technologies to be employed.

Constraint	Impact

Assumptions

Record any assumptions that were made (as opposed to known facts) when conceiving the project. An assumption is a statement that CANNOT be proven to be true, but needs to be taken as being true in order to proceed with the solution. Since the assumption is essential to progress, it is also a dependency that carries some risk because of its indeterminate nature, and the associated risk should be noted in the Business Risks section below.

- Assumption #1
- Assumption #2
- Etc.

Dependencies

Note any major external dependencies the project must rely upon for success, such as specific technologies, third-party vendors, development partners, or other business relationships. Also, identify any other projects that are related to this project in some way or may have a bearing on its outcome.

- Dependency #1
- Dependency #2
- Etc.

Risks

Summarize the major business risks associated with this project or program, such as marketplace competition, timing issues, user acceptance, implementation issues, or possible negative impacts on the business. Estimate the severity of each risk's potential impact and identify any risk mitigation actions that could be taken. This is not the place for the project's overall risk list.

Risk	Mitigation strategy

Timeline Estimate

The following table contains an estimated schedule for the major milestones in the project. Include a list of major project milestones and key deliverables with estimated target dates.

Dates	Milestone	Deliverables

D Suggested Course Outline

THIS BOOK IS DESIGNED TO be used to support a graduate-level course in Agile Project Management. The following is a suggested course outline that can be used with the course.

COURSE OVERVIEW

The course provides an understanding of how new Agile principles and practices are changing the landscape of project management and is designed to give experienced project managers fresh new insight into how to successfully blend Agile and classical plan-driven project management principles and practices in the right proportions to fit any business and project situation.

- It is desirable but not essential for the students to have a project management background as well as a basic foundation of knowledge of Agile principles and practices.

- The course provides a much deeper understanding of both Agile and classical plan-driven project management principles and practices in order to see them as complementary rather than competitive approaches.

   Topics include:

- Agile fundamentals, principles, and practices including the Scrum methodology.

- The impact of Agile on the Project Management profession.

- Understanding Agile at a deeper level and the roots of Agile.

- Agile Project Management Principles and Practices, Tools and Techniques.

- Agile Project Management Principles and Practices.

- Adapting an Agile approach to fit a business environment at an enterprise level.

- Planning and managing an enterprise-level Agile transformation for a business.

- Scaling agile to an enterprise level (including the use of enterprise-level Agile frameworks and Agile Project Management tools).

- Enterprise-level Agile Transformations.

- The course uses case studies and interactive discussions extensively to provide students with a real-world perspective of how to put these concepts into action.

COURSE OBJECTIVES

The general objective of the course can be summed up as giving experienced project managers the tools and skills:

- To assume a major leadership role in applying a new and emerging Agile Project Management approach to large and complex enterprise-level initiatives and

- To develop a more adaptive project management approach that can be applied to almost any project.

The course objectives are to give you a very fresh new outlook on how to blend Agile and more classical plan-driven Project Management principles and practices to dramatically enhance your repertoire of project management capabilities. In addition, you will learn how to adapt these principles and practices to different business environments and how to lead overall Agile business transformations.

The more specific objectives of the course are:

1. Develop new ways of thinking and begin to see Agile principles and practices in a new light as complementary rather than competitive to classical plan-driven project management practices.
2. Gain an understanding of the fundamentals of Agile practices and learn the principles behind the Agile practices at a deeper level in order to understand why they make sense and how they can be adapted as necessary to fit a given situation.
3. Learn how to go beyond the classical notion of plan-driven project management and develop an adaptive approach to project management that blends both Agile and classical plan-driven project management principles and practices in the right proportions to fit a given project and business environment.
4. Understand the potential roles that an Agile Project Manager can play and begin to reshape project management skills around those roles.

5. Learn some of the challenges of scaling Agile to an enterprise level and develop experience in applying these concepts in large, complex enterprise-level environments.

COURSE OUTLINE

Course Outline	Reference	Lesson
Part 1 – Fundamentals of Agile		
1. Introduction, Course Objectives, and Agile Overview		1
1.1. Introductions and Course Objectives		1
1.2. Evolution of the Project Management Profession	Chapter 1	1
1.3. Agile Overview	Chapter 1	1
1.3.1. What Is Agile?	Chapter 1	1
1.3.2. Agile Perception vs. Reality	Chapter 1	1
1.3.3. Agile Project Management Benefits	Chapter 1	1
Assignment: Complete Discussion Topics		
2. Agile Fundamentals		1
2.1. Agile History, Values, and Principles	Chapter 2	1
Assignment: Complete Discussion Topics		
3. Scrum Overview	Chapter 3	1
3.1. Scrum Roles	Chapter 3	1
3.2. Scrum Methodology	Chapter 3	1
3.3. General Scrum/Agile Principles	Chapter 3	1
3.4. Scrum Values	Chapter 3	1
Assignment: Complete Discussion Topics Draft Proposal for Research Paper Due		
4. Overview of Agile Practices		2
4.1. Agile Planning/Requirements Practices & Product Backlog	Chapter 4	2
4.1.1. General Agile Planning Practices	Chapter 4	2
4.1.2. General Agile Requirements Principles and Practices	Chapter 4	2
4.1.3. Product Backlog	Chapter 4	2
4.2. Agile Development, Quality, and Testing Practices	Chapter 5	2
4.2.1. Agile Development Practices	Chapter 5	2
4.2.2. General Agile Quality Management Practices	Chapter 5	2
4.2.3. Agile Testing Practices	Chapter 5	2
Assignment: Complete Discussion Topics		

(continued)

(Continued)

Course Outline	Reference	Lesson
Part 2 – Agile Project Management		
5. Time Boxing, Kanban, Theory of Constraints, Estimation	Chapter 6	3
5.1. Kanban Process Overview	Chapter 6	3
5.2. Theory of Constraints	Chapter 6	3
5.3. Agile Estimation	Chapter 7	3
Assignment: Complete Discussion Topics and Theory of Constraints Assignment		
Complete Estimation Exercise		
6. Agile Project Management Role	Chapter 8	3
6.1. Levels of Agile Implementation	Chapter 8	3
6.2. The Role of an Agile Project Manager	Chapter 8	3
6.2.1. Agile Project Management Shifts in Thinking	Chapter 8	3
6.2.2. Agile and PMBOK	Chapter 8	3
6.2.3. Potential Agile Project Management Roles	Chapter 8	3
Mid-Term ExamFinal Research Paper Proposal Due		
7. Agile Communications Practices & Tools	Chapter 9	4
7.1. Agile Communications Practices	Chapter 9	4
7.2. Agile Project Management Tools	Chapter 9	4
Assignment: Complete Discussion Topics		
8. Learning to See the Big Picture	Chapter 10	4
8.1. Systems Thinking	Chapter 10	4
8.2. Complex Adaptive Systems	Chapter 10	4
Assignment: Complete Discussion Topics		
8.3. The Roots of Agile	Chapter 11	4
8.3.1. The Influence of Total Quality Management (TQM)	Chapter 11	4
8.3.2. The Influence of Lean Manufacturing	Chapter 11	4
Assignment: Complete Discussion Topics		
Part 3 – Agile Project Management Planning and Management		
9. Hybrid Agile Models	Chapter 12	5
9.1. What is a Hybrid Agile Model and Why Would You Use It?	Chapter 12	5
9.1.1. What Are the Benefits of a Hybrid Agile Model?	Chapter 12	5
9.1.2. What Is Different About a Hybrid Agile Approach?	Chapter 12	5
9.1.3. Choosing the Right Approach	Chapter 12	5
Assignment: Complete Discussion Topics		

(Continued)

(continued)

(Continued)

(Continued)

Course Outline	Reference	Lesson
Part 6 – Case Studies		
17. Case Studies		13
17.1. "Not-So-Successful" Case Studies	Chapter 26	13
17.2. Valpak	Chapter 27	13
17.3. Harvard Pilgrim Healthcare	Chapter 28	13
17.4. General Dynamics, UK	Chapter 29	13
17.5. Non-Software Case Studies	Chapter 30	13
17.6. Overall Summary	Chapter 31	13
Assignment: Complete Discussion Topics and Final Research Paper Due		
18. Presentation of Student Research Papers		14
Final Exam		15

INDEX